UKCAT
Practice Papers

Volumes One & Two

UniAdmissions

ISBN 978-1-912557-39-4

Published by *RAR Medical Services Limited*
www.uniadmissions.co.uk
info@uniadmissions.co.uk
Tel: 0208 068 0438

UKCAT
Practice Papers

6 Full Papers & Solutions

Matthew Williams
Rohan Agarwal

UniAdmissions

About the Authors

Matthew is **Medical Resources Editor** at *UniAdmissions* and a 4th year medical student at St Catherine's College, Oxford. As the first student from Barry Comprehensive School in South Wales to receive a place on the Oxford medicine course he embraced all aspects of university life, both social and academic. Matt scored in the **top 5% for his UKCAT and BMAT** to secure his offer at the University of Oxford.

He has been part of the UniAdmissions team since 2014 – tutoring several applicants successfully into Oxbridge and Russell group universities. His work has been published in international scientific journals and he has presented his research at conferences across the globe. In his spare time, Matt enjoys playing rugby and golf.

Rohan is the **Director of Operations** at *UniAdmissions* and is responsible for its technical and commercial arms. He graduated from Gonville and Caius College, Cambridge and is a fully qualified doctor. Over the last five years, he has tutored hundreds of successful Oxbridge and Medical applicants. He has also authored twenty books on admissions tests and interviews.

Rohan has taught physiology to undergraduates and interviewed medical school applicants for Cambridge. He has published research on bone physiology and writes education articles for the Independent and Huffington Post. In his spare time, Rohan enjoys playing the piano and table tennis.

Introduction

The Basics
The UK Clinical Aptitude Test (UKCAT) is the 2-hour written aptitude exam taken by students applying for Medicine or Dentistry.

It is a highly time pressured exam that does not test things which you typically learn at school. Instead the UKCAT requires very little prior knowledge but rather the ability to apply basic reasoning skills to practical and often novel situations. In this respect simply remembering solutions taught in class or from past papers is not enough.

However, fear not, despite what people say, you can actually prepare for the UKCAT! With a little practice you can train your brain to manipulate and apply learnt methodologies to novel problems with ease. The best way to do this is through exposure to as many past/specimen papers as you can.

But the UKCAT consortium do not release past papers...
Well hopefully that is where this book comes in. It contains six unique mock papers written by expert Oxbridge medical tutors at *UniAdmissions*. Having successfully gained places at Oxbridge and other competitive medical schools, our tutors are intimately familiar with the UKCAT and its associated admission procedures. So, the novel questions presented to you here are of the correct style and difficulty to continue your revision and stretch you to meet the demands of the UKCAT.

General Advice

Start Early
It is much easier to prepare if you practice little and often. **Start your preparation well in advance**; ideally 10 weeks but at the latest within a month. This way you will have plenty of time to complete as many papers as you wish to feel comfortable and won't have to panic and cram just before the test, which is a much less effective and more stressful way to learn. In general, an early start will give you the opportunity to identify the complex issues and work at your own pace.

Prioritise
Some questions in sections can be long and complex – and given the intense time pressure you need to know your limits. It is essential that you don't get stuck with very difficult questions. If a question looks particularly long or complex, mark it for review and move on. You don't want to be caught 5 questions short at the end just because you took more than 3 minutes in answering a challenging multi-step question. If a question is taking too long, choose a sensible answer and move on. Remember that each question carries equal weighting and therefore, you should adjust your timing in accordingly. With practice and discipline, you can get very good at this and learn to maximise your efficiency.

Positive Marking

There are no penalties for incorrect answers; you will gain one for each right answer and will not get one for each wrong or unanswered one. This provides you with the luxury that you can always guess should you absolutely be not able to figure out the right answer for a question or run behind time. Since each question provides you with 4 to 6 possible answers, you have a 16-25% chance of guessing correctly. Therefore, if you aren't sure (and are running short of time), then make an educated guess and move on. Before 'guessing' you should try to eliminate a couple of answers to increase your chances of getting the question correct. For example, if a question has 5 options and you manage to eliminate 2 options- your chances of getting the question increase from 20% to 33%!

Avoid losing easy marks on other questions because of poor exam technique. Similarly, if you have failed to finish the exam, take the last 10 seconds to guess the remaining questions to at least give yourself a chance of getting them right.

Practice

This is the best way of familiarising yourself with the style of questions and the timing for this section. Although the exam will essentially only test GCSE level knowledge, you are unlikely to be familiar with the style of questions in all sections when you first encounter them. Therefore, you want to be comfortable at using this before you sit the test.

Practising questions will put you at ease and make you more comfortable with the exam. The more comfortable you are, the less you will panic on the test day and the more likely you are to score highly. Initially, work through the questions at your own pace, and spend time carefully reading the questions and looking at any additional data. When it becomes closer to the test, **make sure you practice the questions under exam conditions**.

Repeat Questions

When checking through answers, pay particular attention to questions you have got wrong. If there is a worked answer, look through that carefully until you feel confident that you understand the reasoning, and then repeat the question without help to check that you can do it. If only the answer is given, have another look at the question and try to work out why that answer is correct. This is the best way to learn from your mistakes, and means you are less likely to make similar mistakes when it comes to the test. The same applies for questions which you were unsure of and made an educated guess which was correct, even if you got it right. When working through this book, **make sure you highlight any questions you are unsure of**, this means you know to spend more time looking over them once marked.

A word on calculators

Whilst you are permitted to use the simple onscreen calculator in the exam, be warned that it is very slow and easy to make mistakes on the touch screen. Thus, it is essential that you have strong numerical skills. For instance, you should be able to rapidly convert between percentages, decimals and fractions. You will seldom get questions that would require calculators, but you would be expected to be able to arrive at a sensible estimate. Consider for example:

Estimate 3.962 x 2.322;

3.962 is approximately 4 and 2.323 is approximately 2.33 = 7/3.

Thus, $3.962 \times 2.322 \approx 4 \times \frac{7}{3} = \frac{28}{3} = 9.33$

Since you will rarely be asked to perform difficult calculations, you can use this as a signpost of if you are tackling a question correctly. For example, when solving a physics question, you end up having to divide 8,079 by 357- this should raise alarm bells as calculations in the UKCAT are rarely this difficult.

A word on timing...

"If you had all day to do your exam, you would get 100%. But you don't."

Whilst this isn't completely true, it illustrates a very important point. Once you've practiced and know how to answer the questions, the clock is your biggest enemy. This seemingly obvious statement has one very important consequence. **The way to improve your score is to improve your speed.** There is no magic bullet. But there are a great number of techniques that, with practice, will give you significant time gains, allowing you to answer more questions and score more marks.

Timing is tight throughout – **mastering timing is the first key to success**. Some candidates choose to work as quickly as possible to save up time at the end to check back, but this is generally not the best way to do it. Often questions can have a lot of information in them – each time you start answering a question it takes time to get familiar with the instructions and information. By splitting the question into two sessions (the first run-through and the return-to-check) you double the amount of time you spend on familiarising yourself with the data, as you have to do it twice instead of only once. This costs valuable time. In addition, candidates who do check back may spend 2–3 minutes doing so and yet not make any actual changes. Whilst this can be reassuring, it is a false reassurance as it is unlikely to have a significant effect on your actual score. Therefore, it is usually best to pace yourself very steadily, aiming to spend the same amount of time on each question and finish the final question in a section just as time runs out. This reduces the time spent on re-familiarising with questions and maximises the time spent on the first attempt, gaining more marks.

It is essential that you don't get stuck with the hardest questions – no doubt there will be some. In the time spent answering only one of these you may miss out on answering three easier questions. If a question is taking too long, choose a sensible answer and move on. Never see this as giving up or in any way failing, rather it is the smart way to approach a test with a tight time limit. With practice and discipline, you can get very good at this and learn to maximise your efficiency. It is not about being a hero and aiming for full marks – this is almost impossible and very much unnecessary (even universities that demand high UKCAT scores e.g. Kings will regard any score higher than 750 as exceptional). It is about maximising your efficiency and gaining the maximum possible number of marks within the time you have.

Top tip! In general, students tend to improve the fastest in abstract reasoning and slowest in section verbal reasoning; the other sections usually fall somewhere in the middle. Thus, if you have very little time left, it's best to prioritise abstract reasoning.

Use the Options:

Some questions may try to overload you with information. When presented with large tables and data, it's essential you look at the answer options so you can focus your mind. This can allow you to reach the correct answer a lot more quickly. Consider the example below:

The table below shows the results of a study investigating antibiotic resistance in staphylococcus populations. A single staphylococcus bacterium is chosen at random from a similar population. Resistance to any one antibiotic is independent of resistance to others.

Calculate the probability that the bacterium selected will be resistant to all four drugs.

A   1 in 10^6
B   1 in 10^{12}
C   1 in 10^{20}
D   1 in 10^{25}
E   1 in 10^{30}
F   1 in 10^{35}

Antibiotic	Number of Bacteria tested	Number of Resistant Bacteria
Benzyl-penicillin	10^{11}	98
Chloramphenicol	10^9	1200
Metronidazole	10^8	256
Erythromycin	10^5	2

Looking at the options first makes it obvious that there is **no need to calculate exact values**- only in powers of 10. This makes your life a lot easier. If you hadn't noticed this, you might have spent well over 90 seconds trying to calculate the exact value when it wasn't even being asked for.

In other cases, you may actually be able to use the options to arrive at the solution quicker than if you had tried to solve the question as you normally would. Consider the example below:

A region is defined by the two inequalities: $x - y^2 > 1 \ and \ xy > 1$. Which of the following points is in the defined region?

A.   (10,3)
B.   (10,2)
C.   (-10,3)
D.   (-10,2)
E.   (-10,-3)

Whilst it's possible to solve this question both algebraically or graphically by manipulating the identities, by far **the quickest way is to actually use the options**. Note that options C, D and E violate the second inequality, narrowing down to answer to either A or B. For A: $10 - 3^2 = 1$ and thus this point is on the boundary of the defined region and not actually in the region. Thus the answer is B (as 10-4 = 6 > 1.)

In general, it pays dividends to look at the options briefly and see if they can be help you arrive at the question more quickly. Get into this habit early – it may feel unnatural at first but it's guaranteed to save you time in the long run.

Keywords

If you're stuck on a question; pay particular attention to the options that contain key modifiers like "**always**", "**only**", "**all**" as examiners like using them to test if there are any gaps in your knowledge. E.g. the statement "arteries carry oxygenated blood" would normally be true; "All arteries carry oxygenated blood" would be false because the pulmonary artery carries deoxygenated blood.

Manage your Time:

It is highly likely that you will be juggling your revision alongside your normal school studies. Whilst it is tempting to put your A-levels on the back burner falling behind in your school subjects is not a good idea, don't forget that to meet the conditions of your offer should you get one you will need at least one A*. So, time management is key!

Make sure you set aside a dedicated 90 minutes (and much more closer to the exam) to commit to your revision each day. The key here is not to sacrifice too many of your extracurricular activities, everybody needs some down time, but instead to be efficient. Take a look at our list of top tips for increasing revision efficiency below:

1. Create a comfortable work station
2. Declutter and stay tidy
3. Treat yourself to some nice stationery
4. See if music works for you → if not, find somewhere peaceful and quiet to work
5. Turn off your mobile or at least put it into silent mode
6. Silence social media alerts
7. Keep the TV off and out of sight
8. Stay organised with to do lists and revision timetables – more importantly, stick to them!
9. Keep to your set study times and don't bite off more than you can chew
10. Study while you're commuting
11. Adopt a positive mental attitude
12. Get into a routine
13. Consider forming a study group to focus on the harder exam concepts
14. Plan rest and reward days into your timetable – these are excellent incentive for you to stay on track with your study plans!

Keep Fit & Eat Well:

'A car won't work if you fill it with the wrong fuel' - your body is exactly the same. You cannot hope to perform unless you remain fit and well. The best way to do this is not underestimate the importance of healthy eating. Beige, starchy foods will make you sluggish; instead start the day with a hearty breakfast like porridge. Aim for the recommended 'five a day' intake of fruit/veg and stock up on the oily fish or blueberries – the so called "super foods".

When hitting the books, it's essential to keep your brain hydrated. If you get dehydrated you'll find yourself lethargic and possibly developing a headache, neither of which will do any favours for your revision. Invest in a good water bottle that you know the total volume of and keep sipping through the day. Don't forget that the amount of water you should be aiming to drink varies depending on your mass, so calculate your own personal recommended intake as follows: 30 ml per kg per day.

It is well known that exercise boosts your wellbeing and instils a sense of discipline. All of which will reflect well in your revision. It's well worth devoting half an hour a day to some exercise, get your heart rate up, break a sweat, and get those endorphins flowing.

Sleep

It's no secret that when revising you need to keep well rested. Don't be tempted to stay up late revising as sleep actually plays an important part in consolidating long term memory. Instead aim for a minimum of 7 hours good sleep each night, in a dark room without any glow from electronic appliances. Install flux (https://justgetflux.com) on your laptop to prevent your computer from disrupting your circadian rhythm. Aim to go to bed the same time each night and no hitting snooze on the alarm clock in the morning!

Revision Timetable

Still struggling to get organised? Then try filling in the example revision timetable below, remember to factor in enough time for short breaks, and stick to it! Remember to schedule in several breaks throughout the day and actually use them to do something you enjoy e.g. TV, reading, YouTube etc.

	8AM	10AM	12PM	2PM	4PM	6PM	8PM
MONDAY							
TUESDAY							
WEDNESDAY							
THURSDAY							
FRIDAY							
SATURDAY							
SUNDAY							
EXAMPLE DAY	School				AR	QR	VR

Getting the most out of Mock Papers

Mock exams can prove invaluable if tackled correctly. Not only do they encourage you to start revision earlier, they also allow you to **practice and perfect your revision technique**. They are often the best way of improving your knowledge base or reinforcing what you have learnt. Probably the best reason for attempting mock papers is to familiarise yourself with the exam conditions of the UKCAT as they are particularly tough.

Start Revision Earlier

Thirty five percent of students agree that they procrastinate to a degree that is detrimental to their exam performance. This is partly explained by the fact that they often seem a long way in the future. In the scientific literature this is well recognised, Dr. Piers Steel, an expert on the field of motivation states that *'the further away an event is, the less impact it has on your decisions'*.

Mock exams are therefore a way of giving you a target to work towards and motivate you in the run up to the real thing – every time you do one treat it as the real deal! If you do well then it's a reassuring sign; if you do poorly then it will motivate you to work harder (and earlier!).

Practice and perfect revision techniques

In case you haven't realised already, revision is a skill all to itself, and can take some time to learn. For example, the most common revision techniques including **highlighting and/or re-reading are quite ineffective** ways of committing things to memory. Unless you are thinking critically about something you are much less likely to remember it or indeed understand it.

Mock exams, therefore allow you to test your revision strategies as you go along. Try spacing out your revision sessions so you have time to forget what you have learnt in-between. This may sound counterintuitive but the second time you remember it for longer. Try teaching another student what you have learnt, this forces you to structure the information in a logical way that may aid memory. Always try to question what you have learnt and appraise its validity. Not only does this aid memory but it is also a useful skill for UKCAT section 3, Oxbridge interview, and beyond.

Improve your knowledge

The act of applying what you have learnt reinforces that piece of knowledge. A question may ask you to think about a relatively basic concept in a novel way (not cited in textbooks), and so deepen your understanding. Exams rarely test word for word what is in the syllabus, so when running through mock papers try to understand how the basic facts are applied and tested in the exam. As you go through the mocks or past papers take note of your performance and see if you consistently under-perform in specific areas, thus highlighting areas for future study.

Get familiar with exam conditions

Pressure can cause all sorts of trouble for even the most brilliant students. The UKCAT is a particularly time pressured exam with high stakes – your future (without exaggerating) does depend on your result to a great extent. The real key to the UKCAT is overcoming this pressure and remaining calm to allow you to think efficiently.

Mock exams are therefore an excellent opportunity to devise and perfect your own exam techniques to beat the pressure and meet the demands of the exam. **Don't treat mock exams like practice questions – it's imperative you do them under time conditions.**

> ***Remember!*** It's better that you make all the mistakes you possibly can now in mock papers and then learn from them so as not to repeat them in the real exam.

Things to have done before using this book

Do the ground work

➤ Read in detail: the background, methods, and aims of the UKCAT as well logistical considerations such as how to take the UKCAT in practice. A good place to start is a UKCAT textbook like *The Ultimate UKCAT Guide* (flick to the back to get a free copy!) which covers all the groundwork but it's also worth looking through the official UKCAT site (www.ukcat.ac.uk).

➤ Get comfortable rapidly converting between percentages, decimals, and fractions.

➤ Practice developing logical arguments and structuring essays with an obvious introduction, main body, and ending.

➤ These are all things which are easiest to do alongside your revision for exams before the summer break. Not only gaining a head start on your UKCAT revision but also complimenting your year 12 studies well.

➤ Discuss scientific problems with others - propose experiments and state what you think the result would be. Be ready to defend your argument.

Ease in gently

With the ground work laid, there's still no point in adopting exam conditions straight away. Instead invest in a beginner's guide to the UKCAT, which will not only describe in detail the background and theory of the exam, but take you through section by section what is expected. *The Ultimate UKCAT Guide* is the most popular UKCAT textbook – you can get a free copy by flicking to the back of this book.

When you are ready to move on to mock papers, take your time and puzzle your way through all the questions. Really try to understand solutions. A mock paper question won't be repeated in your real exam, so don't rote learn methods or facts. Instead, focus on applying prior knowledge to formulate your own approach.

If you're really struggling and have to take a sneak peek at the answers, then practice thinking of alternative solutions, or arguments. It is unlikely that your answer will be more elegant or succinct than the model answer, but it is still a good task for encouraging creativity with your thinking. Get used to thinking outside the box!

Accelerate and Intensify

Start adopting exam conditions after you've completed the official mock papers. Don't forget that **it's the time pressure that makes the UKCAT hard** – if you had as long as you wanted to sit the exam you would probably get 100%.

Try and expose yourself to as many sample questions as possible, although these can be hard to come by outside of this book. In any case, choose a question bank and proceed with strict exam conditions. Take a short break and then mark your answers before reviewing your progress. For revision purposes, as you go along, keep track of those questions that you guess – these are equally as important to review as those you get wrong.

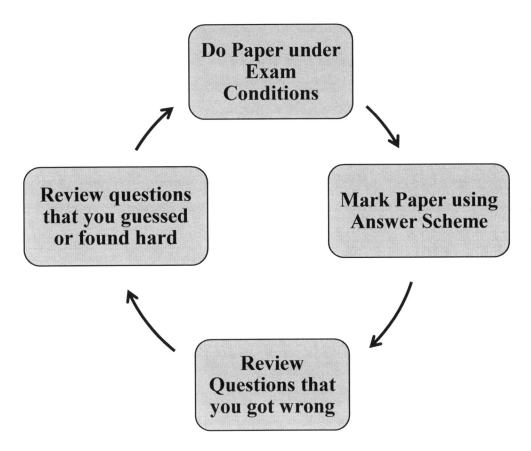

Once you've exhausted all the specimen and sample questions, move on to tackling the unique mock papers in this book. In general, you should aim to complete one to two mock papers every night in the ten days preceding your exam.

UKCAT sections overview

Verbal Reasoning

This is the first subtest, assessing your ability to read and reason novel information presented to you. As with all sections of the UKCAT it is very important to only consider the information which the test question presents to you – do not draw on information that you have learnt previously even if you believe it to be more accurate than what is presented in the exam!

In the verbal reasoning section, you will be presented with 11 texts, each followed by 4 questions. Meaning in total you have 21 minutes to answer 44 questions. There are generally only two questions formats:
  ➢ **Type 1:** you will be asked to draw conclusions or other inferences from the passage.
  ➢ **Type 2:** you will be asked to decide if a statement follows logically from the passage. You may answer either "true", "false", or "can't tell".

Decision Making

This sub test assesses your ability to utilise logical methods to reach a valid conclusion or decision when presented with novel arguments or statistics. This can include information in the form of charts, graphs, tables, diagrams, or lists. In total you have 31 minutes to answer 29 questions. Questions come in two forms:
  ➢ **Type 1:** select one correct answer from the options.
  ➢ **Type 2:** work through a list of 5 statements and respond with either "yes", or "no".

Quantitative Reasoning

This section of the exam assesses your numerical skills and ability to apply them to solve novel problems. The test assumes sufficient knowledge to achieve a good pass in GCSE maths. Questions may present you with tables, graphs, charts, or number arrays, with each data set followed by 4 (sometimes linked) questions. You will have 24 minutes to solve 36 questions. There is only one type of questions for this sub-section:
  ➢ **Type 1:** choose the most suitable answer from the options.

Abstract Reasoning

This section assesses your ability to recognise and match patterns, whilst ignoring irrelevant or distracting material. You will be required to generate and test your own hypotheses in solving the questions. You will have 13 minutes to solve 55 questions which can be 1 of 4 types:
  ➢ **Type 1:** when presented with two sets of shapes you must decide whether a given test shape belongs to either "set A", "set B", or "neither".
  ➢ **Type 2:** select the shape which would occur next in a given series.
  ➢ **Type 3:** determine which shape completes a given statement.
  ➢ **Type 4:** when given two sets of shapes, you must decide from a list options which shape belongs to a given set.

Situational Judgement

This final section of the UKCAT assesses your ability to understand real world situations. You will need to identify which factors in the given scenarios are important and how you would appropriately deal with them. In total you have 69 questions to answer in 26 minutes. You do not require any medical or procedural knowledge for this section.

Questions are generally of 2 types:
  ➢ **Type 1:** consider the appropriateness of a given action.
  ➢ **Type 2:** consider the importance of a given factor in making a decision.

How to use this Book

If you have done everything this book has described so far then you should be well equipped to meet the demands of the UKCAT, and therefore **the mock papers in the rest of this book should ONLY be completed under exam conditions**.

This means:

➤ Absolute silence – no TV or music
➤ Absolute focus – no distractions such as eating your dinner
➤ Strict time constraints – no pausing half way through
➤ No checking the answers as you go
➤ Give yourself a maximum of three minutes between sections – keep the pressure up
➤ Complete the entire paper before marking
➤ Mark harshly

In practice this means setting aside 2 hours and 15 minutes in an evening to find a quiet spot without interruptions and tackle the paper. Completing one mock paper every evening in the week running up to the exam would be an ideal target.

➤ Tackle the paper as you would in the exam.
➤ Return to mark your answers, but mark harshly if there's any ambiguity.
➤ Highlight any areas of concern.
➤ If warranted read up on the areas you felt you underperformed to reinforce your knowledge.
➤ If you inadvertently learnt anything new by muddling through a question, go and tell somebody about it to reinforce what you've discovered.

Finally relax… the UKCAT is an exhausting exam, concentrating so hard continually for two hours will take its toll. So, being able to relax and switch off is essential to keep yourself sharp for exam day! Make sure you reward yourself after you finish marking your exam.

How to calculate your scaled UKCAT score for each section:

Your exact score in the UKCAT will depend on the difficulty of the questions your test consists of however; you can use the following formula to work out your result for each subsection: *300 + (% correct x 600)*

Scoring Tables

Use these to keep a record of your scores from mock papers – you can then easily see which paper you should attempt next (always the one with the lowest score).

VR	1st Attempt	2nd Attempt	3rd Attempt
Mock A			
Mock B			
Mock C			
Mock D			
Mock E			
Mock F			

DM	1st Attempt	2nd Attempt	3rd Attempt
Mock A			
Mock B			
Mock C			
Mock D			
Mock E			
Mock F			

QR	1st Attempt	2nd Attempt	3rd Attempt
Mock A			
Mock B			
Mock C			
Mock D			
Mock E			
Mock F			

AR	1st Attempt	2nd Attempt	3rd Attempt
Mock A			
Mock B			
Mock C			
Mock D			
Mock E			
Mock F			

SJT	1st Attempt	2nd Attempt	3rd Attempt
Mock A			
Mock B			
Mock C			
Mock D			
Mock E			
Mock F			

Mock Paper A

Section A: Verbal Reasoning

Passage 1

Atomic Structure

Although the existence of atoms has been suggested since ancient Greece, the modern understanding of atomic structure is the product of the hard work of many dedicated scientists. In the early 1800s, John Dalton's experiments with chemical reactions led him to theorise that matter is composed of tiny individual units. These individual units of matter were later understood to be what we call atoms. In 1897, the English Physicist J. J. Thomson discovered that atoms contain both positive and negative electrical charges by performing experiments with cathode ray tubes. Thomson visualised the negative charges as immersed within positively charged material. Although his model of atomic structure is now considered incorrect, it was the most accurate picture of an atom at that time.

In 1909, Ernest Rutherford, a physicist at the University of Manchester, performed his famous alpha particle experiment and concluded that atoms have a positively charged nucleus at their centre. Danish physicist Niels Bohr theorized in 1912 that negatively charged electrons orbit the positively charged nucleus in fixed paths, not unlike planets around the sun. In the 1920s, several scientists modified Bohr's ideas of the orbital paths of electrons by demonstrating that electrons do not behave like other particles and that their location as they orbit the nucleus is difficult to predict.

The rest of the 20th Century saw the discovery of neutrons, which carry a neutral charge, and quarks, which make up all protons, neutrons and electrons. By the 1990s, no less than six different types of quarks had been identified. New discoveries in atomic structure will surely be made in the future.

1. The existence of atoms:
A. is a recent idea.
B. is an old idea.
C. is an early 19th century idea.
D. is a late 19th century idea.

2. Which of the following statements is true according to the above passage:
A. John Dalton proved that matter is composed of tiny individual units.
B. J.J. Thomson proved atoms were made up of negative charges immersed within positively charged material.
C. Niels Bohr discovered that electrons orbit in fixed paths around the nucleus.
D. That atoms do not contain a negatively charged nucleus was discovered in the 20th century.

3. Which of the following statements about quarks is true:
A. There are over six types of quarks.
B. Quarks are made up of neutrons, electrons and protons.
C. The discovery of quarks has settled discussions on atomic structure.
D. Quarks are even smaller than atoms.

4. Which of the following statements is INCORRECT:
A. Electrons' behaviour is dissimilar to other particles' behaviour.
B. Electrons' behaviour is predictable.
C. Electrons contain quarks.
D. Electrons go around a nucleus.

Passage 2

The Space Race

After the Second World War, Western Europe and the United States found themselves at odds with the Soviet Union. Although they never went to war directly, the two sides were in a constant race to develop advanced weapons and other technology.

In 1957, the Soviets launched the first man-made satellite, Sputnik 1, into space. The successful launch and orbit of Sputnik 1 caused the United States to attempt a satellite launch two months later. The rocket carrying the American satellite failed to launch and exploded on the launch pad. Two months after that, the U.S. did successfully launch a satellite into earth's orbit, and the rivalry between the Soviet and American space programs began.

The next goal was to successfully launch a human into outer space, which the Soviets did first in April 1961 when Yuri Gagarin completed almost two hours orbiting the earth. A month later, an American astronaut was successfully launched into space, although not into orbit. U.S. President Kennedy then began campaigning for a space program aimed at reaching the moon faster than the Soviets.

The mid-1960s were spent with both the Americans and Soviets achieving multiple-day manned space flights, the manual manoeuvring of spacecraft, and spacewalks. In 1967, both sides began a series of manned space flights whose ultimate goal was to land on the moon.

The first humans to orbit the moon were American astronauts in December, 1968. The Soviets never achieved manned lunar orbit, as their space program was plagued with difficulties. In June of 1969 American astronaut Neil Armstrong became the first human to set foot on the moon, effectively defeating the rival Soviet lunar program.

After the race to the moon was over, both sides began to focus on developing space stations, as they would be less expensive than lunar programs and would provide valuable opportunities for research. The Soviets launched the first space station, Salyut 1, into orbit in 1971, followed by the American space station Skylab in 1973. In 1975, a joint space mission involving Soviet and American crews in cooperation was the symbolic end of the rivalry.

5. Which of the following statements is correct:
A. Rivalry throughout WWII between the US and the Soviet Union continued after the war's end.
B. The aftermath of WWII caused tensions between the Soviet Union and Western forces including the US and Western Europe.
C. The developments in each other's technology sparked fear and tension between the US and the Soviet Union.
D. The passage does not state the reasons for the US and the Soviet Union being at odds.

6. According to the passage:
A. The US's space programme was inferior to the Soviet Union's.
B. Developments in space technology fuels developments in weapons technology.
C. The Soviet Union was initially more successful in the Space Race.
D. The US won the Space Race.

7. Which of the following statements is supported by the above passage?
A. The Soviet Union's poverty crippled their attempts to achieve the landing of the first man on the moon.
B. The Soviet Union's civil unrest crippled their attempts to achieve the landing of the first man on the moon.
C. Neil Armstrong was the first man to achieve a spacewalk.
D. The moon-landing had political elements to it.

8. According to the above passage, the 1975 space mission was preceded by:
A. Multiple humans orbiting the moon.
B. Political truce between the Soviet Union and the US.
C. Multiple men on the moon.
D. Much valuable research had been achieved by the 1973 space mission.

Passage 3

Childcare

Childcare is a very important service for most families, whether it be day care, preschool, primary school, secondary school, or simply babysitting. Parents generally cherish their children above all other things, so they do not normally hand their children over to just anyone. Parents need to trust that their children will be physically and emotionally safe while they are apart. Furthermore, if the childcare situation is at a school, parents need to trust that their children are receiving an appropriate education, or they may make other arrangements.

Childcare professionals, babysitters, teachers and school staff should make every effort to ensure a safe setting for the children in their care. This normally includes meeting the physical needs of children by providing clean nappies, bathrooms, food, water, shelter, and sleep if necessary. Children also have emotional and intellectual needs that should be met by adults speaking with them, allowing them to make decisions, helping them overcome challenges, giving them advice, allowing them to make friends, and making them feel respected.

People that care for children should also monitor them for signs of emotional and physical abuse. If a teacher, babysitter, or childcare professional were to observe bruises on a child or to notice the child acting unusually aggressive or depressed, he or she should consider that the child may be having serious problems at home. If a teacher, babysitter, or childcare professional confirms that a child is likely experiencing physical or emotional abuse, he or she should contact an appropriate authority to investigate the situation.

9. Parents:
A. All hold childcare as a very important service.
B. All cherish their children above most things, at all times.
C. May be discerning when it comes to childcare providers.
D. Above all, need to trust that their kids will be emotionally safe when apart from them.

10. Which of these is NOT stated as something to consider when caring for someone else's child:
A. Nutrition.
B. The child's feelings.
C. A tidy appearance.
D. The child's feeling respected.

11. People who care for children should know:
A. Bruises mean the child is experiencing serious problems at home.
B. Unusual aggression means the child is being abused.
C. Unusual depressions means the child is being abused.
D. There is a procedure to follow if they suspect child abuse.

12. Which of the following statements is NOT supported by the above passage:
A. People involved in child care include educators.
B. Childcare services include educational establishments, holiday camps and babysitting.
C. People who care for children should make sure the children in their care are getting water.
D. People who care for children should watch out for signs of abuse.

Passage 4

Australia

The first European contact with Australia was apparently with Dutch explorers in the 17th Century, but the English were the first to explore and colonize the enormous island. English settlers arrived in Australia in 1770 and soon English became the dominant language, as opposed to the roughly 250 language groups found in the country before this immigration, and Christianity the dominant religion. Initial settling was through penal transportation, the moving of criminals to the Oceanian land. This practice was also used in the Victorian era, and the fictional character of Sweeney Todd is said to have been sent to this land 'on a trumped up charge'.

The 19th Century saw the development of the territory into a modern economic and political force, making it an important part of the British Empire. During the 20th Century, Australia gained independence from Great Britain and became a major world power.

Most Australians now have European ancestry and Native Australians, also known as Aboriginal Australians, currently make up just a small fraction of the Australian population. Australian society saw little variation for millennia, as the Aborigines inhabited it for at least 40,000 years, but in the last 250 years it has undergone sweeping changes.

13. Australians of European descent may well be able to trace their ancestry back to an 18th century criminal.

A. True.
B. False.
C. Can't tell.

14. Dutch is *not* the dominant language in Australia, because:

A. The English stole the land from them.
B. English is easier to learn than Dutch.
C. The Dutch had the first European contact with Australia, but did not settle there.
D. The natives hated the Dutch.

15. According to the above passage, Australian independence:

A. Improved the conditions of Aboriginal Australians.
B. Decreased the wealth of Great Britain.
C. Embarrassed the English.
D. Resulted in the creation of a new major world power.

16. Which of the following is true according to the passage:

A. Many were sent to Australia on trumped-up charges.
B. Natives have reproduced less since the advent of English immigration to Australia.
C. Immigrants to Australia have all reproduced to a great extent.
D. A quarter of a millennium can be enough time to change a country's demographic situation.

Passage 5

Louis Pasteur

Louis Pasteur, a French scientist, is known as the father of microbiology. Pasteur's work demonstrated once and for all that microorganisms such as mould and bacteria, along with insect larvae, were not the product of spontaneous generation. Spontaneous generation was the ancient and medieval idea that certain types of life could appear on their own, without coming from a parental source, such as mould growing in standing water or maggots growing in uneaten food. Pasteur theorized that microorganisms were responsible for the appearance of these phenomena and proved himself correct in one of his most famous experiments. In 1859, he sterilized several containers and filled them with beef broth. In containers with long, curved necks, the beef broth remained uncontaminated because microbes could not reach the broth within. If a container were opened or turned so that microbes and other particles could reach the broth, its contents became contaminated with bacteria in a matter of hours.

The work of Pasteur and other biologists led to modern food preservation techniques. Canned food was a new development in Pasteur's day and his advancements in microbiology helped perfect and promote the process. Pasteur developed the practice of heating wine and beer to a temperature of 50 – 60 degrees Celsius to prevent spoilage. The practice was later expanded to include milk and other beverages. Boiled wine, beer or milk would be ruined, but if they were heated to this lower temperature, many of the microbes died and the structure of the beverage remained intact. This process was later termed "Pasteurization" after Louis Pasteur.

17. The idea of spontaneous generation:
A. Stated organisms could come into being from nothing.
B. Stated organisms could be utterly different from their parents, like maggots being born from uneaten parents.
C. Started in the middle ages.
D. Ended with the middle ages.

18. Which of the following is Pasteur NOT stated to be in the above passage:
A. European.
B. Male.
C. A genius.
D. A biologist.

19. Pasteur's discoveries are stated to have helped modern food production because:
A. It made food more palatable, as micro-organisms are known to be foul-tasting.
B. It made food more attractive, as micro-organisms like mould are unsightly.
C. It stopped food from smelling unpleasant.
D. It helped to keep food from going off.

20. According to the passage, when treating milk, beer and wine through pasteurisation:
A. The fluid becomes ruined by heat.
B. The fluid becomes completely free of microbes.
C. The fluid maintains its structure.
D. The fluid is 100% guaranteed not to cause any sickness through consumption.

Passage 6

Agricultural Reform

The 18th Century saw many advances in agriculture, which spurred Europe into the Industrial Revolution and spawned the modern societies now known by most of the world. Farming had been common in Europe for thousands of years, but few innovations had taken hold. When crops grow in a field, they remove the nutrients and fertilisers from the land. Ancient and Medieval farmers knew to leave fields fallow for a growing season or two to allow the soil to regain its fertility, and this practice remained largely unchanged. In the 1700s, however, British farmers found that if they grew other crops on their unused fields, nutrients returned to the soil faster than if the fields were left fallow.

Clover restores fields well; after growing clover in a field for a season, farmers could replant and grow on that field the next season with success. We now know that this is because clover absorbs nitrogen from the atmosphere and returns it to the soil as a natural fertilizer. Turnips were commonly planted in hitherto unused fields because they would also return fields to fertility. We now know that this is because the deep roots of turnips collect nutrients and bring them to the topsoil where they can be reached by the roots of other crops when the fields are replanted.

21. For thousands of years, farming in Europe was done in roughly the same way.
A. True.
B. False.
C. Can't tell.

22. There was a link between agriculture and industry in 1700s Europe.
A. True.
B. False.
C. Can't tell.

23. What did 18th Century British farmers know that Medieval ones did not:
A. That crops sap nutrients from the land.
B. That crops sap fertilisers from the land.
C. Certain plants replenish nutrients.
D. Leaving fields fallow allows them to recover.

24. According to the passage, turnips and clovers
A. Improve topsoil.
B. Collect nutrients from the atmosphere.
C. Collect nutrients from the earth.
D. Provide an alternative to leaving land fallow.

Passage 7

The Industrial Revolution

The increased production of food in the 18th Century allowed Europe to support a higher population than ever before. Advances in agricultural technology also allowed farms to be operated by fewer labourers than ever before, creating a surplus of labour. Many of these labourers found work in mills, mines and eventually factories, and their manpower helped drive the Industrial Revolution. The excess population in Europe also helped fuel European colonial empires, as a ready supply of colonists appeared to fill the New World.

The cities of Europe soon experienced overpopulation, which brought the problems of disease, poverty and crime. Orphanages and workhouses were common in the 19th Century, as many of the poor lived in squalor. Factories and mills were generally unsafe places to work and workplace accidents were common. The steady supply of labour meant that employers could overwork and underpay their workers. Churches and charities began caring for the poor and campaigning for better living and working conditions. Eventually, laws were enacted across the western world that limited work hours and child labour, promoted workplace safety and public health, and established the minimum wage.

25. Increased food production meant:
A. There were fewer jobs for farm labourers.
B. More people could be fed.
C. More people moved to the colonies.
D. Mills improved.

26. The Industrial revolution was predominately responsible for colonisation.
A. True.
B. False.
C. Can't tell.

27. The passage states:
A. The Industrial revolution was a positive event.
B. The Industrial revolution is an event to be regretted.
C. Things became a lot simpler with the advent of the Industrial Revolution.
D. Higher populations can come with problems.

28. Employers in the above passage are described as possibly:
A. Benevolent.
B. Anxious.
C. Exploitative.
D. Sadistic.

Passage 8

The Ice Age

For much of the last 100,000 years, Earth's climate was colder than it is now and enormous ice sheets covered large parts of North America, Europe and Asia. The Scandinavian Ice Sheet covered what is now Scotland and northern England, as well as what are now Scandinavia and northern Russia. What are now Canada and the northern United States were covered by the Laurentide Ice Sheet. Humans inhabited many of these areas before the ice sheets formed, but were forced out by the cooling climate. By around 12,000 years ago, much of this ice had melted and sea levels had risen to their present state.

As the Earth warmed, the ice sheets slowly receded, creating many of the landforms now present in North America, Europe and Asia. Sand, gravel and rocks of various sizes were all carried northward with the receding ice, and were re-deposited as melting ice formed new rivers and lakes. When melting ice flowed heavily from an ice sheet, a new river could be formed as the flowing water dug a trench through the earth. When a large piece of ice broke off from an ice sheet and was left behind to melt, a lake or pond could be formed. This is how many of the lakes in northern Europe and Canada were created, such as Osterseen in Germany, Loch Fergus in Scotland, and Wilcox Lake in Canada.

As the ice sheets shrank, humans moved back into the newly inhabitable areas of the arctic. Asians crossed into the Americas, and sea levels rose to separate Siberia from Alaska and the British Isles from Europe.

29. For 100,000 years Earth's climate was colder than it is now.
A. True.
B. False.
C. Can't tell.

30. Which of the following is NOT true:
A. Scotland is covered by the Scandinavian Ice Sheet.
B. The same ice sheet covered what were northern Russia and Scotland.
C. Multiple ice sheets existed.
D. Ice sheets were very large.

31. Loch Fergus was created:
A. By heavy-flowing melting ice.
B. By sand, gravel and rocks being deposited by melting ice.
C. By non-flowing melting ice.
D. By medium-flowing melting ice.

32. The British Isles was once part of mainland Europe:
A. True.
B. False.
C. Can't tell.

Passage 9

Germany

The Federal Republic of Germany (or the *Bundesrepublik Deutschland)* is the most populous country in the European Union and has Europe's largest economy. It also has the fourth largest economy in the world, giving it an influential role in geopolitics. The German language, also widely spoken in Switzerland and Austria, has over 100 million native speakers and some 80 million speakers who learned it as a foreign language. Many Europeans have migrated to Germany to take advantage of job opportunities that are available in its strong economy.

The European Union's Freedom of Movement for Workers principle (described in Treaty on European Union Article 39) means that anyone from an EU country (a state that is a member of the European Union) can seek and gain work in any other EU country, without experiencing discrimination due to their citizenship, excepting the people of Croatia. Accordingly, in 2014 many immigrants from Poland, Romania and Bulgaria came to Germany looking for employment. Croatian citizens also attempted to find paid work in the country, despite their exclusion from the above treaty. There were also those from the Middle East, who were refugees from ongoing conflicts in the region, who sought asylum within Germany's boarders.

33. Germany is a monarchy.
A. True.
B. False.
C. Can't tell.

34. According to the passage, the German language:
A. Is spoken as a mother tongue by over 100 million people.
B. Is spoken as a mother tongue by over 180 million people.
C. Is the dominant language in Austria and Switzerland.
D. Is the second most important language in geopolitics.

35. The German economy:
A. Is strong because of the influx of immigrant workers.
B. Ensures everyone in Germany is well-off.
C. Is in the top ten biggest global economies.
D. Ensures jobs for migrant workers.

36. According to the above passage, Croatians have more right to German jobs than Middle Eastern migrants.
A. True.
B. False.
C. Can't tell.

Passage 10

Thomas Hobbes

Born in 1588 in Malmesbury, Wiltshire, with a clergyman father, Thomas Hobbes was an English political philosopher and political scientist. His birth was linked to acts of war: he was born prematurely, when his mother heard of the approaching Spanish Armada invasion. This instance led to him pithily saying his parent 'gave birth to twins: myself and fear.'

In 1651, he published the book *Leviathan*. The following is an excerpt from this text, which he wrote in France, and is in reference to the English Civil War and its effects on society:

"In such condition, there is no place for industry; because the fruit thereof is uncertain; and consequently no culture of the earth; no navigation, nor use of the commodities that may be imported by sea; no commodious building; no instruments of moving and removing such things as require much force; no knowledge of the face of the earth; no account of time; no arts; no letters; no society; and which is worst of all, continual fear, and danger of violent death; and the life of man, solitary, poor, nasty, brutish, and short."

37. *Leviathan* considers biblical stances.
A.   True.
B.   False.
C.   Can't tell.

38. Hobbes:
A.   Hated the Spanish.
B.   Hated war from birth.
C.   Was born in a condition of stress.
D.   Was born to be a clergyman.

39. The passage states that the seventeenth century:
A.   Saw the birth of this great political mind.
B.   Saw the birth of a great philosophical mind.
C.   Saw the product of a political-philosophical mind.
D.   Saw the depression of Thomas Hobbes.

40. Hobbes' speaks on:
A.   The necessity of war.
B.   The destruction of the enemy through war.
C.   The shutting down of society.
D.   The problems with the predominate illiteracy of the poor of his time.

Passage 11

Emily Davison

The first-wave feminism movement saw many passionate women fighting for their cause, and sacrificing much in pursuit of equal rights for both sexes. An example of a feminist who paid much for her beliefs in Emily Davison, who campaigned in Britain and had experienced multiple run-ins with the law before her death.

On nine separate occasions she was arrested and thrown into jail. She continued protesting even from a cell, by refusing to eat. Feminist hunger strikes often ended in force-feeding, a horrible process where a tube is passed through the mouth (or, occasionally, the nose) into the stomach so food can be poured directly into the prisoner's body. The inmate would be held down whilst this happened. Davison endured this invasive treatment 49 times.

In 1913, at the Epsom Derby, she ran in front of King George's horse and ended up trampled. It has been debated what exactly her intentions were, with many arguing it was not a suicidal act. Some believe that analysis of the newsreel supports the notion that Davison was trying to attach a scarf to the King's horse's bridle, and that her behaviour was more a publicity stunt than a conscious sacrifice of her life. If this is so, then she paid the ultimate price for her beliefs and died the way she lived: campaigning for male and female equality.

41. Equal rights did not exist at all in the 20th century.
A. True.
B. False.
C. Can't tell.

42. According to the passage, feminists fought for:
A. Superior rights for women.
B. Equal rights for different sexualities.
C. The vote.
D. Men to have the same rights as women.

43. Hunger strikes:
A. Discredited the feminist cause.
B. Were a waste of time.
C. Were countered with awful treatment.
D. Eventually died out.

44. Emily Davison was:
A. Suicidal.
B. Willing to sacrifice personal comfort.
C. A genius.
D. A pre-eminent figure in first wave feminism.

END OF SECTION

Section B: Decision Making

1. Chocolates come in boxes of 6, 9 and 20. What is the largest number of chocolates you CANNOT buy using the above combinations?

    A. 19   B. 27   C. 35   D. 43   E. 52

2. "It is best to start medical school aged 18."

    Which statement gives the best supporting reason for this statement?

    A. Medical schools are unlikely to admit people under 18 years old.
    B. At 18, students have reached the right level of maturity to enter medical school.
    C. Students cannot live away from their parents before they are 18.
    D. You must be an adult before starting medical school.

3. Dr Smith is only able to prescribe drugs. All antidepressants are drugs. Carbamazepine is not an antidepressant. Most home remedies are drugs. Place true and false next to the statements below.

    A. Dr Smith can prescribe antidepressants.
    B. Dr Smith can prescribe carbamazepine.
    C. Carbamazepine is not a drug.
    D. Dr Smith can prescribe all home remedies.
    E. Most home remedies are antidepressants.

4. The probability Lucas misses the bus to school on a sunny day is 0.3 and on a rainy day is 0.2. The probability he carries an umbrella on any day is 0.4. If he misses the bus, Lucas walks to school. Last week, it rained on the last three days of the school week. Lucas thinks that last week, he was more likely to walk to school in the rain without an umbrella than to walk in the sun with an umbrella. Is he correct?

    A. Yes
    B. No, there is an equal chance of both occurrences in the last week
    C. No because the probability of getting the bus when it is raining is greater than catching the bus when it is sunny
    D. No, because the probability that he carries an umbrella on any given day is 0.4

5. All musicians play instruments. All oboe players are musicians. Oboes and pianos are instruments. Karen is a musician. Which statement is true?

    A. Karen plays two instruments.
    B. All musicians are oboe players.
    C. All instruments are pianos or oboes.
    D. Karen is an oboe player.
    E. None of the above

6. All of James's sons have brown eyes and all of his daughters have blue eyes. His wife has just become pregnant with a boy. Which statement is most likely to be correct?

    A.   The baby will have brown eyes.
    B.   James' wife has blue eyes.
    C.   Brown eyes are more likely than blue eyes.
    D.   Blue eyes are equally as likely as brown eyes.
    E.   None of the above.

7. Millie and Ben play a game. There is a stack of pennies and each player takes it in turn to remove one, two or three pennies each turn. The person who takes the last penny wins. If Millie starts the games, how many pennies will she need to start the game with to guarantee a win?

    A. 4 pennies.   B. 8 pennies   C. 13 pennies.   D. 16 pennies.

8. "Obesity is a growing problem, therefore there should be a tax on high calorie foods." Which option is the best argument against the above statement?

    A.   Those from low income backgrounds will be hit hardest from this tax.
    B.   Cost does not affect choice of food.
    C.   You are more likely to be overweight if you are rich.
    D.   Many high calorie foods are healthy

9. B is right of A. C is left of B. D is in front of C. E is in front of B. Where is D is relation to E?

    A. D is behind E.
    B. E is behind D.
    C. D is to the right of E.
    D. D is to the left of E.
    E. E is to the left of D.

10. Arnold, Carrie and Eric are arguing about the number of cars their father owns. Arnold says "Dad owns at least four cars", Carrie says "No, he owns less than four cars", and Eric says "he owns at least one car". If only one of them is tell the truth, how many cars does their father own?

    A. 0   B. 1   C. 3   D. 4   E. 5

11. "Increased traffic is bad for your health." Which statement provides the best evidence for this conclusion?

    A. Traffic congested increases carbon monoxide in the environment to harmful levels.
    B. Sitting in traffic reduces the amount of time for people to exercise.
    C. Towns with higher road tax have healthier people.
    D. Pollution from cars causes acid rain.
    E. Some traffic congestion is not hazardous to health

12. Fred has a drawer full of socks. There are 1 red, 2 green, 4 blue and 10 orange. In the dark, he cannot distinguish colours. What is the least number of socks he has to pick to ensure he has three matching pairs?

    A. 10   E. 11
    B. 8
    C. 3
    D. 9

13. Gabby is older than Maria. Maria's older sister Olivia is older than Gabby. Gabby wins more often than Olivia. Olivia's boyfriend Tom loses the most often. All four play cards. Which statement is true?

    A. Maria is the youngest.
    B. Olivia wins more than Maria.
    C. Gabby hates playing cards.
    D. Maria wins more than Tom.
    E. Tom is the oldest.

14. Three rats are placed in a maize that is in the shape of an equilateral triangle. They pick a direction at random and walk along the side of a triangle. Sophie thinks they are less likely to collide than not. Is she correct?

    A. Yes, mice naturally keep away from each other.
    B. No. They are more likely to collide than not.
    C. No. they are equally likely to collide than not collide.
    D. Yes, the probability they collide is 0.25

15. Jane says, "Plums are not sweets. Some plums are sweet. All sweets are tasty." Which of the below statements is most in keeping with Jane's thoughts?

    A. Some sweets are plums
    B. Some plums are not tasty.
    C. Some plums are tasty
    D. No plum is tasty
    E. Some plums are not sweet

16. In Leeds, a survey is done on a school. Strawberry ice cream is liked by 8 children, chocolate ice cream is liked by 5 children and vanilla ice cream is liked by 4 children. Three children like all flavours and no children like chocolate and strawberry, or vanilla and strawberry. Only one child likes chocolate and vanilla. Two children in the survey don't like ice cream. How many children took the survey?

    A. 15   B. 12   C. 10   D. 19

17. In Newcastle, a survey is done on a school. 4 children like all types of ice cream. 4 children like only strawberry and vanilla, and 12 like chocolate. However, 4 like vanilla ice cream only. Only one child likes everything but strawberry. Which Venn diagram represents this information?

A.

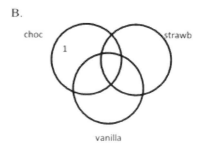

B.

C.

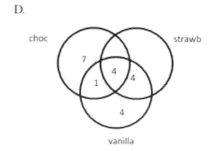

D.

18. Tasha went shopping. She says that every dress she bought was blue and she bought every blue dress she saw. Which statement is true?

A. Blue dresses were the only dresses Tasha saw while shopping.
B. While Tasha was shopping, she bought only blue dresses.
C. In the area that Tasha shopped, there were only blue dresses being sold.
D. All of the dresses that Tasha saw she bought.
E. Tasha did not see any other dress while she was shopping.

19. In a game of kissing catch, everyone put their names in a hat. Each player takes turns to draw a name. If they pick their own name, it is placed back and they draw another. If the last person to pick draws their own name, everyone starts again. John, Kelly and Lisa play and pick in alphabetical order. Which statement is true?

A. Jon has a one in three chance of picking and kissing Lisa
B. Kelly is more likely to pick John than picking Lisa
C. John has an equal chance of kissing Lisa or Kelly
D. Lisa should reverse the order is she wants to increase her chances of kissing John

20. Mary is showing photos to her daughter. She points to a woman and says "her brother's father is the only son of my grandfather." How is the woman in the photo related to Mary's daughter?

A. Cousin   B. Mother   C. Sister   D. Daughter   E. Aunt

21. "If vaccinations are now compulsory because society has decided that they should be forced, then society should pay for them." Which of the following statements would weaken the argument?

   A. Many people disagree that vaccinations should be compulsory.
   B. The cost of vaccinations is too high to be funded locally.
   C. Vaccinations are supported by many local communities and GPs.
   D. Healthcare workers do not want vaccinations.

22. Tim is going to the doctor for a blood test today. He says that he knows he will be in pain today. What assumption has Tim made?

   A. Using a needle will cause pain
   B. The doctor will have a hard time finding Tim's vein.
   C. He has had pain when he visited the doctor before so it must always happen
   D. Tim will have a bruise after his blood is taken
   E. The doctor will need repeated attempts to get blood

23. William, Xavier and Yolanda race in a 100m race. All of them run at a constant speed during the race. William beats Xavier by 20m. Xavier beats Yolanda by 20m. How many metres does William beat Yolanda?

   A. 30m   B. 36m   C. 40m   D. 60m   E. 64m

24. Chris is shorter than Ellen. Jane is shorter than Mark who is shorter than Ellen. Ellen and Jane are shorter than Naomi. Who is the tallest?

   A. Chris   B. Ellen   C. Jane   D. Mark   E. Naomi

25. Diane, Erica and Harry have Ferraris. Michael and Harry have Fords. Chris just bought an Audi. All the girls have Mercedes, except Lily who has a Volkswagen. Michael, Erica and Chris have BMWs. Who has the most cars?

   A. Erica   B. Chris   C. Harry   D. Lily   E. Michael

26. Watermelon is 99% water. Penny has 100 grams of watermelon. After drying in the sun, the shrivelled watermelon is 98% water. What is the weight of the watermelon now?

   A. 98g   B. 75g   C. 68g   D. 50g   E. 49g

27. Jon, Emmanuel and Saigeet are in a Rubik's cube solving competition. They need to solve the cube in less than 30 seconds to qualify. Jon and Saigeet solve faster than Emmanuel. Emmanuel's best time is 32.1 seconds. Which statement must be correct?

   A. Only Saigeet qualifies   C. Emmanuel   doesn't   D. Only Jon Qualifies
   B. No one qualifies   qualify   E. Jon and Saigeet qualify

28. All doctors are handsome. Some doctors are popular. Francis is handsome, and Oscar is popular. Choose a correct statement.

    A. A doctor can be popular and handsome
    B. Oscar is handsome
    C. Some popular people are handsome
    D. Francis is a doctor
    E. Oscar is popular with doctors.

29. There are four houses on a street. Lucy, Vicky and Shannon live in adjacent houses. Shannon has a black dog named Chrissie, Lucy has a white Persian cat and Vicky has a red parrot that shouts obscenities. The owner of a four legged pet has a blue door. Vicky has a neighbour with a red door. Either a cat or bird owner has a white door. Lucy lives opposite a green door. Vicky and Shannon are not neighbours. What colour is Lucy's door?

    A. Green
    B. Red
    C. White
    D. Blue
    E. Cannot tell

END OF SECTION

Section C: Quantitative Reasoning

Data Set 1

The following graph describes the travel of a car and bike along a road. Study the graph, then answer the following six questions.

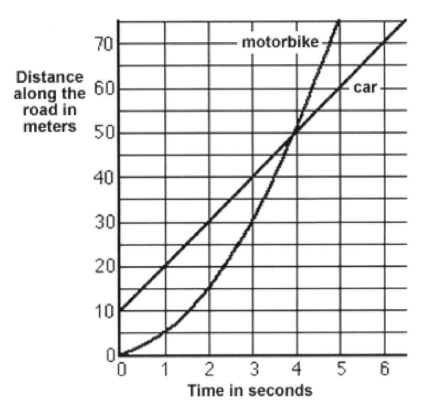

1. After 1 second, what is the speed of the car, in metres per second?

A. 5 m/s   C. 12 m/s   E. 30 m/s
B. 10 m/s   D. 20 m/s

2. Approximately, what is the highest speed the motorcycle reaches?

A. 20 m/s   C. 30 m/s   E. 100 m/s
B. 25 m/s   D. 50 m/s

3. What is the peak acceleration of the car, in m/s²?

A. 0 m/s²   C. 2 m/s²   E. 10 m/s²
B. 1 m/s²   D. m/s²

4. The motorbike accelerates more quickly than the car

A. True   C. True – but not initially   E. Can't tell
B. False   D. True – but only initially

Data Set 2

The following tables describe the cost of making international telephone calls. The cost of any given telephone call is calculated by adding together the connection charge, the duration charge plus a surcharge if applicable. The connection charge is only paid if the call is answered.

Connection charge between two countries (in UK pence):

	UK	France	USA	China	Australia
UK	-	25	47	52	68
France	25	-	51	54	78
USA	47	51	-	43	56
China	52	54	43	-	45
Australia	68	78	56	45	-

Cost per minute for international calls (in UK pence). All calls are rounded up to the nearest minute for calculation purposes. Increasing the duration of the call does not make previous minutes cheaper – only those minutes above any threshold are subject to the lower rate.

	1 – 10 mins	11 – 20 mins	21 – 60 mins	Over 60mins
Peak	42	37	34	28
Off-peak	25	18	16	15

Peak time is recorded as between 0800 and 1800 in the country making the call.

A surcharge of 88 pence is payable on calls over an hour, off peak only. A different surcharge of 10 pence is placed on international calls from Europe to Australia if the phone isn't answered.

5.   What is the total cost of a seven-minute peak time call from the UK to France?

A. £ 0.67   B. £ 2.94   C. £ 3.19   D. £ 3.29   E. £ 3.77

6.   What is the cost of a call at 0930 local time from France to Australia, that rings for 63 seconds but is unanswered?

A. £ 0.00   B. £ 0.10   C. £ 0.58   D. £ 0.84   E. £ 1.62

7.   What percentage of the overall cost of a call (of duration 463 seconds made at 1543 hours local time from USA to China) is represented by fixed (i.e. non-duration dependent) charges?

A. 0%   B. 8%   C. 11%   D. 13%   E. 15%

8.   What is the total cost of a 75-minute call, from China to France at 1935 local time?

A. £ 11.79   B. £ 12.67   C. £ 13.49   D. £ 14.37   E. £ 21.45

9.   What is the total cost of a 763 second call, from France to Australia at 1215 local time?

A. £ 5.31   B. £ 5.59   C. £ 6.09   D. £ 6.19   E. £ 7.19

Data Set 3

The following graph describes the price of gold. The next two questions refer to this graph.

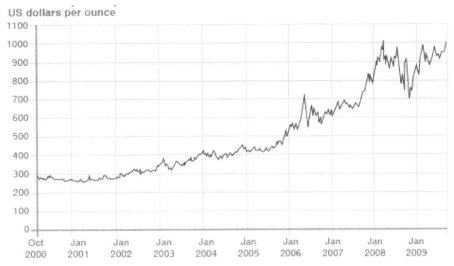

Assume where required that the current conversion rate is 1 USD = 0.68 GBP
One ounce is the equivalent of 28g

10. What is the approximate percentage change in gold value between January 2004 and January 2007?

A. 63%   B. 145%   C. 155%   D. 165%   E. 195%

11. What was the approximate total cost (in GBP) of 50g of gold in January 2004?

A. £456   B. £486   C. £714   D. £767   E. £20,000

Data Set 4

The following table shows the nutritional composition of three different single-portion pre-prepared meals.
Read it, and then answer the following questions.

Meal	Energy content / Kcal	Sugar content / g	Total mass / g
Lasagne	427	19	450
Chicken curry	783	24	600
Beef noodles	722	35	475
Ratatouille	359	14	320

12. How much more sugar per unit mass does the Beef Noodle dish contain relative to the Lasagne?

A. 16.0 g/kg   C. 31.5 g/kg   E. 37.1 g/kg
B. 21.5 g/kg   D. 34.3 g/kg

13. Which dish has the least proportion of its energy content provided by sugar?

A. Lasagne   C. Beef Noodles   E. More information required
B. Chicken Curry   D. Ratatouille

Data Set 5

Five respondents were asked to estimate the value of three bottles of wine, in pounds sterling.

Respondent	Wine 1	Wine 2	Wine 3
1	13	16	25
2	17	16	23
3	11	17	21
4	13	15	14
5	15	19	29
Actual retail value	8	25	23

14. What is the mean error margin in the guessing of the value of wine 1?

A. £4.80   B. £5.60   C. £5.80   D. £6.20   E. £6.40

15. Which respondent guessed most accurately on average?

A. Respondent 1   C. Respondent 3   E. Respondent 5
B. Respondent 2   D. Respondent 4

Data Set 6

A sweet shop stocks a range of different products. A new popular product is released, which the shopkeeper is keen to stock. However in order to do this, he must discontinue one of his current lines to create shelf space. The amount of shelf space required for this is 0.2m. The data below show sales figures for four different products currently stocked.

Product	Gobstopper	Bubblegum	Everton mints	Jelly beans
Cost per unit	22p	35p	45p	50p
Sale price	40p	50p	90p	65p
Number sold per week	150	180	300	420
Shelf space taken	0.2m	0.1m	0.2m	0.2m

16. What is the total weekly profit from these four items?

A. £189.00   C. £252.00   E. £472.50
B. £225.00   D. £295.50

17. What is the total value of sales for Gobstoppers and Everton mints combined, minus the total purchase price of one week's supply of Bubblegum?

A. £99.00   C. £267.00   E. £330.00
B. £162.00   D. £290.00

18. Based on the information available to you, which of these items would you recommend the shopkeeper replace with the new product?

A. Gobstopper   C. Everton mints   E. Gobstopper and
B. Bubblegum   D. Jelly beans   Bubblegum

Data Set 7

The population of Country A is 40% greater than the population of Country B.
The population of Country C is 30% less than the population of Country D (which is has a population 20% greater than Country B).

19. Given that the population of Country A is 45 million, what is the population of country D?

A. 32.1 million people
B. 35.8 million people

C. 36.6 million people
D. 38.6 million people

E. 39.0 million people

20. The population of Country A is still 45 million. If Country B introduced a new health initiative costing $ 45 per capita, what would be the total cost?

A. $ 1.35 bn
B. $ 1.45 bn

C. $ 1.50 bn
D. $ 1.55 bn

E. $ 1.65 bn

21. The population of Country C now changes to 25 million. The ratios are still preserved. Assuming that 52% of the population are female and 28% of the population are aged under 18, how many adult men are there in Country D?

A. 8.6 million
B. 10.3 million

C. 12.3 million
D. 17.1 million

E. 25.7 million

Data Set 8

The table below displays the costs associated with recruiting skilled workers in different industries in euros.

Training sector	Trade and industry	Civil service	Liberal professions	Crafts and skilled trades	Agri-culture
Application process This includes:	1,525	1,168	1,157	664	536
Advertising costs	576	502	337	231	183
Application process (personnel costs)	568	640	562	395	352
External consultants	381	26	258	38	0
Continuing training during the familiarisation period This includes:	1,048	1,029	183	329	376
Lost working hours during continuing training	447	431	75	139	168
Cost of continuing training courses	600	598	107	190	208
Difference in productivity during the familiarisation period	2,798	2,183	1,660	1,902	1,399
Total personnel recruitment costs	**5,370**	**4,380**	**3,001**	**2,895**	**2,311**

22. What is the difference in advertising costs per position between the most expensive and least expensive? What is the difference between the most expensive and the least expensive advertising cost?

A. €271   B. €288   C. €319   D. €363   E. €393

23. In which industry is there the greatest cost due to lost working hours relative to the total overall recruitment cost?

A. Trade and industry
B. Civil service

C. Liberal professions
D. Crafts and skilled trades

E. Agriculture

Data Set 9

The table below shows crime data for some types of crime from the town of Westwich over a three-year period.

Crime code	2011	2012	2013
X632	2,350	2,453	2,670
X652	3,821	3,663	3,231
Y321	230	210	?
Y632	456	490	432
Y115	321	?	431
Y230	763	754	714

24. Data for offence Y321 is missing for 2013, however you are told that the rate is 10% lower than in 2011. What is the rate of crime for Y321 in 2013?

A. 189   B. 195   C. 199   D. 207   E. 210

25. Data for offence Y115 is missing for 2012. You are informed that of the four "Y code" crimes recorded here, there were less than 1,837 in 2012. How many Y115 offences were committed?

A. 373   B. 383   C. 388   D. 393   E. 399

26. Which crime experienced the biggest percentage reduction from 2012 to 2013?

A. X632   B. X652   C. Y321   D. Y632   E. Y230

Data Set 10

This is a bus timetable taken from a route in Southern England. Use the data in the table to answer the following questions.

Mondays to Fridays / Saturdays

		Sch						Sch					
Petersfield Tesco	—	—	0825	1025	1225	1425	1525	1620	1810	—	1025	1325	1625
Petersfield Square	0700	0735	0830	1030	1230	1430	1530	1625	1815	0730	1030	1330	1630
Petersfield Station ▭	0702	0737	0832	1032	1232	1432	1532	1627	1817	0732	1032	1332	1632
Stroud The Seven Stars	0707	0742	0837	1037	1237	1437	1537	1632	1822	0737	1037	1337	1637
East Meon All Saints Church	0717	0752	0847	1047	1247	1447	1547	1642	1832	0747	1047	1347	1647
West Meon The Thomas Lord	0723	0758	0853	1053	1253	1453	1553	1648	1838	0753	1053	1353	1653
Bramdean The Fox Inn	0730	0805	0900	1100	1300	1500	1600	—	1845	0800	1100	1400	1700
Cheriton Cheriton Hall	0736	0811	0906	1106	1306	1506	1606	—	1851	0806	1106	1406	1706
New Alresford Perins School	0744	0819	0914	1114	1314	1514	1614	—	1859	0814	1114	1414	1714
Itchen Abbas Trout Inn	0755	—	0925	1125	1325	—	1625	—	—	0825	1125	1425	1725
Kings Worthy Cart and Horses	0802	—	0932	1132	1332	—	1632	—	—	0832	1132	1432	1732
Winchester City Road 🚲	0808	—	0938	1138	1338	—	1638	—	—	0838	1138	1438	1738
Winchester Bus Station	0811	—	0941	1141	1341	—	1641	—	—	0841	1141	1441	1741

No service on Sundays or Public Holidays

Mondays to Fridays / Saturdays

		Sch	SH					Sch						
Winchester Bus Station	—	—	0900	1100	1300	1400	—	1650	1750	0900	1200	1500	1750
Winchester City Road 🚲	—	—	0904	1104	1304	1404	—	1654	1754	0904	1204	1504	1754
Kings Worthy Cart and Horses	—	—	0911	1111	1311	1411	—	1701	1801	0911	1211	1511	1801
Itchen Abbas Trout Inn	—	—	0918	1118	1318	1418	—	1708	1808	0918	1218	1518	1808
New Alresford Perins School	0727	0729	0929	1129	1329	1429	1529	1719	1819	0929	1229	1529	1819
Cheriton Cheriton Hall	0735	0737	0937	1137	1337	1437	1537	1727	1827	0937	1237	1537	1837
Bramdean The Fox Inn	0741	0743	0943	1143	1343	1443	1543	1733	1833	0943	1243	1543	1833
West Meon The Thomas Lord	0748	0750	0950	1150	1350	1450	1550	1740	1840	0950	1250	1550	1840
East Meon All Saints Church	0755	0757	0957	1157	1357	1457	1557	1747	1847	0957	1257	1557	1847
East Meon Primary School	0757	—	—	—	—	—	—	—	—	—	—	—	—
Stroud The Seven Stars	0807	0807	1007	1207	1407	1507	1607	1757	1857	1007	1307	1607	1857
Petersfield Station ▭	0812	0812	1012	1212	1412	1512	1612	1802	1902	1012	1312	1612	1902
Petersfield Square	0814	0814	1014	1214	1414	1514	1614	1804	1904	1014	1314	1614	1904
Petersfield Tesco	0817	0817	1017	1217	1417	1517	1617	1807	—	1017	1317	1617	—

Liable to change from 11 June 2012

27. On a Monday, how long does the 1225 to Winchester Bus Station take to travel between East Meon All Saints Church and Itchen Abbas Trout Inn?

A. 13 minutes   C. 38 minutes   E. 60 minutes

B. 25 minutes   D. 47 minutes

28. If you were at Cheriton, Cheriton Hall at 1321 on a Tuesday, how long would you have to wait until the next bus to Winchester City Road?

A. 45 minutes   C. 105 minutes   E. 165 minutes

B. 65 minutes   D. 145 minutes

Data Set 11

Tables 1 and 3 show data relating to the cultivation of corn grain. Interpret the tables and answer the subsequent questions. (A hectare is the equivalent of 10000m^2).

Table 1. Monthly precipitation during the growing season, 1985-1989.

| | Precipitatiion (mm) | | | | | |
	1985	1986	1987	1988	1989	Long-Term Average
May	69	89	38	15	124	66
June	56	119	64	5	84	89
July	51	71	66	61	46	71
August	104	99	127	97	175	76
September	89	203	102	94	150	64
Total	369	581	397	272	579	366

Table 3. Corn grain yield, seed moisture content, and stand count at harvest for three erosion classes of Marlette soils.

Degree of Erosion	1985	1986	1987	1988	1989	Mean
	Yield (kg/ha)					
Slight	6,770a	8,150a	8,400a	3,510a	9,910a	7,340a
Moderate	5,580ab	7,340a	8,590a	3,820a	9,910a	7,090a
Severe	5,080b	5,960b	6,840b	1,820b	9,280a	5,830b
	Seed moisture content (%)					
Slight	22.6a	22.7a	26.5ab	26.0a	31.6a	25.9a
Moderate	24.8b	26.0a	25.9a	25.1a	30.4a	26.4ab
Severe	24.7b	25.9a	27.2b	31.3b	31.6a	28.2b
	Stand count (plants/ha)					
Slight	43,600a	59,400a	52,000ab	53,600a	57,500a	53,200a
Moderate	33,900a	48,800b	54,500a	46,800ab	57,700a	48,400ab
Severe	33,700a	32,800c	49,000b	34,700b	56,500a	41,300b

29. What is the percentage difference between the July 1987 precipitation levels and the long-term average for the month?

A. 5%   B. 6%   C. 7%   D. 8%   E. 9%

30. In what year was the overall mean seed moisture content the lowest?

A. 1985   B. 1986   C. 1987   D. 1988   E. 198

31. What was the difference in corn grain yield between slightly eroded Marlette soil in 1986 and moderately eroded Marlette soil in 1989, expressed in kg/hectare?

A. 0   B. 1330   C. 1640   D. 1760   E. 1920

32. In what month was there the highest average precipitation, taking the years 1985 – 1987 inclusive?

A. May   B. June   C. July   D. August   E. September

33. What sized area of ground did each plant occupy, in lightly eroded Marlette soil in 1986?

A. 1262 cm^2 C. 1684 cm^2 E. 2012 cm^2
B. 1384 cm^2 D. 1836 cm^2

Data Set 12

The following graph plots a child's length and weight up to the age of 36 months.

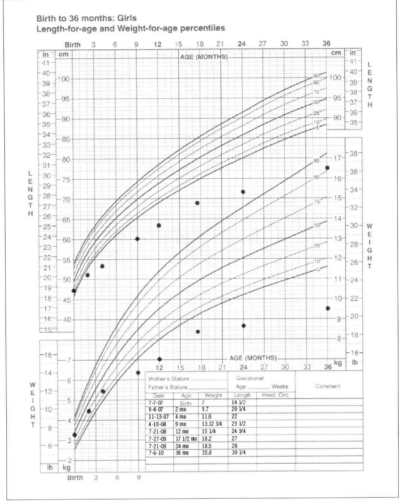

Figure 1 – The patient's weight and length from birth to age 36 months are shown. A deceleration of growth is apparent at age 9 months.

34. After two months, on which centile did the child's weight fall?

A. 75th   B. 5th   C. 10th   D. 25th   E. 50th

35. At the 24 month check, what was the child's actual length, in cm?

A. 38.5   B. 70.5   C. 71.5   D. 72.5   E. 73.5

36. What is the mean weight, in pounds (lbs.), of the final three measurements recorded on the chart?

A. 8.5 lbs   B. 9.5 lbs   C. 18.6 lbs   D. 19.2 lbs   E. 21.2 lbs

END OF SECTION

Section D: Abstract Reasoning

For each question, decide whether each test shape fits best with Set A, Set B or with neither.

For each question, work through the test shapes from left to right as you see them on the page. Make your decision and fill it into the answer sheet.

Answer as follows:
A = Set A
B = Set B
C = Neither

Set 1:   Set A   Set B

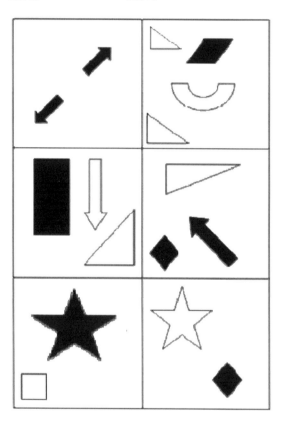 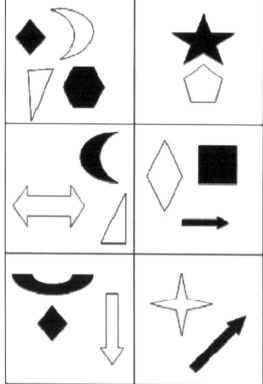

Questions 1-5:

  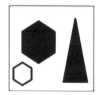

Set 2:   Set A   Set B

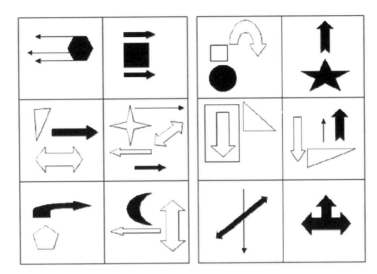

Questions 6-10:

Set 3:   Set A   Set B

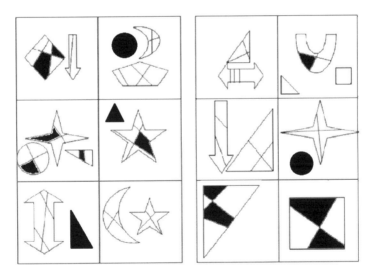

Questions 11-15:

Set 4:   Set A   Set B

Questions 16-20:

Set 5:   Set A   Set B

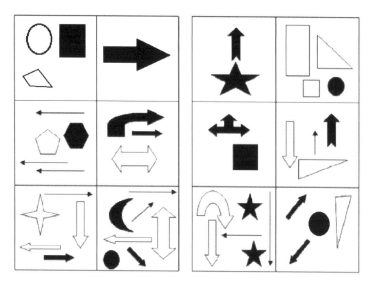

Questions 21-25:

Set 6:   Set A   Set B

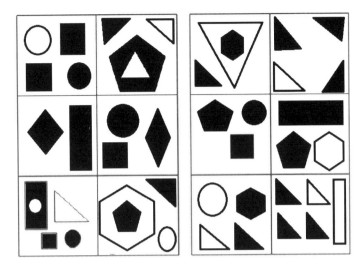

Questions 26-30:

Set 7:   Set A   Set B

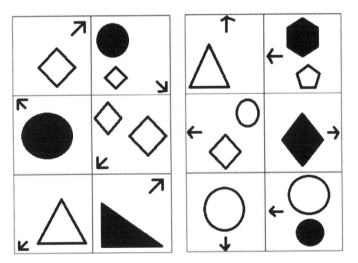

Questions 31-35:

Set 8:

Set A   Set B

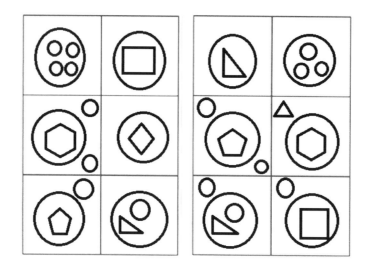

Questions 36-40:

Set 9:

Which answer completes the series?

Question 41:

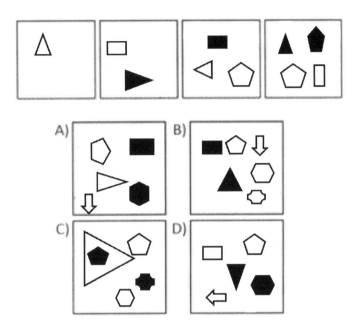

Question 42:

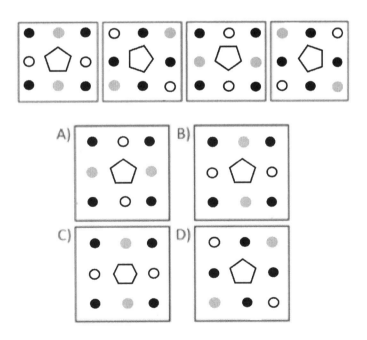

Question 43:

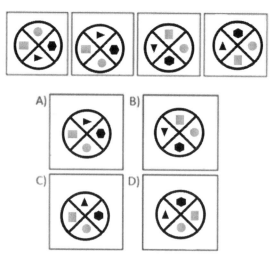

Question 44:

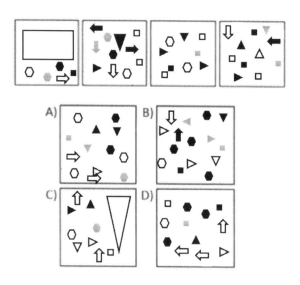

Question 45:

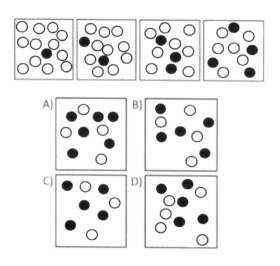

Set 10:

Which answer completes the statement?

Question 46:

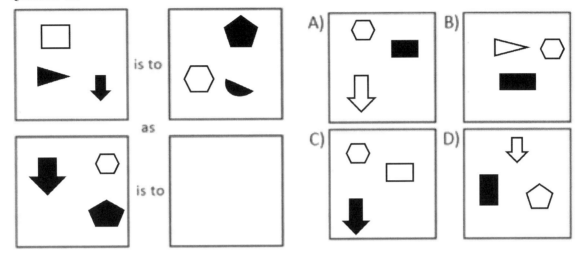

Question 47:

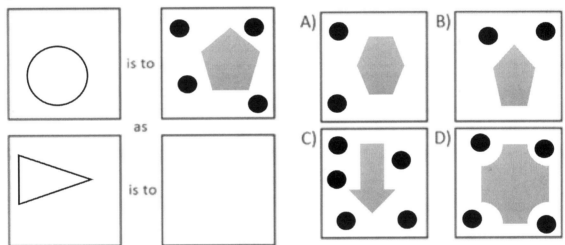

Question 48:

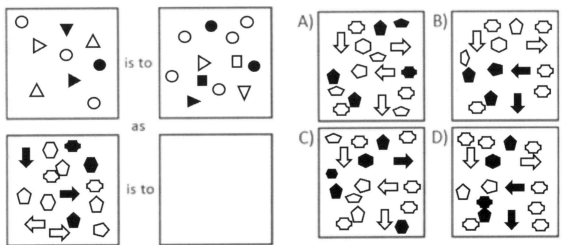

Question 49:

Question 50:

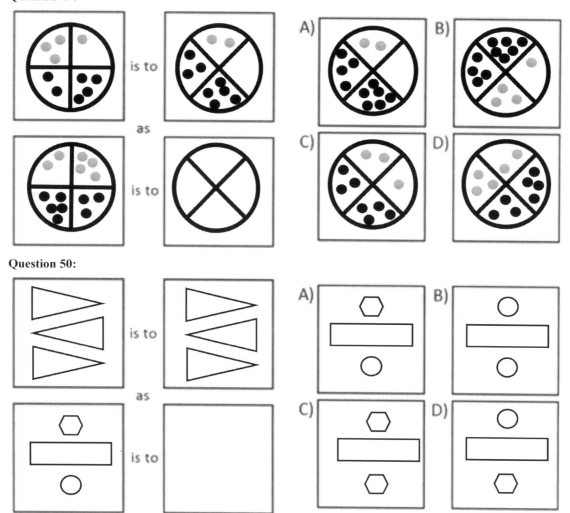

Set 11:

Which of the four response options belongs to either set A or set B?

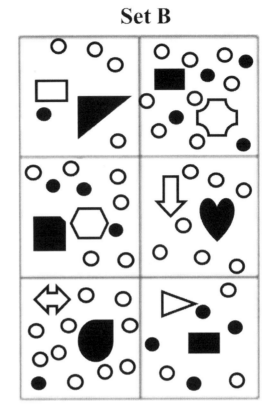

Question 51:   **Question 52:**

Set A?

Question 53:

Set A?

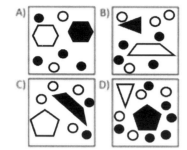

Question 54:

Set B?

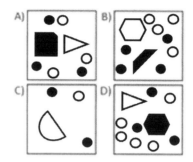

Question 55:

Set B?

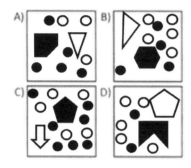

END OF SECTION

Section E: Situational Judgement Test

Read each scenario. Each question refers to the scenario directly above. For each question, select one of these four options. Select whichever you feel is the option that best represents your view on each suggested action.

For questions 1 – 30, choose one of the following options:

A   A highly appropriate action
B   Appropriate, but not ideal
C   Inappropriate, but not awful
D   A highly inappropriate action

Scenario 1

Emily is a third year medical student. She has been looking forward to the summer holidays for a long time because she has booked a once-in-a-lifetime trip to Tanzania with a group of her best friends from her course. She extended her overdraft to be able to pay for this trip, and has spent a lot of time planning it. Unfortunately, the end-of-year exam results have just come out, and she has only passed 2 out of 3 exams. She only failed by 2%, but must re-sit this final exam in order to continue with her studies. The re-sit is scheduled to be only two days after her return from Tanzania.

How appropriate are the following responses to Emily's dilemma?
1.   Go on her trip and cram her revision into the two days before the exam – she only just failed anyway.
2.   Cancel the entire trip and let her friends go without her.
3.   Ask her friends if they can all reschedule for the following year.
4.   Take revision with her to Tanzania – it'll be an active holiday but maybe she can study in her free time.
5.   Go for part of the trip but return early to ensure sufficient time for revision.

Scenario 2

Sanjay is a first year medical student who has recently joined the university football team. The word has just got out that this year's sports tour will be abroad. Sanjay is desperate to go as he has heard this is the best week of the academic year, and he wants to have a more active role within the football club. Regrettably, tour will be especially expensive this year and Sanjay is short on funds, having already spent most of his student loan and increased his overdraft.

How appropriate are the following actions to Steve's problem?
6.   Ask his other football friends to chip in as they all want him to be there too.
7.   Further increase his overdraft to the limit (this would be difficult to pay back).
8.   Ask his parents to contribute, despite their own financial hardships.
9.   Forget about tour this year – there is always next year when it is likely to be cheaper.
10.   Get a part-time job to pay for the tour, even though this will mean sacrificing some of his medical studies.

Scenario 3

Ade is heading into his second year at university. He lived at home with his parents throughout his first year, as the proximity of his home to the university meant that he was not eligible for student accommodation. He struggled to make close friends in his first year and often felt isolated from the other students who all lived together. He has recently seen an advertisement on the year group Facebook page, where several members of his year are looking for another flatmate. He is keen to apply, but is unsure whether his parents would be able to provide financial aid.

How appropriate are the following responses by Ade?

11.   Apply to the advertisement without telling his parents.
12.   Continue living at home but find other ways to enjoy the student experience – e.g. joining a society.
13.   Carry on exactly as he is.
14.   Demand his parents pay for him to move out as they are denying him the real student experience.
15.   Discuss his concerns with his parents and come up with a financial solution together – e.g. getting a part-time job.

Scenario 4

Gillian has recently been attending communication classes, and has learnt about the importance of having a good manner when talking to patients. That afternoon, she attends a clinic and is shocked to observe the doctor's rude and indifferent manner towards several patients. These patients leave feeling ignored and upset.

How appropriate are the following actions from Gillian?

16.   Tell the doctor what she thinks of his rude manner.
17.   Inform her supervisor and get their advice on what to do.
18.   Do nothing.
19.   Talk to the patients and direct them on how to make a complaint.
20.   Report the doctor to the dean of education.

Scenario 5

Ethel is a 3rd year medical student observing a prostate cancer clinic. She witnesses her consultant break the bad news of a terminal prognosis to a patient, who clearly struggles to comprehend his situation. Suddenly, the consultant receives a phone call and leaves the room, leaving Ethel alone with the upset patient.

How appropriate are each of the following responses by Ethel in this situation?

21.   Say nothing. Leave the patient to his thoughts.
22.   Offer words of consolation and support.
23.   Leave the room and the patient alone in order to look for the consultant.
24.   Share what little information about prostate cancer she knows.
25.   Tell the patient everything will be Okay.
26.   Ask the patient if he has any questions or concerns.

Scenario 6

Larissa is a third year medical student in her second term of clinical training. Whilst observing in A&E, she is asked by a nurse to perform an ABG (Arterial blood gas). Although she has learnt many of the clinical skills expected at her stage of training, an ABG is a fourth year skill and she has never attempted one before. However, the patient needs the procedure done as soon as possible and no other members of staff are immediately available. She is keen to learn and help out.

How appropriate are the following actions by Larissa in this situation?

27.   Refuse to carry out the procedure. It is not worth putting the patient at risk.

28.   Wait until a member of staff is available to assist.

29.   Go ahead and do the procedure. She has learnt similar skills such as taking blood – how different can it really be?

30.   Inform the patient that this would be her first time doing an ABG, and only proceed if they consent.

31.   Avoid the embarrassment of telling the nurse she can't do it, and rush off pretending to go to a lecture.

For questions 31 – 69, decide how important each statement is when deciding how to respond to the situation?

A   **Very important**

B   **important**

C   **Of minor importance**

D   **Not important at all**

Scenario 7

Romario is a final year medical student. He and his friends have carefully planned a trip to the Caribbean – their last hurrah before graduation. However, his consultant has just offered him a chance to participate in a study that is very likely to be published. Romario has been concerned as he is yet to have any publications to his name, but all of his friends have several. Unfortunately, the study will be carried out at the same time as the Caribbean trip.

How important are the following factors for Romario's decision?

32.   The likelihood of the study resulting in a publication – there is no guarantee.

33.   This is his last year as a student and a publication will help him when it comes to job applications.

34.   This is his last year as a student and the last chance to go away with his university friends.

35.   Rejecting the offer may look bad to his consultant.

36.   Cancelling the holiday may be letting his friends down.

Scenario 8

Mahood has been offered a one-off teaching session with a senior consultant on the High Dependency Unit (HDU) for respiratory patients. He is very keen to attend as this is a great opportunity to see lots of interesting cases. However, in the preceding days, Mahood comes down with the flu. Whilst he feels able to continue with his day, he is unsure whether to attend the teaching as it could put the patients at risk. He has been told to stay at home if he is unwell, but he is very eager to attend this prestigious teaching.

How important are the following factors for Mahood in deciding what to do?

37.   He only has the flu and it seems quite mild to him.

38.   This is a one-off teaching session with a top doctor.

39.   He has been told to stay at home if he is unwell.

40.   This is his only chance to see the interesting cases on HDU.

41.   This unit is clearly for very sick patients.

Scenario 9

Theodore is a third year medical student who is captain of his university rugby team. It is his first term on a clinical placement and he is keen to impress his consultant, who will be responsible for his final grade for the year. However, Theodore has noticed that his consultant teaching sessions will always be on a Wednesday afternoon, which clashes with his rugby matches.

How important are the following factors for Theodore in deciding on what to do?

42. He is the team captain – they need him to be present at the matches.
43. His consultant will determine his final grade.
44. His final grade will determine whether he can progress to fourth year.
45. Wednesday afternoons are the only time the consultant can make teaching.
46. His rugby team have the chance to win the championships this year.

Scenario 10

Malaika, a fourth year medical student, has been invited to attend an extra clinic with a senior doctor in a field she is interested in pursuing as a career. Unfortunately, she has not yet finished an important essay that is due in the next day.

How important are the following factors for Malaika to consider in deciding on what to do?

47. How long it will take her to finish the essay.
48. The importance of the essay towards her final mark.
49. She might not learn that much in the clinic.
50. Her reputation with the doctor.
51. Whether or not Malaika will be able to attend another clinic with the doctor.

Scenario 11

Jean and Franklin are medical students and clinical partners. Jean notices that Franklin frequently arrives looking very untidy, and some of their doctors have started to comment on his unprofessional appearance. However, Jean is worried that Franklin would take it personally and get offended if she mentioned something to him.

How important are the following factors for Jean in deciding on what to do?

52. Their reputation with the doctors.
53. Jean's friendship with Franklin.
54. Mentioning his appearance may hurt Franklin's feelings.
55. Their daily contact with patients.
56. Franklin's appearance may reflect badly on Jean as they are clinical partners.

Scenario 12

Rory and Priya have been working on a project together based on an interesting patient they have seen. They divided the work between them so each is responsible for different sections. Rory has been going into the hospital early every day in order to look at the patient's notes and construct his part of the project. However, he notices that Priya has been secretly taking sections of the patient's notes home with her each evening and bringing them back the next morning.

How important are the following factors to Rory's situation?

57. The project must be completed on time.
58. It risks breaking confidentiality for Priya to take identifiable patient notes home.
59. The overnight staff may not be able to access the patient's notes if Priya has taken them home.
60. A member of staff might find out.
61. Rory and Priya will receive a joint mark for their project.
62. Rory has no responsibility towards how Priya works on her part of the project.

Scenario 13

Horatio and Nelson are medical students on a surgical placement. They have been invited to observe an interesting operation, but have been advised to stand back and not to touch any of the sterile equipment. Horatio sees Nelson accidently touch the sterile trolley out of the corner of his eye. Nelson does not say anything, and the procedure is about to begin.

How important are the following factors for Horatio in deciding what he should do?

63. Nelson would be very embarrassed if Horatio told on him.
64. The risk to the patient being operated on with unsterile equipment
65. Nelson only very briefly touched the trolley.
66. The inconvenience for all of the surgical staff if they have to bring out a new sterile trolley.
67. They may be asked to leave the theatre.
68. The procedure may not be able to be carried out if the equipment is contaminated.
69. Nelson is scrubbed in a sterile surgical gown.

END OF PAPER

Mock Paper B

Section A: Verbal Reasoning

Passage 1

Rosa Parks

American Rosa Parks, born Rosa Louise McCauley, met her husband at protests fighting for justice for the Scottsboro boys, a group of young men who were falsely accused and convicted of rape because they were African-Americans. Her vested interest in civil rights had led to her marriage, and would lead to her becoming one of the most recognised names in black history. Montgomery, Alabama, USA had passed a city ordinance in 1900 concerning racial segregation on busses. Conductors were vested in the power to assign seats to their passengers, in order to achieve the sectioning off of black people from white people on the vehicle. Though the law stated no-one should have to give up a seat or move when a bus was crowded, in practice the custom evolved that conductors would get African-Americans to vacate their seat when there were no white-only ones left.

On the 1st of December, 1955, Parks got on a bus after a full day of work. She sat at the front of the 'coloured section' on the bus, a section whose size was determined by a movable sign. At one point during the journey, the driver noticed several white people were standing, and moved the sign back a row, telling the seated black passengers to get up. Three others stood from their seats, but Rosa did not: she instead slid along to a seat next to the window. She refused to give up, and was arrested for this refusal. She said that, contrary to popular belief, she did not remain seated because she was physically tired, but 'tired of giving in'. By not giving in, she made history.

1. Bus segregation was not the only concern for a black person in Rosa Parks' lifetime.

A. True.
B. False.
C. Can't tell.

2. Parks's actions were preceded by:

A. Over half a century of racist law demanding black people vacate seats for white people.
B. Conductors treating African-Americans as inferiors.
C. Her previous arrest for her involvement in the Scottsboro boy protests.
D. Years of tiring hard work being answered by unpleasant bus journeys.

3. A movable sign on a Montgomery bus could NOT:

A. Determine how many seats were for white people.
B. Determine how many black people were allowed to sit.
C. Determine how many African-Americans would be made to stand.
D. Distinguish between the ages, genders and circumstances of black or white customers.

4. Rosa Parks was arrested:

A. For refusing to move.
B. For being black.
C. Wrongly.
D. For not conforming to the behaviours of other black passengers.

Passage 2

Urania Cottage

19th Century social reformers like the author Charles Dickens often desired to help prostitutes leave behind their lives of sex work. Angela Burdett-Coutts, a philanthropist of considerable financial means, approached the novelist in 1846 to talk to him about a project to help save sex workers from their profession. The novelist at first tried to dissuade her from this idea, but ended up persuaded himself, and set up a refuge in Urania Cottage, a building in Shepherd's Bush, London. Unlike other establishments for 'fallen women', which were harsher in their treatments of ex-prostitutes, Dickens and Coutts planned a kind approach to the women who sought help from them.

Dickens sent out an advert to gain the attention of sex workers:

"If you have ever wished (I know you must have done so, sometimes) for a chance of rising out of your sad life, and having friends, a quiet home, means of being useful to yourself and others, peace of mind, self-respect, everything you have lost, pray read... attentively... I am going to offer you, not the chance but the certainty of all these blessings, if you will exert yourself to deserve them. And do not think that I write to you as if I felt myself very much above you, or wished to hurt your feelings by reminding you of the situation in which you are placed. God forbid! I mean nothing but kindness to you, and I write as if you were my sister."

5. Charles Dickens:

A. Wrote on the condition of prostitutes.
B. Used prostitutes.
C. Was romantically engaged with Angela Burdett-Coutts.
D. Wanted to improve working conditions for prostitutes.

6. Dickens:

A. Supported Coutts's idea from the start.
B. Initially opposed Coutts's idea.
C. Greatly admired Coutts.
D. Patronised Coutts.

7. The new refuge was to be:

A. Punishing.
B. Neutral.
C. Heavily religious.
D. Sympathetic.

8. According to the advert, Dickens saw sex workers as:

A. Women who have suffered irredeemable loss.
B. Women who had purposely chosen the way of the devil.
C. Women immensely inferior to himself.
D. Women who could achieve lovely things.

Passage 3

Moral Relativism

Despite the good intentions of those who would reform policies, such people must always be aware of their audience and ask if those they would help actually wish to receive this 'help'. Though it is tempting to project personal ideas of happiness and value onto another, it is not always correct. Human beings do not all hold the same desires or ideals, and what is misery for one person could be contentment for another. The phrase 'one man's trash is another man's treasure' may spring to mind.

A good example of this can be found in an 1858 British incident involving a letter in the newspaper, *The Times*. An unnamed 'Unfortunate' had written two columns on being a prostitute that had gained the attention of Angela Burdett-Coutts. Coutts then asked Charles Dickens to find out the writer, so that she may reach out and help this 'unfortunate' herself.

Unfortunately, neither of the well-meaning people had read the two columns in full. By the end of her letter, the prostitute stated she was happy doing what she did, and rather angered by those who would actively try to get rid of her means of earning. She scorned the type of person who would seek to redeem her:

"You the pious, the moral, the respectable, as you call yourselves ... why stand you on your eminence shouting that we should be ashamed of ourselves? What have we to be ashamed of, we who do not know what shame is?"

9. The above passage states:

A. We should all mind our own business.
B. A good deed is its own punishment.
C. Good deeds are not necessarily considered 'good' by all.
D. Other people's views on morality are wrong.

10. When the above passage states 'one man's trash is another man's treasure', they specifically refer to the fact that:

A. Poor people will enjoy things rich people do not.
B. Prostitutes love their profession, whereas non-prostitutes would not.
C. Dickens demands has higher expectations of happiness than sex workers do.
D. There is no universal route to happiness for all humans.

11. According to the passage, the letter accused do-gooders like Coutts of:

A. Prudishness.
B. Making working conditions worse.
C. Depriving people of work.
D. Conservatism.

12. The excerpt from the letter:

A. Personally attacks Coutts and Dickens.
B. Is from the 18th century.
C. Questions reformers' definition of piety.
D. States do-gooders should be ashamed of *their* actions.

Passage 4

Dystopias in Fiction

Literature and popular culture have seen a great range of dystopias. George Orwell's *1984* depicts a surveillance state where every individual is being constantly watched and assessed, where the 'truth' may be destroyed and rewritten by those in the Ministry of Truth and where those who don't toe the line may be subject to horrendous torture. The novel sees children informing on their parents and love being sacrificed to fear. Considering the severity of all of this, it may seem odd that the reality TV show 'Big Brother' took both its name and concept from such a frightening piece of fiction, replicating a house where one can never find privacy. One wonders what Orwell would have made of his work being used in this populist way. Another oddity is the fact that the awful torture chamber of the book, Room 101, has been used as the title of a BBC comedy programme where people discuss their pet hates - not, as in the novel, their greatest fear.

Other dystopian works include the graphic novel *V for Vendetta* written by Alan Moore, *A Clockwork Orange* written by Anthony Burgess and episodes of Charlie Brooker's *Black Mirror* TV series. The literal translation of dystopia is 'not-good-place' and many audiences are very much drawn to nightmarish imaginings of awful places and societies.

13. *V for Vendetta* and *A Clockwork Orange* both describe horrible future realities.

A. True.
B. False.
C. Can't tell.

14. George Orwell would be pleased with the way *1984* has inspired TV.

A. True.
B. False.
C. Can't tell.

15. Which of the following is NOT described as occurring within *1984*:

A. Constant camera surveillance.
B. Honesty being undermined.
C. Children turned against parents.
D. Love being destroyed.

16. Which of the following is true according to the passage:

A. *Black Mirror* voices the same concerns previously brought up in *A Clockwork Orange* and *1984*.
B. Audiences universally love dystopias.
C. Dystopias help us feel good about things as they are.
D. Horrible fictional realities do not necessarily repel people.

Passage 5

Oscar Wilde on Art

Oscar Wilde wrote the following as a preface to the only novel he wrote, *The Picture of Dorian Grey*:

'The artist is the creator of beautiful things. To reveal art and conceal the artist is art's aim. The critic is he who can translate into another manner or a new material his impression of beautiful things.

The highest as the lowest form of criticism is a mode of autobiography. Those who find ugly meanings in beautiful things are corrupt without being charming. This is a fault.

Those who find beautiful meanings in beautiful things are the cultivated. For these there is hope. They are the elect to whom beautiful things mean only beauty.

There is no such thing as a moral or an immoral book. Books are well written, or badly written. That is all.

The nineteenth century dislike of realism is the rage of Caliban seeing his own face in a glass.

The nineteenth century dislike of romanticism is the rage of Caliban not seeing his own face in a glass...No artist desires to prove anything. Even things that are true can be proved. No artist has ethical sympathies. An ethical sympathy in an artist is an unpardonable mannerism of style....

All art is quite useless.'

17. The above excerpt is cited as being written by:

A. A playwright.
B. A writer of fiction.
C. A Victorian-era man.
D. A comedian.

18. According to Wilde, the critic:

A. Is a beast.
B. Should find beautiful meanings in ugly things.
C. Should avoid all things that are not beautiful.
D. Can be corrupt or cultivated.

19. Beautifully written books:

A. Are immoral.
B. Are moral.
C. Are the highest form of art.
D. Are simply beautifully written.

20. The book to follow this preface, according to the writer himself:

A. Is well written.
B. Has no purpose.
C. Will have no impression of the creator.
D. Will be translated into further beauty by critics.

Passage 6

G.K. Chesterton on Eugenics

The following is from G.K. Chesterton's 1922 'Eugenics and Other Evils':

'It is not really difficult to sum up the essence of Eugenics: though some of the Eugenists seem to be rather vague about it. The movement consists of two parts: a moral basis, which is common to all, and a scheme of social application which varies a good deal. For the moral basis, it is obvious that man's ethical responsibility varies with his knowledge of consequences. If I were in charge of a baby (like Dr. Johnson in that tower of vision), and if the baby was ill through having eaten the soap, I might possibly send for a doctor. I might be calling him away from much more serious cases, from the bedsides of babies whose diet had been far more deadly; but I should be justified. I could not be expected to know enough about his other patients to be obliged (or even entitled) to sacrifice to them the baby for whom I was primarily and directly responsible. Now the Eugenic moral basis is this; that the baby for whom we are primarily and directly responsible is the babe unborn. That is, that we know (or may come to know) enough of certain inevitable tendencies in biology to consider the fruit of some contemplated union in that direct and clear light of conscience which we can now only fix on the other partner in that union. The one duty can conceivably be as definite as or more definite than the other. The baby that does not exist can be considered even before the wife who does. Now it is essential to grasp that this is a comparatively new note in morality.'

21. G.K. Chesterton accuses Eugenicists of being:

A.   Too blunt.
B.   Too pithy.
C.   Imprecise in their definition of their belief.
D.   Crude in their definition of anti-Eugenicists.

22. Chesterton acknowledges one might criticise the guardian of the soap-eating baby for:

A.   Allowing the baby to consume soap.
B.   Not treating the baby himself.
C.   Having a baby with no self-preservation instincts.
D.   Taking up a doctor's valuable time for a case that may not be as serious as others.

23. 'The babe unborn' can be contemplated because:

A.   Of the inherent value of human life.
B.   Our ability to predict the products of a union through science.
C.   Babies are essentially just like their mothers.
D.   Babies are all the same.

24. The 'new note in morality' described in the passage is:

A.   A belief that the living babies of others are more important than yours.
B.   A belief that we should use biology to stop births of undesirable babies.
C.   That non-existent babies can be more important than existing women.
D.   People should have as many babies as possible.

Passage 7

Asylums in the 19th Century

The following is an extract from a column from Fanny Fern, describing a visit to an American insane asylum in the 19th century:

'It is a very curious sight, these lunatics – men and women, preparing food in the perfectly-arranged kitchen. One's first thought, to be sure, is some possibly noxious ingredient that might be cunningly mixed in the viands; but further observation showed the impossibility of this under the rigid surveillance exercised. As to the pies, and meats, and vegetables, in process of preparation, they looked sufficiently tempting to those who had earned a good appetite like ourselves, by a walk across the fields. Some lunatic-women who were employed in the laundry, eyed me as I stood watching them, and, glancing at the embroidery on the hem of my skirt, a little the worse for the wet and dust of the road, exclaimed, "Oh, fie! A soiled skirt!" In fact, I almost began to doubt whether our guide was not humbugging us as to the real state of these people's intellects; particularly as some of them employed in the grounds, as we went out, took off their hats, and smiled and bowed to us in the most approved manner.'

25. It is curious for the author to see men in a kitchen.

A. True.
B. False.
C. Can't tell.

26. The writer wonders whether the inmates of the asylum:

A. May burn themselves working in a kitchen.
B. Should be kept away from sharp objects like knives.
C. Would be distressed by food preparation.
D. May poison the asylum's food.

27. The laundry-women demonstrate they are women of their trade through:

A. Busily washing the sheets.
B. Cleaning the writer's skirt.
C. Commenting on the dirtiness of clothing.
D. The fine state of everyone's apparel in the asylum.

28. The writer doubts that the inmates are insane, specifically as:

A. They all demonstrate good intelligence.
B. Some demonstrate good courtesy.
C. They are not violent.
D. They speak good English.

Passage 8

Self-immolation

The following is an excerpt from a Victorian era newspaper column:

I RECENTLY witnessed one of the most extraordinary and horrid scenes ever performed by a human being – namely, the self-immolation of a woman on the funeral pile of her husband. The dreadful sacrifice has made an impression on my mind, that years will not efface...

Yesterday morning, at seven o'clock, this woman was brought in a palanquin to the place of sacrifice. It is on the banks of the Ganges, only two miles from Calcutta. Her husband had been previously brought to the river to expire. His disorder was hydrophobia – think of the agony this must have occasioned him. He had been dead for twenty-four hours, and no person could prevail on the wife to save herself. She had three children, whom she committed to the care of her mother. A woman, called to be undertaker, was preparing the pile. It was composed of bamboo, firewood, oils, resin, and a kind of flax, altogether very combustible. It was elevated above the ground, I should say twenty inches, and supported by strong stakes. The dead body was lying on a rude couch, very near, covered with a white cloth. The eldest child, a boy of seven years, who was to light the pile, was standing near the corpse. The woman sat perfectly unmoved during all the preparation, apparently at prayer, and counting a string of beads which she held in her hand. She was just thirty years old; her husband twenty-seven years older.

29. Which of the following terms could NOT be used to correctly describe the Indian wife in the passage:

A. A daughter.
B. A mother.
C. A wife.
D. Alive.

30. The agony of the husband is due to:

A. The painfulness of his condition.
B. The knowledge of his imminent death.
C. The fear he held of water.
D. The knowledge of his wife's imminent sacrifice.

31. Which of the following is the seven year old boy NOT described to do:

A. Enable his mother's suicide.
B. Burn his father's body.
C. Become an orphan.
D. Become the support of his younger siblings.

32. Which of the following is definitely true:

A. The wife was in prayer during the preparation of the pile.
B. The wife was immobilised during the preparation of the pile.
C. The wife was three decades the junior of her husband.
D. The wife chose to leave her children.

Passage 9

Advice for Boys

The following is an excerpt from a Victorian era newspaper:

A WORD TO BOYS. – Who is respected? It is the boy who conducts himself well, who is honest, diligent, and obedient in all things. It is the boy who is making an effort continually to respect his father, and obey him in whatever he may direct to be done. It is the boy who is kind to other little boys, who respects age, and who never gets into difficulties and quarrels with his companions. It is the boy who leaves no effort untried to improve himself in knowledge and wisdom every day – who is busy and attentive in trying to do good acts towards others. Show us a boy who obeys his parents, who is diligent, who has respect for age, who always has a friendly disposition, and who applies himself diligently to get wisdom and to do good towards others, and if he is not respected and beloved by everybody, then there is no such thing as truth in this world. Remember this, boys, and you will be respected by others, and will grow up and become useful men.

33. A boy must be and do several things before he can be 'respected'.

A. True.
B. False.
C. Can't tell.

34. Girls, unlike boys, cannot hope to be 'respected' for qualities like honesty, diligence and obedience.

A. True.
B. False.
C. Can't tell.

35. A boy should avoid unnecessary confrontation.

A. True.
B. False.
C. Can't tell.

36. Respected boys will one day be valuable men.

A. True.
B. False.
C. Can't tell.

Passage 10

Renaissance Revenge Tragedy

Renaissance Revenge tragedies were very popular dramas, often filled with sexuality and murder. Shakespeare himself wrote two revenge tragedies, *Titus Andronicus* (set in ancient Roman times) and *Hamlet*, and though he is often considered the greatest English playwright (if not, writer) that ever lived, he has a sense of wickedness that thrilled through his dark plays. For example, Titus Adronicus involves an act of familial cannibalism, as Tamora's two sons (who had raped and mutilated Titus' daughter) are killed and baked into a pie, to be served to their mother (who had committed a series of deeds to destroy the family and happiness of Titus). Other such tragedies involve unusual scenes, like a Duke being killed by a poisoned skull he believes to be a living woman (*The Revenger's Tragedy*) or a man biting off his own tongue so that he will not reveal secrets when he is tortured (*The Spanish Tragedy*).

The avengers in revenge tragedies often die themselves: having killed a wrong-doer, they become murderers themselves, and often have lost their previous moral high-ground. Morality dictates that they must be served with a bloody end. Though the plays do not protect innocents, who may be raped or murdered, the genre does not allow its killers to go unpunished - even when the killers are also the protagonists.

37. Shakespeare was not above writing in a populist genre.

A. True.
B. False.
C. Can't tell.

38. Only Shakespeare named his revenge tragedies after characters in the play.

A. True.
B. False.
C. Can't tell.

39. Revenge tragedies are NOT described as involving:

A. A sexually abused woman.
B. Deadly bones.
C. Self-mutilation.
D. Suicide.

40. Revenge tragedies are described as:

A. All set in European places like Italy and Spain.
B. Poems and plays involving tales of avengers.
C. Following an ethical code.
D. Obsessed with cannibals.

Passage 11

Cats have been domesticated for thousands and thousands of years, but as yet there has been no conclusive explanation for what started off this trend of making felines pets.

One theory for the origins of cat domestication lies in the animal's ability to kill rodents. Washington University's Wesley Warren suggests that originally people would welcome the predators, as they would destroy pests that would otherwise eat up grain harvests. The human may then have offered, as a reward, food, to make sure the cat would stay.

One can see in this a possible beginning in human's affectionate relationship with felines: having done such a good job in controlling rodents, an owner would be pleased with their pet and perhaps even grateful. It may well be this very practical use of the creature that led to our modern culture of keeping the animal as a pampered pet.

To this day, domestic cats, with their simple gut suited for raw meat, are mostly carnivorous. Their bodies have another feature suited to this diet of flesh: a rough tongue. The tongue's textured surface helps the cat take every last morsel of meat from the bone of some other animal.

Of course, even if the cat's owner attempts to limit their intake of meat, a wily feline can still hunt for their own prey, perhaps outside the home, even if they are 'domesticated'.

41. There is no longer such a thing as a wild cat.

A. True
B. False
C. Can't tell

42. Which of the following statements is NOT supported by the passage:

A. Cats may have been traditionally used as hunters.
B. Cats are currently used for their killing ability.
C. Cats are anatomically suited for a diet of meat.
D. Cats may not exclusively rely on their owners for food.

43. Which of the following statements is supported by the passage:

A. Wesley Warren is a student of Washington University.
B. Wesley Warren is an acclaimed professor of Washington University.
C. Wesley Warren is affiliated with an academic institution.
D. Wesley Warren is a man.

44. According to the passage, humans:

A. All love cats.
B. Love only those that do something for them in return.
C. Have as a race conducted relationships with another species for millennia.
D. All hate rats.

END OF SECTION

Section B: Decision Making

1. "All chemistry exams are written. All written exams are flammable. Some written exams are double sided."
    If the above is true which of the below statements are also true?
    >   I.   Some chemistry papers are double sided.
    >   II.   All flammable exams are chemistry exams
    >   III.   All double sided exams are flammable

    A. I only
    B. II only
    C. III only
    D. II and III
    E. All of the above

2. The Government is starting a new health initiative to sponsor athletic shoes, in the hopes of improving the fitness of the public by encouraging more walking. Outdoor activities are cheaper than gym memberships. What can be concluded from this?

    A. Athletics is the best form of exercise
    B. Outdoor training is better than indoor training
    C. Gym memberships are a poor investment
    D. New shoes will mean fewer people go to the gym
    E. Walking is a good form of exercise

3. There are 30 medical students in a classroom, discussing their subjects. Everyone states a preference for at least one subject. 6 medical students like anatomy only. 14 medical students like pharmacology, and 19 like biochemistry. 5 like all three. 3 hate pharmacology but like the others. No one enjoys pharmacology and anatomy. How many medical students like pharmacology and biochemistry.

    A. 7   B. 4   C. 0   D. 6   E. 3

4. "Medical knowledge and supplies are urgently needed around the world. People in developing countries need better medical care." Which statement supports these?

    A. The majority of people in the world have never seen a doctor
    B. People in developing countries are not getting minimum medical aid
    C. Doctors are selfish and only care about money
    D. Insurance companies have increased their prices
    E. Developed countries have a lot of medical supplies

5. Which diagrams shows the relationship between women, mothers and doctors?

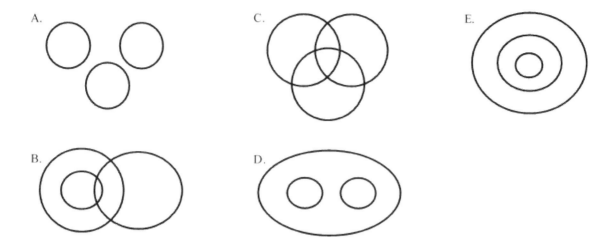

A.

C.

E.

B.

D.

6. A survey is down on pets owned in a town.

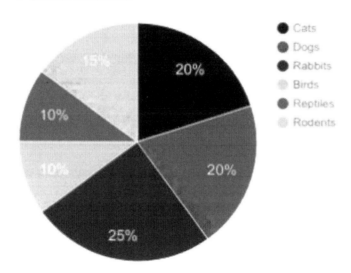

Points scored

- Cats
- Dogs
- Rabbits
- Birds
- Reptiles
- Rodents

20%
20%
25%
10%
10%
15%

Which statement is correct?

A. There are 25% fewer rodents than cats.
B. It is more expensive to feed dogs than birds.
C. Only 65% of pets have four legs
D. Reptiles and rodents must be sold together
E. Rabbits are less popular than dogs.

7. The chart shows the profits between three competing shops: Jake's, Lauren's and Nathan's, over the course of 5 years.

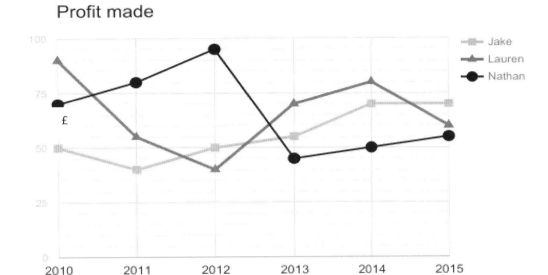

Which statement is correct?

A. The greatest overall loss of profits was 2012 to 2013.
B. Jake has the greatest percentage change in profit at the end of 5 years.
C. Lauren made more than Nathan for 2 years.
D. Everyone made the same amount of money from 2010 to 2012

8. Scores are taken from a recent exam period.

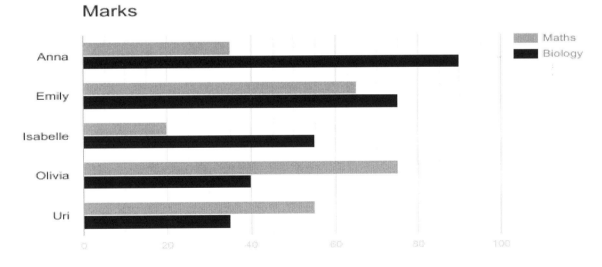

Which statement is correct?

A. Anna beats Olivia in biology by double the number of marks that Oliva beat Anna in Maths
B. Maths was a harder exam than Biology
C. Emily has the highest average score
D. Isabelle and Uri combined have the highest Biology score
E. It is an all girls' school.

9. Millie's Marvellous Cars is reviewing their yearly sales and profit.

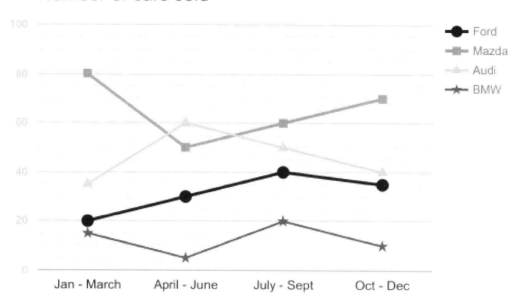

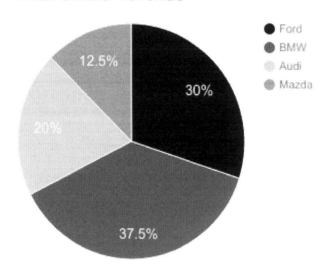

Which statement is true?

A. Millie did not make money selling Audi cars
B. All the cars increased in value throughout the year
C. BMW and Ford sold fewer cars in the second half of the year than Audi did in the first half
D. Mazda cars are cheaper than BMWs

10. The table shows the proportion of female students and swimmers in a school. There are 200 students in Y10, Y11 and Sixth Form. 80% of students are in Sixth form and there are an equal number of Y10 and Y11 students.

	Female	Swimmers
Year 10	0.3	0.65
Year 11	0.7	0.5
Sixth form	0.45	0.3
Total	0.46	0.355

Which statement is true?

A. The percentage of swimmers in the school that are in Y11 is 65%
B. The total number of swimmers in the school is 82
C. There are the same number of male students in the Sixth Form as there are in Y11.
D. You are more likely to be a male swimmer than a female non swimmer.
E. There are twelve times as many girls in Sixth form as in Y10.

11. Queenie, Rose, and Susanna are playing a game where the loser gives half her money to the other players, who share it equally. They play three games and each loses a game, in alphabetical order. At the end of last game, each girl has 32p. If Queenie started with the least money, who had the most money?

A. They all started with the same amount
B. Rose
C. Susanna
D. Rose and Susana started with the same amount

12. Communism is the only truly just social system where everybody is equal.

Which of the following statements **most strongly** argues against this point?

A. Communism must fail because humans inherently strive for personal gain.
B. Communism has so far always failed to produce functioning state systems.
C. Communist societies always fail economically.
D. Capitalism is better than Communism.

13. In a shooting competition, an individual is awarded 5 points for hitting a target. Missing a target results in no points being added, but there is no penalty for missing. Each shooter fires 5 shots at the target. 5 Shooters take part in the competition.

Pat achieves 15 points.
Lauren misses 4 targets, but does not finish last.
Peter hits 4 targets.
Dave achieves 5 more points than Lauren.

Which of the following statements is correct?

A. Joe finishes last.
B. Dave finishes second
C. Pat is a poor shot.
D. Dave shoots better than Pat.

14. Mr Smith is a surgeon. One day, Mr Smith and his wife decide to sort through his wardrobe. He has a total of 20 shirts, of which ¼ striped. Mrs Smith doesn't like Mr Smith's pink shirt. Mr Smith has 3 blue shirts. He only wears his white shirts with black trousers.

Select if the following statements are true or false from the information provided.

   A. Mrs Smith is a surgeon.
   B. Mr Smith has black shirts.
   C. Mr Smith only wears black trousers with white shirts
   D. Mr Smith owns white shirts.

15. Explorers in the US in the 18th Century had to contend with a great variety of obstacles ranging from natural to man-made. Natural obstacles included the very nature and set up of the land, presenting explorers with the sheer size of the land mass, the lack of reliable mapping as well as the lack of paths and bridges. On a human level, challenges included the threat from outlaws and other hostile groups. Due to the nature of the settling situation, availability of medical assistance was sketchy and there was a constant threat of diseases and fatal results of injuries.

Which of the following statements is correct with regards to the above text?

   A. Medical supply was good in the US in the 18th Century.
   B. The land was easy to navigate.
   C. There were few outlaws threatening the individual.
   D. Crossing rivers could be difficult.

16. People who practice extreme sports should have to buy private health insurance.

Which of the following statements most strongly supports this argument?

   A. Exercise is healthy and private insurance offers better reward schemes.
   B. Extreme spots have a higher likelihood of injury.
   C. Healthcare should be free for all.
   D. People that practice extreme sports are more likely to be wealthy.

17. A group of scientists investigates the role of different nutrients after exercise. They set up two groups of averagely fit individuals consisting of the same number of both males and females aged 20 – 25 and weighing between 70 and 85 kilos. Each group will conduct the same 1hr exercise routine of resistance training, consisting of various weighted movements. After the workout they will receive a shake with vanilla flavour that has identical consistency and colour in all cases. Group A will receive a shake containing 50 g of protein and 50 g of carbohydrates. Group B will receive a shake containing 100 g of protein and 50 g of carbohydrates. All participants have their lean body mass measured before starting the experiment.

Which of the following statements is correct?

   A. The experiment compares the response of men and women to endurance training.
   B. The experiment is flawed as it does not take into consideration that men and women respond differently to exercise.
   C. The experiment does not consider age.
   D. The experiment mainly looks at the role of protein after exercise.

18. When considering the reproductive behaviour of animals, different factors have to be considered. On one hand, there are social structures within the local animal population that play a role. In herd animals for example, it is common for only the lead male or female to reproduce in order to ensure that offspring have optimal genetic material. In addition to that, different animals reproduce during different times of the year in order to optimise survival chances for their offspring based on different gestation periods. Animals with longer periods of gestation, for example, are more likely to reproduce in the spring so that the young are born in late summer, allowing them to grow during the fall months in order to prepare for the winter months. In areas where there are less marked differences between winter and summer months, this behaviour is less relevant.

Which of the following statements is correct?

A. In herd animals lead animals are always the only individuals reproducing.
B. Animal reproductive behaviour is complex.
C. Animals with longer gestation periods are more likely to deliver in spring in colder climates.
D. All of the above.

19. A restaurant re-prints their menus and decides to display their range of beers according to alcohol content.

The Monk's Brew is brewed in Germany and has an alcohol content of 4.5%.
Fisher's Ale has the least alcohol content at 3.8% and is brewed using triple filtered spring water.
Knight's Brew has more alcohol than Fisher's, but less than Monk's.
Ferrier Ale has 5% alcohol, but is not the strongest beer.

Which of the following is true?

A. Brewer's Choice is the strongest beer.
B. German beer is the weakest.
C. Spring water makes for particularly strong beers.
D. Fischer's Ale is not the weakest beer.

20. Jack sets up a vegetable garden. He starts out with 10 potato plants. For every potato plant, he plants twice as many carrots. He also plants twice as many tomatoes as he plants carrots. In the end, he also plants 10 heads of salad.

Which of the following is true?

A. There are a total of 70 plants
B. He plants more potatoes than heads of salad.
C. There must be 20 tomato plants.
D. There are four times as many tomatoes than heads of salad.

21. A group of scientists investigate the prevalence of allergies in the population. In order to do that, they use medical files of those treated for allergies or diagnosed with allergies. They find that over the last 30 years the overall prevalence of allergies has increased. This is particularly the case for food intolerances and allergies to inhaled substances such as dust and pollen, which represent the majority of all allergies documented. The scientists also find that there is a proportionally larger increase in younger individuals being diagnosed with allergies than there was 30 years ago. The scientists find that 30% of diagnoses are being made within the first 5 years of live, 20% of diagnoses are being made in the second 5 years of life and the remaining 50% are being made between age 10 and 50.

Which of the following statements is true?

A. Contact allergies are very common.
B. The majority of diagnoses are being made between 6 and 10 years.
C. The scientists used unreliable data.
D. Allergy prevalence in the young has increased over the past 3 decades.

22. Rowing is a complex sport that requires a variety of different skills in order to perform well. On one hand, a good rower must possess physical strength in order to move the heavy boats as well as his own body effectively through the water. The more power a rower can exert on the oars, the more his boat will accelerate. Due to the length of rowing races of 2000m, rowers also need a high degree of cardiovascular fitness and muscular endurance in order to be able to perform well for the duration of their race. The final aspect of rowing is not a physical one but a mental one, requiring the individual to perform complex motor patterns with technical perfection in order to reduce opposing forces on the motion of his boat and in order to maintain constant and even acceleration.

Which of the following is true?

A. Rowing has little technical skill involved.
B. Physics plays no role in rowing.
C. Rowing requires a combination of physical and mental abilities.
D. Being strong is the most important aspect of rowing.

23. A group of scientists conduct a study into the density of fast food restaurants in different income neighbourhoods and relate this to the incidence of obesity. After collecting their data they find that the density of fast food restaurants in low-income neighbourhoods is higher than in high-income neighbourhoods. They also find that there are comparatively more fast food restaurants in the vicinity of schools than there is in the vicinity of housing areas. With regards to obesity, the study finds that with increasing density of fast food restaurants, the incidence of obesity increases.

Which of the following statements is true?

A. The incidence of obesity is higher in low-income neighbourhoods.
B. The most fast food restaurants exist around housing areas.
C. Places that serve fast food can hardly be called restaurants.
D. In order to tackle the obesity crisis, fast food needs to be more expensive.

24. Jason is unsure of what to get his little sister for her birthday, so he conducts a poll amongst his friends.

5 friends think that new ear rings are a good idea.
Out those that think earrings are a good idea, 3 also support a necklace and 2 also support shoes.
2 other friends also think that shoes are a good idea, but one of those also thinks a handbag is a good idea.
3 friends only support the idea of a dress.

Which of the following most accurately describes the distribution of opinion?

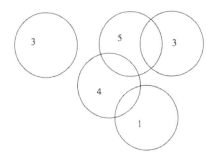

A.

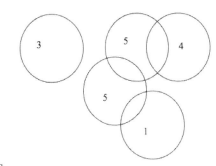

C.

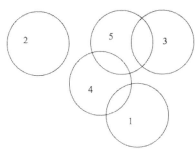

B.

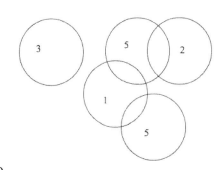

D.

25. Parents with high incomes should pay more school fees so the extra money can be used to subsidise the fees of poorer children.

Which of the following arguments supports this statement the most?

A.   By law, everybody has to go to school.
B.   Children from poorer backgrounds are more likely to undergo poorer education due to financial constraints.
C.   Wealthy families have a social responsibility to support others.
D.   It is not justifiable to charge wealthier families for being more successful.

26. Antibiotic resistance is an increasingly important problem for modern medicine. There are several key mechanisms with which bacteria aim to evade the action of antibiotics including the mutation of target structures or the production of lytic enzymes that destroy antibiotic molecules before they can act on the bacteria. This represents an important resistance mechanism. It is also common for bacteria to develop mutations in important transporter molecules that reduce the entry of antibiotic molecules into the cell or increase their extrusion from the cell. The worst case scenario is that one lineage of bacteria has developed multiple mechanisms of resistance that allow it to counteract the action of many different types of antibiotics.

Which of the following is correct?

A. Antibiotic resistance is of little relevance today.
B. Bacteria rarely develop lytic enzymes that destroy antibiotic molecules.
C. Bacteria have no way to reduce uptake of antibiotic molecules.
D. Antibiotic resistance can be caused by more than one mechanism simultaneously.

27. Stan collects stamps. He has stamps from Europe. He also has a half a dozen from Asia, and 3 from Canada. Select whether each statement is true or false.
A. Stan has no stamps from North America.
B. Stan may have stamps from Germany.
C. There are fewer stamps from Asia in Stan's collection than from Canada.
D. Stan may have stamps from France.

28. A survey asks a group of universities are asked for what they consider the most important quality in students applying for medicine.

> Dependability receives 28% of the votes.
> Intelligence receives 10% less of the votes.
> Communication skills receives 5% of the votes.
> Mental flexibility receives 10% more than communication skills.

Communication skills				

Which of the following is correct?

A. Intelligence is less important than mental flexibility.
B. Mental flexibility is the most important.
C. Dependability is not the most important skill in a prospective medical student.
D. Empathy receives less than 30% of votes.

29. Sugar should be taxed like alcohol and cigarettes.

Which of the following arguments most supports this claim?

A. Sugar can cause diabetes.
B. Sugar has a high addictive potential and is associated with various health concerns.
C. High sugar diets increase obesity.
D. People that eat a lot of sugar are more likely to start abusing alcohol.

END OF SECTION

Section C: Quantitative Reasoning

SET 1

Anne is looking to buy food from a takeaway. A sample of the menu is shown:

Item	Cost	Item	Cost
Plain Rice	£1.55	Chicken Satay	£4.30
Egg Fried Rice	£2.00	Kung Po Chicken	£3.40
Special Fried Rice	£5.00	Sichuan Pork	£5.60
Special Curry	£4.30	Roast Pork	£6.00
Chicken Curry	£4.00	Prawn Chow Mein	£5.50

1. Anne wishes to purchase 2 portions of egg fried rice, 1 special fried rice, a special curry, roast pork and a portion of Kung Po Chicken. What proportion of her bill comes from rice (to 2 s.f)?

A. 1:2.3   B. 1:2.5   C. 1:3.2   D. 1:4.2

2. The shop offers a set menu whereby someone can buy 2 portions of rice (either plain or egg fried) and any two different main meals for £10. Suppose Hannah buys the most expensive options. What percentage of the bill does she save?

A. 32%   B. 36%   C. 40%   D. 44%

3. The shop charges £2.50 delivery on orders under £25 and 10% of the cost of the order otherwise. If Anne buys two set menus consisting of Special Curry and Chicken Curry, each with Egg Fried Rice, how much has she saved on delivery relative to if she bought the items separately?

A. £0.00   B. £0.30   C. £0.60   D. £1.00

4. The shop wishes to offer a new deal, when a customer spends more than £40, they get free delivery. Anne is planning to spend £39.40 and wants delivery. What percentage of her bill does she save by purchasing an extra portion of egg fried rice?

A. 2.3%   B. 4.5%   C. 6.8%   D. 9.2%

SET 2

Bob is looking to insure his car. He gets the following three quotes:

Company	Cost
Red Flag	£25 a month, 10% discount for 5+ years with no claims, 2 free months a year for 10+ years with no claims (the final two deals accumulate together).
Chamberlain	£350 a year, 3% discount for each year with no claims.
Meerkat Market	£300 a year, but for every year with no claims £10 is saved.
Munich	£250 a year, no discounts.

5. Bob has 4 years with no claims. Which is the cheapest company for him to go with?

A. Red Flag   B. Chamberlain   C. Meerkat Market   D. Munich

6. What percentage of the Red Flag normal cost is saved by a driver with 11 years without claims?

A. 12.5%   B. 25.0%   C. 37.5%   D. 40.0%

7. Laura is insured with Chamberlain. She has 3 years no claims discount and is thinking about switching to Meerkat Market. How much money would she save by switching?

A. £58.50   B. £62.00   C. £68.50   D. £72.00

8. Munich decides to introduce a system whereby if two people buy a policy from them, they save 10%. Suppose that Delia and Elliot have 3 and 5 year no claims discounts respectively. Delia buys her insurance from Meerkat market and Elliot buys his from Chamberlain. Which of the following proportions gives the total new cost:old cost?

A. 0.79:1   B. 0.88:1   C. 1:1.14   D. 1:1.32

SET 3

The figure gives details of the number of students in a school who play any of the four games – cricket, basketball, football and hockey.

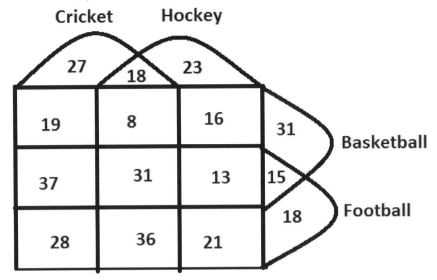

The total number of students in the school is 500.

9. How many of the students play at most one game?

A. 99   B. 159   C. 258   D. 341

10. What is the number of students who play either cricket or basketball, but not football?

A. 92   B. 119   C. 126   D. 138

11. How many students play at least three games?

A. 75   B. 125   C. 175   D. 225

12. The number of students playing at most one game exceeds those playing at least two games by

A. 8   B. 16   C. 18   D. 36

13. How many students do not play cricket, football or hockey?

A. 160   B. 180   C. 190   D. 200

14. By what number did the number of students playing none of the games **outnumber** the number of students playing exactly one game?

A. 20   B. 30   C. 60   D. 90

SET 4

The table below shows the daily number of passengers (in thousands ('000)) using trains as a means of transportation from 1998–2005.

Year	Number (in '000)
1998	200
1999	150
2000	300

For the year 2000, the following table shows the daily break-down of the proportion of passengers using different transportation modes:

Mode of Transport	Percentage
Taxis	8.33
Trains	41.67
Buses	33.33
Private Cars	16.67
Total	**100**

15. In 2000, how many people used a private car as a daily means of transportation?

A. 120,000   B. 180,000   C. 280,000   D. 320,000

16. If the same proportional break-down in 2000 is also applicable for 1999, the number of persons using taxis as a daily means of transport in 1999 is closest to:

A. 20,000   B. 30,000   C. 40,000   D. 50,000

17. If in 1998, one quarter of the total passengers travelled by train, what is the percentage change in the total number of daily passengers from 1998 to 2000?

A. 10% increase   B. 10% decrease   C. 20% increase   D. 20% decrease

18. From 1998 to 2000, the number of passengers using taxis as a means of transportation increased by 50%. What is the percentage of passengers using daily taxis in 1998, given that the total population using public transport was 0.8 million in that year?

A. 5%   B. 10%   C. 15%   D. 20%

19. If the ratio of number of people using buses as a means of transport in 1998, 1999 and 2000 is 8:5:6 and the bus fares in these three years are in the ratio of 2:3:4, then find the total amount collected in 1998 and 1999 together, given that fare in 2000 was £2 per person and the buses operate for 358 days a year.

A. £22,196,000   B. £211,360,000   C. £221,960,000   D. £235,020,000

20. If the same break-down as for 2000 is also applicable for 1999, then the number of persons using buses and private cars combined as a means of transport in 1999 is closest to:

A. 140,000   B. 160,000   C. 180,000   D. 200,000

SET 5

The following are the details of different steps involved in opening a fully functional computer centre.

S. No.	Work	Work Code	Duration (Weeks)	Other works that should be completed before the work
1.	Planning by architect	A	2	-
2.	First painting of roof	B	1	-
3.	First painting of walls	C	1	B
4.	Partitioning the workspace	D	2	A, B
5.	Networking	E	4	A
6.	Air-conditioning	F	6	D
7.	Electrical work	G	3	B
8.	False ceiling	H	4	E, F, J
9.	False flooring	I	1	L
10.	Fire safety systems	J	1	G
11.	Final painting of ceiling	K	2	H
12.	Final painting of walls	L	2	K, C

Note: In any week one or more activities are in progress.

21. The earliest time by which the false ceiling work can start is:

A. 8th week   B. 9th week   C. 10th week   D. 11th week

22. If the work is to be done at the earliest, then the latest by which the fire safety systems work can start is:

A. 8th week   B. 9th week   C. 10th week   D. 11th week

23. If not more than one activity can be undertaken at any given time, then what is the minimum possible time in which all of the above-mentioned works can be completed?

A. 25 weeks   B. 29 weeks   C. 31 weeks   D. 33 weeks

24. What can be the maximum time gap between the start of the air-conditioning work and the start of the final painting of walls, if no time is wasted in between?

A. 17 weeks   B. 19 weeks   C. 21 weeks   D. 23 weeks

25. If the work is to be done at the earliest, then the latest the air-conditioning work can start is:

A. 5th week   B. 6th week   C. 7th week   D. 8th week

SET 6

The production of 5 food grains is listed in the following table:

Food Grain	Year 1989–90	Year 1999–2000
A (Rice)	75	90
B (Wheat)	50	70
C (Coarse Cereals)	35	30
D (Total Cereals)	160	190
E (Pulses)	15	13

Note:

Food Grains = Total Cereals + Pulses

Total Cereals= Total of A, B and C

26. The production target for food grains from 1999-2000 was 20% more than that of 1989-90, but this target was missed. What was the percentage deficit of production in 1999-2000 relative to the target production?

A. 1.3%   B. 2.3%   C. 3.3%   D. 4.3%

27. The percentage decrease in the production of pulses from 1989-90 to 1999-2000 is equal to the percentage increase in the price of the pulses during the same period. If the average price of pulses is £1.80 per kg in 1989-90, what is the price in 1999-2000?

A. £2.00   B. £3.00   C. £4.00   D. £5.00

28. What is the percentage of total cereal production in 1999-2000 relative to total cereal production in 1989-90?

A. 112.5%   B. 118.75%   C. 143.95%   D. 145.25%

SET 7

Four countries (USA, Kenya, Russia and Australia) participated in a 4 × 400 metre relay race. This is a race around a 400m athletic track with each team comprising four runners, each of whom completes 1 complete lap of 400 metres before handing a baton over to the next runner, who starts at the moment the previous runner finishes their lap. The winning team is the team whose final runner finishes their lap first.

The following bar graph gives the average speeds at which runners from the four teams ran to complete their laps.

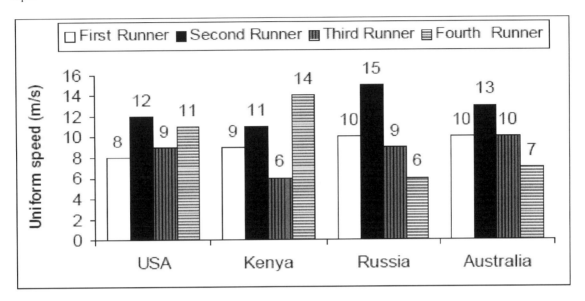

29. Which country won the event?

A. USA   B. Australia   C. Russia   D. Kenya

30. Which country finished last in the event?

A. USA   B. Australia   C. Russia   D. Kenya

31. Which of the following statements is <u>definitely</u> true?

A. A runner from Australia was never behind the corresponding numbered runner from USA.
B. The third runners from Russia and Kenya never met during the course of the event.
C. The second runner from Kenya and Russia definitely met once during the event.
D. A runner from Kenya was always behind the corresponding numbered runner from Russia.

32. Taking all four countries into account, which runners were the slowest on average?

A. First Runners   B. Second Runners   C. Third Runners   D. Fourth Runners

SET 8

The table below shows data for export and import for Simuland, a hypothetical country.

Year	Exports (Million USD)	Imports (Million USD)
1996	100	90
1997	120	130
1998	130	110
1999	140	180
2000	190	220
2001	160	150

Trade Surplus = Exports – Imports   **Trade Deficit = Imports – Exports**

33. During the year 1999, the average cost of exports was USD 7000 per tonne and that of imports was USD 6000 per tonne. By what percentage is the total weight of exports less than the total weight of imports in 1999?

A. 30.33%   B. 31.33%   C. 32.33%   D. 33.33%

34. The percentage decrease in trade surplus from 2001 to 2002 is the same as during the period 1998 to 2001 overall. Given that imports in 2002 increased by 20%, what is the value of exports in 2002 in Million USD?

A. 175   B. 180   C. 185   D. 190

35. If imports grow at a rate of 10% per year from 2001 onwards, what is the approximate total import value in 2004?

A. 100 Million USD
B. 150 Million USD

C. 200 Million USD
D. 250 Million USD

36. What is the approximate compound annual growth rate (CAGR) for exports during 1996 to 2001?

A. 5%   B. 10%   C. 15%   D. 60%

END OF SECTION

Section D: Abstract Reasoning

For each question, decide whether each test shape fits best with Set A, Set B or with neither.

For each question, work through the test shapes from left to right as you see them on the page. Make your decision and fill it into the answer sheet.

Answer as follows:

A = Set A

B = Set B

C = Neither

Set 1:

Set A

Set B

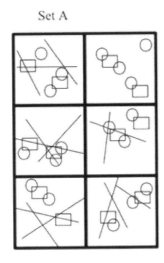
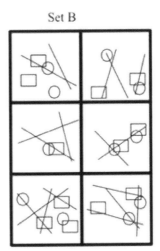

Questions 1- 5

Set 2:

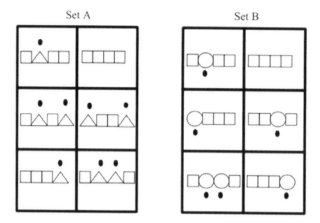

Set A   Set B

Questions 6 - 10

Set 3:

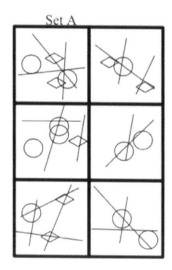 

Set A   Set B

Questions 11 - 15

Set 4

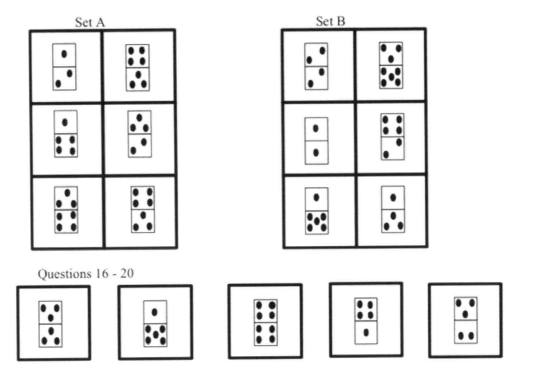

Questions 16 - 20

Set 5

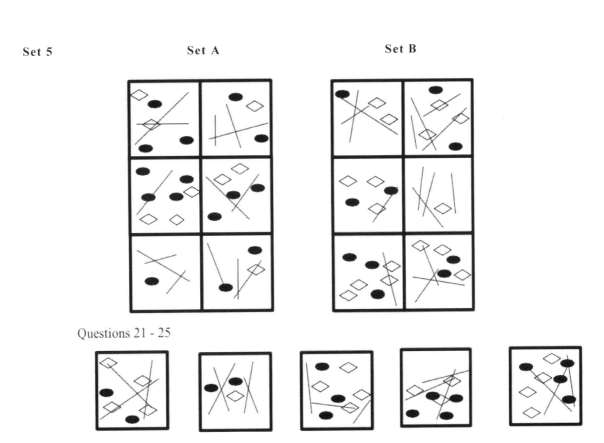

Questions 21 - 25

Set 6

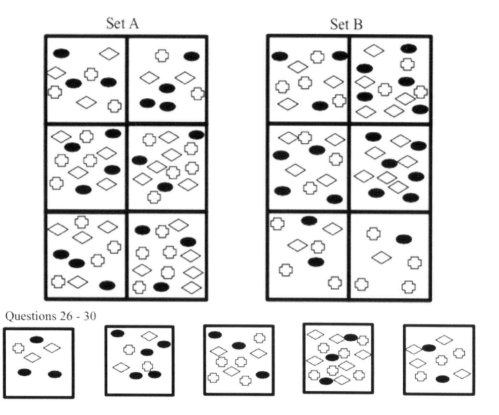

Set A   Set B

Questions 26 - 30

Set 7

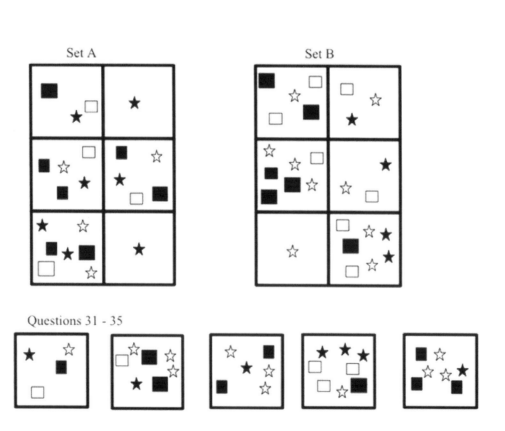

Set A   Set B

Questions 31 - 35

Set 8

Set A

Set B

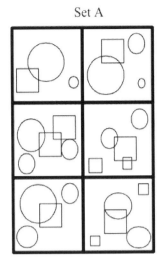

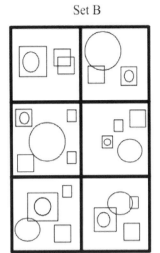

Questions 36 – 40

Set 9

Which answer completes the series?

Question 41:

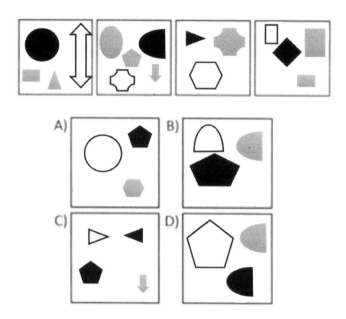

Question 42:

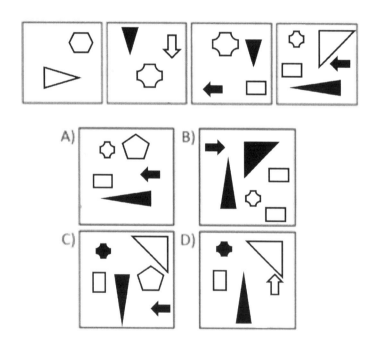

Question 43:

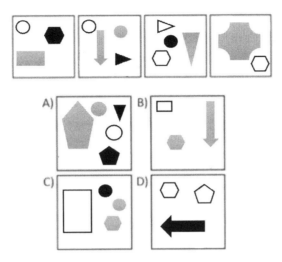

Question 44:

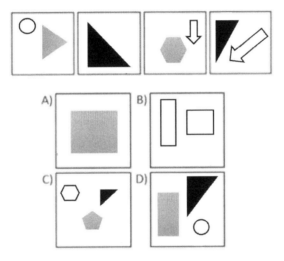

Question 45:

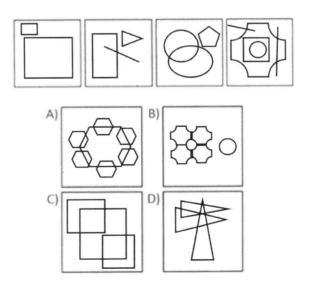

Set 10
Which answer completes the statement?

Question 46:

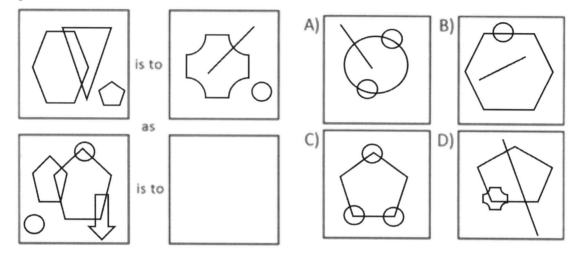

Question 47:

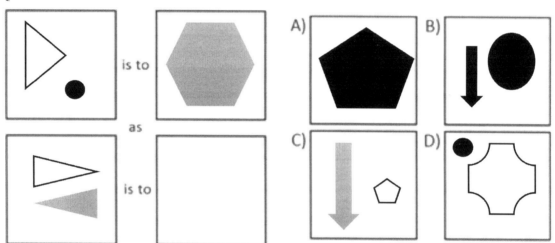

Question 48:

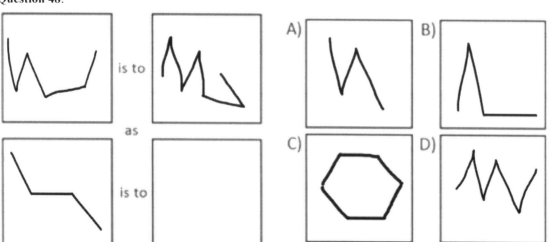

Question 49:

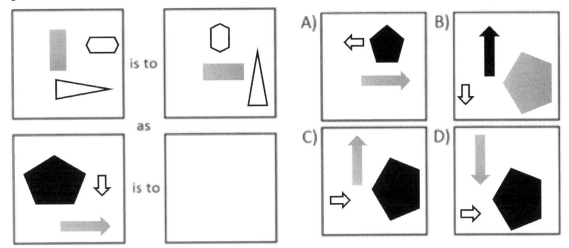

Question 50:

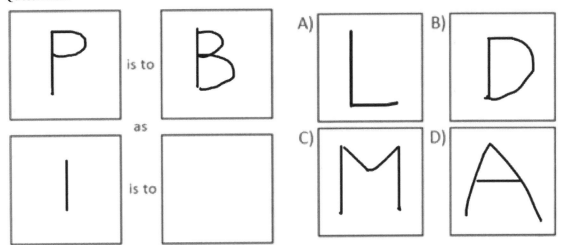

Set 11

Which of the four response options belongs to either set A or set B?

Set A   ## Set B

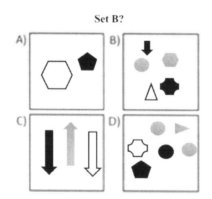

Question 51:

Set B?

Question 52:

Set A?

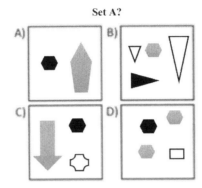

Question 53:

Set A?

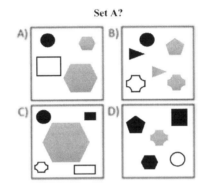

Question 54:

Set B?

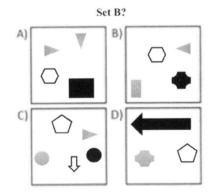

Question 55:

Set B?

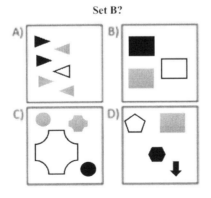

END OF SECTION

Section E: Situational Judgement Test

Read each scenario. Each question refers to the scenario directly above. For each question, select one of these four options. Select whichever you feel is the option that best represents your view on each suggested action.

For questions 1 – 40, choose one of the following options:

A Very appropriate
B Appropriate, but not ideal
C Inappropriate, but not awful
D Very inappropriate

Scenario 1

Arthur is a final year medical student on his GP placement. Having just observed the doctor take a history from a new patient, the doctor quickly steps outside to discuss something with a colleague. In her absence, the patient reveals to Arthur that he has been having suicidal thoughts. These feelings had not been shared with the doctor. The patient pleads with Arthur to not share this information.

How appropriate are the following responses to Arthur's dilemma?
1.   Encourage the patient to share his feelings with the doctor on her return.
2.   When the doctor returns, Arthur should reveal what the patient shared whilst the patient is still present – ignoring the patient's request.
3.   After the patient has left, take the doctor aside and tell her his concerns over the patient. It is worth breaking confidentiality to potentially save a life.
4.   Say nothing.
5.   Don't explicitly inform the doctor, but hint to her that there might be more than meets the eye with this patient.

Scenario 2

Nico is studying for his first year medical exams. He is feeling very unprepared and starting to panic about how much he last left to do. The prospect of failing is very embarrassing to him, and he would rather tell his friends and family that he is too ill to take the exams, rather than face the truth. It is the rules of the medical school that an exam can only not be taken for a valid medical reason with a certified doctor's letter.

How appropriate are the following actions for Nico to take?
6.   Speak to his personal tutor about his concerns.
7.   Refuse to attend the exam.
8.   Make an efficient revision timetable for the few remaining days and do the best he can.
9.   Feign an illness and obtain a sick note.
10.   Bury his head in the sand and ignore how much he has left to do – what will be will be.

Scenario 3

Alan is a fourth year medical student. On his way home from the hospital one afternoon, he notices Mrs Hamilton, a patient whose case he was previously involved in, on the bus.

How appropriate are the following actions for Alan to take?
11.   Politely acknowledge Mrs Hamilton and ask her how she is.
12.   Pretend not to see her and look the other way.
13.   Discuss her medical history openly on the bus.
14.   Get off the bus early without acknowledging her, even though she has clearly spotted him.
15.   Not go out of his way to talk to her.

Scenario 4

Sabrina is a final year medical student. On her way home from hospital one evening, she spots two girls from the year below her. She overhears their conversation, and realises they are openly discussing a patient in detail, including mentioning the patient's name.

How appropriate are the following responses to Sabrina's situation?

16. Ignore the girls and hope no one else recognises the patient's name.
17. Report the girls for breaking patient confidentiality.
18. Inform the girls that they should not be talking about patients publically.
19. Advise the girls to speak more quietly.
20. Pretend she does not recognise the girls.

Scenario 5

Maxine is a second year medical student visiting the hospital for the first time since she started at medical school. The doctor supervising her and her group had previously emailed saying they should prepare for a small assessment. Unfortunately, Maxine is the only one who did not receive the email and is therefore unprepared.

How appropriate are the following responses to Maxine's dilemma?

21. Ask the doctor not to hold the test as she was not informed.
22. Tell the doctor that she did not receive the email.
23. Decide not to turn up to the session.
24. Quickly ask the other students what they have learnt and attempt to understand it before the assessment.
25. Copy one of the other students' work.

Scenario 6

Marcus is a third year medical student attending a clinic with his consultant. He observes the consultant inform a patient that their condition is terminal. However, he does not believe that the patient has fully understood this as he thinks the consultant gave the patient false hope.

How appropriate are the following responses to Marcus' situation?

26. Follow the patient and bluntly inform them of their prognosis.
27. Tell the consultant he believes he gave the patient false hope.
28. Ask the consultant why they chose to deliver the information in this way.
29. Say nothing.
30. Inform the patient's family of the reality of the patient's condition.

Scenario 7

Tunde is a third year medical student on a clinical placement. His consultant informed him a week in advance that their group of students would be having a test. However, Tunde completely forgets to inform the other students in his group, and only remembers the night before the test. The results of the test will contribute towards the final placement mark.

How appropriate are the following responses to Tunde's problem?

31. Pretend to the consultant that he had told all of the students and that they are lying.
32. Inform the consultant the night before of his mistake, and ask to reschedule.
33. Inform the other students and apologise profusely.
34. Deny the consultant told him anything about a test.
35. Decide not to come in to avoid the situation.

Scenario 8

Delilah is a third year medical student on a sexual health rotation. Whilst she is taking the history of a young male patient, she feels like he is being rather inappropriate towards her. At the end of the consultation, the patient asks if she would like to go for a drink with him.

How appropriate are the following responses to Delilah's situation?

36. Go for a drink with him, he is a nice young man after all.

37. Thank the patient politely, but inform him that she is unable to do so.

38. Report the patient for inappropriate behaviour.

39. Become very offended by the patient and storm out of the room.

40. Ask another student to take over from her.

For questions 40 – 69, decide how important each statement is when deciding how to respond to the situation?

    A **Very important**

    B **Important**

    C **Of minor importance**

    D **Not important at all**

Scenario 9

Chad is a second year medical student approaching his end-of-year exams. He also plays football for the medical school, and joins the team in the national inter-university tournament. After they bring home the cup, Chad is approached by a scout for a professional football team, and asked if he would like to trial for them.

How important are the following factors for Chad in deciding what he should do?

41. His end-of-year exams may be compromised if he focuses on football rather than studying.

42. He has only been offered a trial; this may not be worth compromising his exams for.

43. He has been offered a trial, which may lead to bigger things.

44. There are still a few weeks remaining before the exams.

45. His parents don't think being a footballer is an appropriate profession.

Scenario 10

Troy is a medical student at a busy GP surgery. The doctor has been delayed and there is a long queue in the waiting room. Troy enters the waiting room to inform the patients that the waiting time will be two hours. One of the patients becomes very aggressive and starts to verbally abuse Troy.

How important are the following factors for Troy in determining what he should do?

46. The patient may be feeling very unwell or might be in pain.

47. The patient may have something important to go to after his appointment.

48. There are many other patients in the waiting room who have not become aggressive.

49. The doctor was delayed due to an emergency.

50. Troy needs to look professional to patients.

51. Troy is just a student, there are plenty of other staff around.

52. The patient has dementia.

Scenario 11

Leroy and Byrony are fourth year medical students who are in a relationship. Byrony has recently won an award for a poster she submitted, and has been asked to present it at a conference. One evening, Byrony tells Leroy in confidence that some of the work on the poster is not her own, and was copied from elsewhere.

How important are the following factors in determining what Leroy should do?

53. Leroy and Byrony are in a relationship.

54. Byrony only told Leroy her secret because she thought he would keep it to himself.

55. Byrony has plagiarised and could get into serious trouble.

56. The award Byrony has won is very prestigious.

57. If Byrony is caught presenting plagiarised work at the conference, the reputation of the medical school could be damaged.

Scenario 12

Benson is a first year medical student. He has become concerned about how many books he will need to help him in his studies, as he does not think he will be able to afford them. He thinks he may need to extend his overdraft in order to pay for these books, or else get a part-time job.

How important are the following factors in determining what Benson should do?

58. These books will greatly help Benson in his studies.

59. The library is likely to have many of the books he needs.

60. An overdraft can be difficult to pay back.

61. A part-time job may compromise his studies.

62. His parents may be able to contribute towards these books.

63. If he spends all of his money on books, he will not have any left for a sports tour later in the year.

Scenario 13

Albert-Clifford, a medical student, has been keen to do a particular project for a long time. He has even spoken to the project supervisor and begun to outline it. However, he knows that this project is a very popular one and lots of other people want to do it too. Despite putting it down as his first choice, Albert-Clifford is randomly allocated to a different project instead. Bitterly disappointed, he discovers that his friend, Anna-Theresa, has been allocated his chosen project, despite putting it down as her last choice. Albert-Clifford would like to make an official appeal.

How important are the following factors for Albert-Clifford in deciding whether or not to appeal?

64. Anna-Theresa did not put the project down as her first choice.

65. Albert-Clifford was not allocated his first choice of project.

66. Many other people wanted to do this project too.

67. Albert-Clifford already started planning the project.

68. Albert-Clifford has no interest in the project he has been allocated.

69. Albert-Clifford has spoken to the project supervisor.

END OF PAPER

Mock Paper C

Section A: Verbal Reasoning

Passage 1

Modern Considerations in Pigeon Pest Control

For much of the 20th century, controlling intra-city pigeon populations primarily meant killing them. To this day culling remains a common fallback position (the United States Department of Agriculture kills 60,000 pigeons a year in response to complaints). However the trade journal "Pest Control" cautions local authorities that with "millions of bird lovers out there," one must consider "the publicity you would receive if your local paper runs photos of hundreds of poisoned pigeons flopping around on Main Street." In an apparent paradoxical situation whilst we have become less tolerant of the birds, we have also, somehow, grown more concerned for their welfare.

Studies have allowed for the calculation of estimates that a single pigeon dispenses over 11 kilograms of excrement each year. Often this muck must be blasted off hard-to-reach places by climbing teams using steam hoses. Pigeon-related damage in America alone has been estimated to cost $1.1 billion a year. But the full scope of our disdain and prejudice for the birds is impossible to quantify. Marketing circulated by the bird-control industry (a profitable offshoot of the $6.7-billion-a-year pest-control business) describes how pigeons and their faeces can spread more than 60 diseases communicable between humans. However one must consider that whilst this is true, it is not necessarily panic-inducing given the vanishingly small incidences of respiratory infections like cryptococcosis.

1. The bird-control industry profits around $6.7 billion a year by controlling the pigeon population.
A. True
B. False
C. Can't tell

2. Millions of bird lovers out there:
A. Are responsible for an increasing risk of cryptococcosis
B. Must feel partly to blame for the $1.1 billion of pigeon related damage every year in America
C. Outnumber those who are intolerant of pigeons
D. Have made it difficult to use traditional pigeon-control methods

3. The necessity of pigeon control is to reduce the number disease vectors.
A. True
B. False
C. Can't tell

4. In light of our growing concern for pigeon well-being there has seen a significant move away from traditional killing practices.
A. True
B. False
C. Can't tell

Passage 2

Little Wars *H G Wells*

"LITTLE WARS" is the game of kings—for players in an inferior social position. It can be played by boys of every age from twelve to one hundred and fifty—and even later if the limbs remain sufficiently supple—by girls of the better sort, and by a few rare and gifted women. This is to be a full History of Little Wars from its recorded and authenticated beginning until the present time, an account of how to make little warfare, and hints of the most priceless sort for the recumbent strategist....

The beginning of the game of Little War, as we know it, became possible with the invention of the spring breechloader gun. This priceless gift to boyhood appeared some when towards the end of the last century, a gun capable of hitting a toy soldier nine times out of ten at a distance of nine yards. It has completely superseded all the spiral-spring and other makes of gun hitherto used in playroom warfare. These spring breechloaders are made in various sizes and patterns, but the one used in our game is that known in England as the four-point-seven gun. It fires a wooden cylinder about an inch long, and has a screw adjustment for elevation and depression. It is an altogether elegant weapon and it was with one of these guns that the beginning of our war game was made. It was at Sandgate—in England.

5. The author created the game of little wars.

A. True
B. False
C. Can't tell

6. The spring breechloader gun has a 90% toy soldier hit rate at 10 yards.

A. True
B. False
C. Can't tell

7. The four-point-seven gun used by the author adopts the use of a spiral spring mechanism.

A. True
B. False
C. Can't tell

8. The game of little wars:

A. Was invented at Sandgate
B. Requires a spiral spring breechloader
C. Uses wooden ammunition
D. Is an expensive hobby

Passage 3

A New German Shipping Canal

The gates which admit the water into the new canal which is to connect the Baltic with the North Sea have been recently opened by the Emperor William. This canal is being constructed by the German government principally for the purpose of strengthening the naval resources of Germany, by giving safer and more direct communication for the ships of the navy to the North German ports. The depth of water will be sufficient for the largest ships of the German navy. The canal will also prove of very great advantage to the numerous timber and other vessels trading between St. Petersburg, Stockholm, Dantzic, Riga, and all the North German ports in the Baltic and this country. The passage by the Kattegat and Skager Rack is exceedingly intricate and very dangerous, the yearly loss of shipping being estimated at half a million of money. In addition to the avoidance of this dangerous course, the saving in distance will be very considerable. Thus, for vessels trading to the Thames the saving will be 250 miles, for those going to Lynn or Boston 220, to Hull 200, to Newcastle or Leith 100. This means a saving of three days for a sailing vessel going to Boston docks, the port lying in the most direct line from the timber ports of the Baltic to all the centre of England. Considering that between 30,000 and 40,000 ships now pass through the Sound annually, the advantage to the Baltic trade is very apparent.

9. The new German shipping canal was constructed to reduce the distance trading ships had to travel when making port.
A. True
B. False
C. Can't tell

10. It could be hoped that the new German shipping canal will save £500,000 by reducing the chance of ships running aground in Kattegat and Skager Rack.
A. True
B. False
C. Can't tell

11. The main aim of this article is demonstrating that the opening of the new German shipping canal is an example of
A. How military developments often benefit many aspects of society
B. A human engineering solution to reduce the costs of transport
C. A human engineering solution to reduce the loss of life at sea
D. Germany's booming trade across the Baltic Sea

12. The opening of the German shipping canal has benefited the shipping trade more than the navy.
A. True
B. False
C. Can't tell

Passage 4

Darwin's Geological Observations of South America

Of the remarkable "trilogy" constituted by Darwin's writings which deal with the adventures of the "Beagle," the member which has perhaps attracted least attention up to the present time is that which describes of the geology of South America. The actual writing of this book appears to have occupied Darwin a shorter period than either of the other volumes of the series; his diary records that the work was accomplished within ten months, namely, between July 1844 and April 1845; but the book was not actually issued till late in the year following, the preface bearing the date "September 1846." Altogether, as Darwin informs us in his "Autobiography," the geological books "consumed four and a half years' steady work," most of the remainder of the ten years that elapsed between the return of the "Beagle," and the completion of his geological books being, it is sad to relate, "lost through illness!"

Concerning the "Geological Observations on South America," Darwin wrote to his friend Lyell, as follows: "My volume will be about 240 pages, dreadfully dull, yet much condensed. I think whenever you have time to look through it, you will think the collection of facts on the elevation of the land and on the formation of terraces pretty good."

13. The author is discussing one of Darwin's books out of a trilogy that he wrote on the geology of South America

A. True

B. False

C. Can't tell

14. The author believes Darwin's work is boring.

A. True

B. False

C. Can't tell

15. Darwin was delayed 12 months in publishing his Geological Observations on South America due to ill health.

A. True

B. False

C. Can't tell

16. Darwin's first draft of his Geological Observations of South America was greater than 200 pages.

A. True

B. False

C. Can't tell

Passage 5

Cattle

The wealth-producing possibilities of cattle-raising are written into the history, literature and art of every race; and with every nationality riches have always been counted in cattle and corn. We find cattle mentioned in the earliest known records of the Hebrews, Chaldeans and Hindus, and carved on the monuments of Egypt, thousands of years before the Christian era.

Among the primitive peoples wealth was, and still is, measured by the size of the cattle herds, whether it be the reindeer of the frigid North, the camel of the Great Sahara, or herds of whatsoever kind that are found in every land and in every climate. The earliest known money, in Ancient Greece, was the image of the ox stamped on metal; and the Latin word *pecunia* and our own English "pecuniary" are derived from *pecus*—cattle.

Although known to the Eastern Hemisphere since the dawn of history, cattle are not native to the Western Hemisphere, but were introduced into America during the sixteenth century. Cortez, Ponce de Leon, De Soto and the other *conquistadores* from Old Madrid, who sailed the seas in quest of gold, brought with them to the New World the monarchs of the bull ring, and introduced the national sport of Spain into the colonies founded in Peru, Mexico, Florida and Louisiana.

17. The rearing of cattle had a major influence on the development of economy in the Eastern Hemisphere.

A. True

B. False

C. Can't tell

18. The Spanish are the reason that cattle exist in America.

A. True

B. False

C. Can't tell

19. The main conclusion to draw from the second paragraph is that

A. Cattle are valuable

B. Cattle does not necessarily refer to cows

C. Cattle are a historical symbol of attaching wealth to goods

D. Cattle are an integral part of modern economy

20. The Conquistadores landed in Peru, Mexico, Florida and Louisiana.

A. True

B. False

C. Can't tell

Passage 6

Succession of Forest Growths

It is the prevailing and almost universal belief that when native forests are destroyed they will be replaced by other kinds, for the simple reason that the soil has been impoverished of the constituents required for the growth of that particular tree or trees. This I believe to be one of the fallacies handed down from past ages, taken for granted, and never questioned. Nowhere does the English oak grow better than where it grew when William the Conqueror found it at the time he invaded Britain. Where do you find white pines growing better than in parts of New England where this tree has grown from time immemorial?

The question why the original growth is not reproduced can best be answered by some illustrations. When a pine forest is burned over, both trees and seeds are destroyed, and as the burned trees cannot sprout from the stump like oaks and many other trees, the land is left in a condition well suited for the germination of tree seeds, but there are no seeds to germinate. It is an open field for pioneers to enter, and the seeds which arrive there first have the right of possession.

21. The author agrees with the prevailing belief of forest growth succession and the influence of impoverished soil.
A. True
B. False
C. Can't tell

22. White pine grows best in New England.
A. True
B. False
C. Can't tell

23. In conclusion
A. Soil nutrient depletion is not an issue for forest growth unless the population is entirely destroyed
B. The most pioneering species of trees succeed in colonising the remains of forest fires
C. When native forests are destroyed they are replaced by a different species of tree
D. The English oak grows best in England

24. The author believes that the species of seed which arrives first holds possession of the land.
A. True
B. False
C. Can't tell

Passage 7

The Neutral Use of Cables

Eleven submarine cables traverse the Atlantic between 60 and 40 degrees north latitude. Nine of these connect the Canadian provinces and the United States with the territory of Great Britain; two (one American, the other Anglo-American) connect France. Of these, seven are largely owned, operated or controlled by American capital, while all the others are under English control and management. There is but one direct submarine cable connecting the territory of the United States with the continent of Europe, and that is the cable owned and operated by the Compagnie Francais Cables Telegraphiques, whose termini are Brest, France, and Cape Cod, on the coast of Massachusetts.

All these cables between 60 and 40 degrees north latitude, which unite the United States with Europe, except the French cable, are under American or English control, and have their termini in the territory of Great Britain or the United States. In the event of war between these countries, unless restrained by conventional act, all these cables might be cut or subjected to exclusive censorship on the part of each of the belligerent states.

25. In the event of War, America would take control of all submarine cables.

A. True

B. False

C. Can't tell

26. America owns all nine of the submarine cables connecting the United States with Great Britain.

A. True

B. False

C. Can't tell

27. The neutral use of cables

A. Adopts the use of 11 underwater telecommunication cables crossing the Pacific

B. Occurs less than 60 degrees north

C. Presents a significant benefit to belligerent states during times of war

D. Is integral in maintaining peace between Great Britain, America and France.

28. France receives least use of the submarine cables

A. True

B. False

C. Can't tell

Passage 8

Tombs of the First Egyptian Dynasty

For many years various European collections of Egyptian antiquities have contained a certain series of objects which gave archæologists great difficulty. There were vases of a peculiar form and colour, greenish plates of slate, many of them in curious animal forms, and other similar things. It was known, positively, that these objects had been found in Egypt, but it was impossible to assign them a place in the known periods of Egyptian art. The puzzle was increased in difficulty by certain plates of slate with hunting and battle scenes and other representations in relief in a style so strange that many investigators considered them products of the art of Western Asia.

The first light was thrown on the question in the winter of 1894-95 by the excavations of Flinders Petrie in Ballas and Neggadeh, two places on the west bank of the Nile, a little below ancient Thebes. This persevering English investigator discovered here a very large necropolis in which he examined about three thousand graves. They all contained the same kinds of pottery and the same slate tablets mentioned above, and many other objects which did not seem to be Egyptian. It was plain that the newly found necropolis and the puzzling objects already in the museums belonged to the same period. Petrie assumed that they represented the art of a foreign people—perhaps the Libyans—who had temporarily resided in Egypt in the time between the old and the middle kingdoms.

29. The peculiar vases referred to in the text originated from Western Asia.

A. True

B. False

C. Can't tell

30. Flinders Petrie undertook is excavation of Ballas and Neggadeh in December 1895.

A. True

B. False

C. Can't tell

31. For a period after the middle kingdoms Libyans may have resided in Egypt.

A. True

B. False

C. Can't tell

32. Flinders Petrie examined 3,000 bodies during his excavation along the west bank of the Nile

A. True

B. False

C. Can't tell

Passage 9

How to Raise Turkeys

The best feed for young turkeys and ducks is the yolks of hard-boiled eggs, and after they are several days old the white may be added. Continue this for two or three weeks, occasionally chopping onions fine and sometimes sprinkling the boiled eggs with black pepper; then give rice, a teacupful with enough milk to just cover it, and boil slowly until the milk is evaporated. Put in enough more to cover the rice again, so that when boiled down the second time it will be soft if pressed between the fingers. Milk must not be used too freely, as it will get too soft and the grains will adhere together. Stir frequently when boiling. Do not use water with the rice, as it forms a paste and the chicks cannot swallow it. In cold, damp weather, a half teaspoonful of Cayenne pepper in a pint of flour, with lard enough to make it stick together, will protect them from diarrhea. This amount of food is sufficient for two meals for seventy-five chicks. Give all food in shallow tin pans. Water and boiled milk, with a little lime water in each occasionally, is the best drink until the chicks are two or three months old. When loppered, buttermilk may take the place of the boiled milk. Turkeys like best to roost on trees, and in their place artificial roots may be made by planting long forked locust poles and laying others across the forks.

33. Turkeys are best raised on a vegetarian diet.
A. True
B. False
C. Can't tell

34. In conclusion turkeys
A. Are best reared on the ground with dairy and grains
B. Should eat a varied diet dependent upon the time of year
C. Are difficult to raise
D. Require separate items of food and drink to sustain them despite the water and milk in the food paste

35. A mixture of rice and water will cause turkey chicks to choke.
A. True
B. False
C. Can't tell

36. For the first month of their life turkey chicks should primarily eat whole eggs.
A. True
B. False
C. Can't tell

Passage 10

Pneumatic Malting

According to K. Lintner, the worst features of the present system of malting are the inequalities of water and temperature in the heaps and the irregular supplies of oxygen to, and removal of carbonic acid from, the germinating grain. The importance of the last two points is demonstrated by the facts that, when oxygen is cut off, alcoholic fermentation - giving rise to the well-known odour of apples - sets in the cells, and that in an atmosphere with 20 percent of carbonic acid, germination ceases. The open pneumatic system, which consists of drawing in warm air through the heaps spread on a perforated floor, should yield better results. All the processes are thoroughly controlled by the eye and by the thermometer, great cleanliness is possible, and the space requisite is only one-third of that required on the old plan. Since May, 1882, this method has been successfully worked at Puntigam, where plant has been established sufficient for an annual output of 7,000 qrs. of malt. The closed pneumatic system labours under the disadvantages from the form of the apparatus; germination cannot be thoroughly controlled, and cleanliness is very difficult to maintain, while the supply of oxygen is, as a rule, more irregular than with the open floors.

37. The author believes that the present system of malting is flawed due to inefficient removal of carbonic acid from the germinating grain.
A. True
B. False
C. Can't tell

38. The open pneumatic system is more hygienic than the closed.
A. True
B. False
C. Can't tell

39. Germination would not be possible in an atmosphere of 30% carbonic acid.
A. True
B. False
C. Can't tell

40. Pneumatic malting
A. Is the first step in the process of alcoholic fermentation.
B. Is a general method of malting with several subtypes.
C. Has only been adopted in Puntigam
D. Was invented in 1882.

Passage 11

The Anglesea Bridge in Cork

The river Lee flows through the city of Cork in two branches, which diverge just above the city, and are reunited at the Custom House. The central portion of the city is situated upon an island between the two arms of the river, both of which are navigable for a short distance above the Custom House, and are lined with quays on each side for the accommodation of the shipping of the port.

The Anglesea bridge crosses the south arm of the river about a quarter of a mile above its junction with the northern branch, and forms the chief line of communication from the northern and central portions of the city to the railway termini and deep-water quays on the southern side of the river.

The new swing bridge occupies the site of an older structure which had been found inadequate to the requirements of the heavy and increasing traffic, and the foundations of the old piers having fallen into an insecure condition, the construction of a new opening bridge was taken in hand jointly by the Corporation and Harbour Commissioners of Cork.

The new bridge, which has recently been completed, is of a somewhat novel design, and the arrangement of the swing-span in particular presents some original and interesting features, which appear to have been dictated by a careful consideration of the existing local conditions and requirements.

41. The river Lee flows north of Cork

A. True

B. False

C. Can't tell

42. Local conditions make it difficult to engineer river structures in Cork

A. True

B. False

C. Can't tell

43. The Anglesea bridge has always been a swing bridge

A. True

B. False

C. Can't tell

44. Cork train station is south of the northern branch of the river Lee

A. True

B. False

C. Can't tell

END OF SECTION

Section B: Decision Making

1. Magazine Zleda provides biased opinions praising President Trump. Thus nobody should read magazine Zleda. What is the assumption in this statement?

A. Readers do not want to read magazines with bias
B. President Trump has done nothing worth praising
C. Magazines should publish balanced opinions
D. Magazine Zleda is not a good magazine

2. A survey by a university Students' union found that 2 students liked beer, wine and spirits, 13 students liked wine, 4 students liked beer and spirits, 4 liked only spirits and nobody liked only beer and wine. How many students liked beer alone from the 25 who took part in the survey?

A. 11   B. 10   C. 8   D. 6

3. "Humans can live off 3 hours of sleep a night". Which statement would allow us to refute this conclusion?

A. Most people sleep between 6-8 hours a day
B. Scientists recommend more than 5 hours of sleep for a functional immune system
C. Sleeping less than 6 hours a night increases mortality
D. The need for sleep increases as animals reach the top of the food chain

4. "Australia was the latest country to approve gay marriage. Next thing we know, Australians will be marrying monkeys and dogs". Which statement would weaken this argument?

A. Gay marriage is not equivalent to marrying animals
B. Australians are not known to be animal lovers
C. Every individual has a right to decide who they love, no matter their gender
D. Animals and gays are not equal

5. In an accident at a coal factory, 18 people were reported injured or dead. The factory claimed that 28% of the casualties had been mishandling ignition that caused the accident. Which statement is correct?

A. The factory was trying to get out of paying compensations to all the casualties
B. There is not enough information to draw any conclusions
C. 5 individuals were the cause of the accident
D. 72% of the individuals did not deserve to die at the coal factory

6. The probability of a professor turning up to their lecture is 0.92. The probability of at least half the cohort of students attending a lecture is 0.8. The probability that more than half of the students turn up to the lecture, but not all is 0.7. What is the probability a professor not turning up to the lecture, and all the students attending it?

A. 0.129   C. 0.045   D. Not enough information known
B. 0.011

7. Marbles can be bought in bags of 7, 16 and 28 from Master Maena. Pervill claimed that he bought a certain number of marbles from Master Maena, but everybody knew he was lying. How many marbles had he bought?

A. 55   B. 51   C. 63   D. 67

8. During admissions interviews at Lewisons Medical School, 60% of the candidates interviewed are have a chance to get in. 8% of the candidates are considered "gifted" but will not get in. About 17% are "gifted international students" who will get in. Which Venn diagram represents the given data?

A.

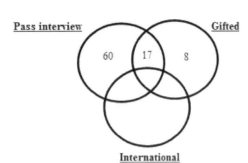

B.

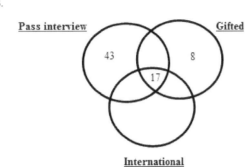

C.

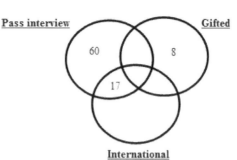

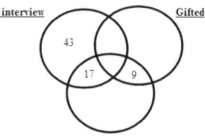

9. In a dance formation, 5 dancers have to make a diamond with the lead dancer in the middle. Jake stands to the left of Macy and Hans, and Amelia stands behind him. Hans stands at the tip of the diamond, with everybody else behind him. Lola stands in front of Amelia and next to Jake. Who is the person in the middle?

A. Amelia   B. Jake   C. Lola   D. Macy

10. Dora has taken up a job as a warehouse assistant. She was sorting through Adidas shoes when the lights went off. There were 4 pairs of Adidas Stan Smiths, 7 pairs of Adidas originals and 3 pairs of Adidas BAIT. What is that probability that she will pick 1 pair of Stan Smiths and 1 pair of BAIT.

A. $\frac{6}{49}$ B. $\frac{3}{49}$ C. $\frac{1}{12}$ D. $\frac{7}{12}$

11. Smoking is the leading cause of lung cancer. The NHS is in debt and cannot afford to treat everyone for free. The NHS should charge smokers for lung cancer treatment. Which statement would support this argument?

A. Children have the most potential and should get preference in free treatment
B. Non-smokers should get free treatment preference
C. Smokers have brought the disease upon themselves
D. All of the above

12. At a conference about reducing anxiety, all the participants were asked to hug one another to relieve stress. 28 hugs were counted in total. How many participants attended the conference?

A. 4   B. 7   C. 8   D. 11

13. A Russian president has never agreed with a British prime minister on trade borders. The American president has always managed to make trade deals with the Russian president, but sometimes fails with the British Prime minister. Which of the following statements are true?

A. Putin and Theresa May will not agree on a trade borders
B. It is likely that Trump will make trade deals with both Putin and Theresa May
C. Trump is the best negotiator amongst the three
D. None of the above

14. Medical students are known to party hard. Most medical students get good grades. Janet gets good grades. Felix is the life of every party

A. Janet is a medical student
B. People can party hard and get good grades
C. Felix gets good grades
D. All medical students party hard and work hard

15. In a 200M medley swimming competition, Jon, Alex and Holly swim at constant speeds. Jon beats Alex by 15m and Holly loses by 15m. At what distance was Holly at when Jon touched the finish line?

A. 156   B. 170   C. 171   D. 185

16. "It is better to do something imperfectly than to do nothing flawlessly". Which of the below options supports this statement?

A. It is impossible to do "nothing" flawlessly
B. We cannot improve by doing nothing
C. Imperfections build the path to success
D. None of the above

17. Doctors are likely to be sued by their patients. Thus, they should hire medicolegal services. Which statement can be an argument against this?

A. If a doctor works towards the greater good, they will not be sued
B. It is expensive to pay for medicolegal services every year
C. Hospitals lawyers can protect doctors
D. Not all doctors are sued by their patients

18. Cybersecurity is a big problem in this day and age. Thus, pornography sites should be banned. What assumption does this conclusion make?

A. Cybersecurity hacks lead to the loss of valuable personal data
B. Pornography sites are a risk to cyber security
C. Pornography sites increase the probability of viewing child pornography
D. The cybersecurity of pornography videos is low

19. Of 130 seafood restaurants in Bali, 63 served shark fin, 28 served lobster, 45 served oysters. 9 restaurants served shark fin and lobster, 9 served lobster and oyster and 12 served shark fin and oyster. How many served all three assuming all restaurant served one of the above?

A. 0   B. 18   C. 24   D. 27

20. Douglas struggles with understanding Venn Diagrams. He was asked to draw one with the following information:

At a swimming pool, children and men above the age of twenty are not allowed to share changing rooms.

Women are allowed to use the same rooms as men above twenty, children or some swim instructors.

Swim instructors are allowed to use any room but some men above twenty have exclusive access and can deny the instructors room use.

Douglas produced the image below.

What does the hexagon represent?

A. Children
B. Men above twenty

C. Swim instructors
D. Women

21. Halep can choose to either dissect the cerebellar cortex, or the brain stem. The probability of completing 85% of her assignment before the class ends is 0.7 if she chooses the cortex. The probability choosing the brain stem and completing 85% of the assignment is 0.3. There is a 0.1 likelihood of her finishing all of her assignment if she chooses the cortex. If she chooses the brain stem, there is a probability of 0.1 that she will finish between 85% and all of the assignment. Which is the better brain region to dissect and why?

A. Brain stem because she is more likely to complete her assignment
B. Brain stem because she is more likely to do 85% of her assignment
C. Cerebellar cortex because she is more likely to complete her assignment
D. Cerebellar cortex because she is more likely to do 85% of her assignment

22. This Dalmatian barks loudly. All dogs that bark loudly are wild. Which conclusion can be drawn?

A. All Dalmatians are wild
B. All dogs are wild
C. Some Dalmatians bark loudly
D. This Dalmatian is wild

23. The cheapest flights from Tokyo to Seoul are provided by Airlines B, F and N. Airline F is the cheapest of the three except on Thursday. Airline B provides extremely low fares for a single flight every Thursday and Saturday. Airline N is the most popular for its good customer service. Which of the following statements is false.

A. Airline B is the cheapest on Thursday
B. Airline F is the cheapest on Saturday, but customer service is not the best
C. Airline N is the cheapest on Saturday
D. Travel with Airline N on Wednesday if you want a cheap flight with good service

24. 83 students indicated their interest to watch the Lion King musical in theatre on Saturday. The probability that any given student will get to watch the play is 0.6. Students are chosen by a random number generator and then informed that there is a 25% chance of them wining a drinks voucher. How many students will watch the play but NOT win the voucher.

A. 50   B. 38   C. 12   D. 63

25. Methane is flammable. Methane is a pollutant. Methane is a greenhouse gas. Greenhouse gases are released in cow faeces.

A. All greenhouse gases are flammable
B. Greenhouse gases are pollutants
C. Methane may increase pollution by causing car fires
D. Methane is released in cow faeces

26. In a game of marbles, Jess and Tim scatter several marbles on the floor. Each can pick 1 or 2 marbles in turn. The person who takes the last marble is the loser. Jess badly wants to win. How many marbles will she need to ensure she defeats Tim, if she starts the game?

A. 4   B. 8   C. 13   D. 16

27. When NASA first sent men to the moon, the probability that Neil Armstrong would place the first step on the moon was 0.7. The probability that Buzz Aldrin would not place the American flag on the moon surface was 0.4. The probability that Armstrong would make the first step, place the flag and then communicate with NASA was 0.14. Either of the two had to do each action. What was the probability that Buzz Aldrin would communicate with NASA

A. 0.2   B. 0.3   C. 0.4   D. 0.5

28. Aspirin is an NSAID (pain reliever). Aspirin causes peptic ulcers. Doctors are recommended not to prescribe Aspirin to patients with gastrointestinal issues. Which of the statements is correct?

A. NSAIDs will always cause peptic ulcers
B. Doctors can prescribe herbal drugs to patients with gastrointestinal pain
C. Patients in pain can be prescribed aspirin
D. None of the above

29. Mike, James, Carrie and Leia decided amongst themselves to cook lunch for the group on one of the four days of the week, Monday, Tuesday Wednesday and Thursday. Each day had a different dish; chicken, fish, lentils and noodles.

James made food on Wednesday. Leia did not food on Thursday. Leia cooked fish on Tuesday. On Monday, chicken was cooked but not by Carrie. Mike did not make noodles and lentils were not cooked on Thursday. Which statement is true about noodles?

A. It was made on Thursday by Carrie
B. It was made on Thursday by Mike
C. It was made on Wednesday by James
D. It was made on Monday by Carrie

END OF SECTION

Section C: Quantitative Reasoning

SET 1

Chris is cooking paella for 10 of his friends from the recipe seen below:

CHRIS' SUPER SECRET PAELLA RECIPE

170g/6oz chorizo, cut into thin slices

110g/4oz pancetta, cut into small dice

2 cloves garlic finely chopped

570ml/1pint calasparra (Spanish short-grain) rice

8 chicken thighs, each chopped in half and browned

110g/4oz fresh or frozen peas

12 jumbo raw prawns, in shells

450g/1lb squid, cleaned and chopped into bite-sized pieces

- Chris' recipe makes 6 portions
- Each portion contains 575kcal which is 25% of Chris' recommended daily kcal intake
- It takes Chris 7 minutes per portion to cook the meal
- The cost per portion is £1.80
- Chris is not eating

1. What quantity of squid does Chris need?

A. 0.45kg   B. 0.6kg   C. 0.75kg   D. 0.9kg   E. 1.35k

2. Chris' friends are due to arrive at 8.30pm. He wants the food to be ready 10 minutes before. What time does Chris need to start preparing the meal?

A. 19:00   B. 19:10   C. 19:15   D. 19:20   E. 19:25

3. If Chris were to eat all 10 portions himself, by how many kcal would he be exceeding his recommended daily intake?

A. 2300   B. 2875   C. 3450   D. 4600   E. 5750

4. Two of Chris' friends are unable to attend. So as not to leave Chris out of pocket his remaining friends agree to split the difference in cost. What is the new cost per portion?

A. £1.95   B. £2.05   C. £2.15   D. £2.25   E. £2.35

5. If Chris had decided to eat with his 10 friends, how many jumbo prawns would he have needed altogether?

A. 2   B. 4   C. 18   D. 20   E. 22

SET 2

The table of admission prices for a cinema is shown below:

	Peak	Off-peak
Adult	£11	£9.50
Child	£7	£5.50
Concession	£7	£5.50
Student	£5	£5

- Peak times include weekends and Friday's from 5pm
- A £2 surcharge is added to any ticket for a film premiere
- A £1 surcharge is added per ticket for the use of 3D glasses
- On Tuesday a special offer means all tickets are charged at the student price

6. How much will it cost for a family of 2 adults, 2 children and 1 concession to watch a premiere 3D film on Saturday evening?

A. £35.50   C. £48   E. £5
B. £43   D. £53

7. How much change would 3 adults receive from a £50 note when viewing a 3D film on a Tuesday?

A. £15   C. £21.50   E. £35
B. £18.50   D. £32

8. Joe buys 4 student tickets for a standard Wednesday afternoon film on his credit card which charges 30% interest. How much does he actually pay?

A. £20   C. £24   E. £28
B. £22   D. £26

9. What is the maximum number of children that could attend an off-peak premiere with £100?

A. 12   B. 13   C. 14   D. 15   E. 16

10. How much money would 1 adult, 2 children and 2 students save by watching a premiere 3D film on a Tuesday instead of a Saturday?

A. £2.50   C. £7.50   E. £12.50
B. £5   D. £10

SET 3

Shown below is a table of figures outlining average energy consumption per property type per year.

	Electricity usage (kWh)	Gas usage (kWh)	Total cost (£)
1 bedroom flat	2,000	0	300
2 bedroom flat	2,000	8,000	700
3 bedroom house	4,750	18,000	1612.50
4 bedroom house	6,000	22,000	2000
Microbusiness	5,000	4,500	635
Small business	15,000	9,000	1770
Medium Business	30,000	12,000	3360

- Gas and electricity are priced differently
- Business and household rates are different

11. What is the cost per kWh for household gas?

A. £0.04
B. £0.05
C. £0.06

D. £0.07
E. £0.08

12. Given that businesses receive gas at a rate of 3p per kWh, what is the difference in price between household and business electricity per kWh?

A. £0.03   B. £0.04   C. £0.05   D. £0.06   E. £0.1

13. If Kate owns both a small business and a 3 bedroom house, what is her expenditure on gas every two years?

A. £900
B. £1170
C. £1350

D. £2340
E. £2700

14. If Luke owns 50% of a medium business in addition to his 1 bedroom flat, what is his monthly energy expenditure?

A. £165
B. £305
C. £440

D. £1980
E. £3660

15. Installing 1m² of photovoltaic cells cuts electricity costs by 1%. Approximately what area of photovoltaic cells would be required to reduce household electricity rates to those of businesses?

A. 0.33m²
B. £0.5m²

C. 0.66m²
D. 33m²

E. 66m²

SET 4

A sports retailer offers discounts on item pricing when bulk buying. Their graph of item prices can be seen below:

- Rugby balls and footballs are sold separately but tennis balls come in packs of 3. (i.e. 3 tennis balls = 1 item)
- Home delivery is charged at £10 within a 15 mile radius and £20 thereafter
- Home delivery is free on orders exceeding £75

16. How much would it cost to collect 50 rugby balls and 60 tennis balls in store?

A. £550   C. £750   E. £980
B. £570   D. £770

17. Alex plans to buy 8 footballs, how many more should she buy to qualify for free delivery?

A. 1   C. 3   E. 5
B. 2   D. 4

18. What is the maximum number of tennis balls that could be delivered to a home 36 miles away on a budget of £75?

A. 11   B. 33   C. 39   D. 45   E. 58

19. What is the difference in price when the quantity of items in an order of 1000 footballs and 500 rugby balls is halved?

A. £2,750   C. £3,250   E. £3,750
B. £3,000   D. £3,500

20. The following year the retailer revokes their free delivery policy and increases all prices by 20%. Now how much does an order of 6,000 tennis balls cost when delivered to a factory 78 miles away?

A. £4,800   C. £4,824   E. £14,420
B. £4,820   D. £14,400

SET 5

The figure shown below is a plan of 4 cattle fields:

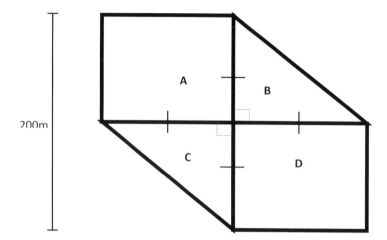

- Each cow legally requires 2m² of space
- A cow drinks 3.5L of water and eats 4kg of pellet food each day
- The farmer's water rate equates to 2p/L and pellet food is purchased at £10 per 20kg
- On average each cow produces 1.5L of milk upon milking
- Milking can only happen every other day

21. What is the total area of all 4 fields?

A. 0.03km²	C. 3km²	E. 30,000km
B. 0.3km²	D. 300km²	

22. How many cattle could be legally housed within field A alone?

A. 50	C. 5,000	E. 50,000
B. 500	D. 10,000	

23. How much money is spent providing the cattle with food and water each day if the maximum legal number of cattle are present?

A. £10,350	C. £35,000	E. £135,000
B. £31,050	D. £86,750	

24. Assuming there is a total of 200 cattle in the fields, what is the maximum volume of milk that could be obtained in a week?

A. 300L	C. 900L	E. 1,500L
B. 600L	D. 1,200L	

25. If the farmer changed to a higher-grade pellet feed instead costing him £2.93 per day, what is the maximum number of cattle that could be housed on a daily budget of £1,000?

A. 288	C. 333	E. 50
B. £315.33	D. 333.33	

SET 6

Shown below is a table of statistics from a race track.

- The statistics shown are for 1 lap of the track
- MPG = miles per gallon
- 1 gallon = 4.5L
- Current fuel prices are 102 pence per litre

CAR	Top Speed (MPH)	Average Speed (MPH)	Fuel consumption (MPG)	Fuel Tank Capacity (L)
A	150	130	54	60
B	180	150	22.5	50
C	165	140	36	60
D	220	180	13.5	40

26. Given that car B completed the lap in 3 minutes exactly, what is the length of the race track?

A. 7.5 miles
B. 9 miles
C. 50 miles
D. 450 miles
E. 540 miles

27. How much does it cost for car C to drive the 10 miles to the race track?

A. 12.75p
B. 87.5p
C. 120p
D. 127.5p
E. 130p

28. How far could car D travel on a full tank of fuel at its average track speed?

A. 105 miles
B. 115 miles
C. 120 miles
D. 125 miles
E. 130 miles

29. If car A was 1.8 miles in front of car B when both cars simultaneously hit top speed; how long would it take for car B to overtake car A–assuming both cars remain travelling at top speed?

A. 0.06 seconds
B. 0.6 seconds
C. 60 seconds
D. 0.06 hours
E. 0.6 hours

30. On a track 8 miles in length, how many whole laps could car A complete on a full tank of petrol assuming that fuel consumption remains constant?

A. 9
B. 90
C. 405
D. 720
E. 3

SET 7

Average annual house prices are shown below.

House prices (£) in South Wales 2010 -2015

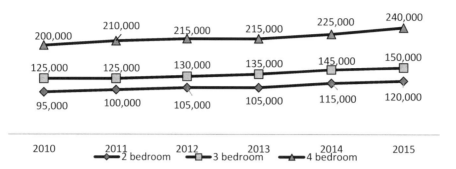

31. What is the percentage rise in 4 bedroom house prices from 2010 to 2015?

A. 17%   C. 23%   E. 29%
B. 20%   D. 26%

32. A couple wants to upgrade from a 2 bedroom to a 4 bedroom house in 2014. They are able to save £5,000 a year for the first 10 years and £10,000 a year thereafter; from what year would they needed to have started saving in order to make their desired move?

A. 1992   C. 1998   E. 2003
B. 1996   D. 2001

33. If house prices rose a staggering 25% between 2009 and 2010, how much would a 4-bedroom house have cost in 2009?

A. £40,000   C. £120,000   E. £200,000
B. £80,000   D. £160,000

34. If an investor buys a 2,3 and 4 bedroom house in 2010, then sells the: 2 bedroom after 2 years, the 3 bedroom after 3 years and the 4 bedroom after 4 years how much profit does he make?

A. £5,000   C. £25,000   E. £45,000
B. £15,000   D. £35,000

35. If a house can only ever be let for 20% of its initial purchase price per annum, for how many years must a 2010, 3 bedroom house be let in order to receive the same profit as if it were sold 4 years later?

A. 0.6   C. 3.6   E. 5.8
B. 0.8   D. 5.6

36. A street has 2, 3, and 3 2 bedroom, 3 bedroom and 4 bedroom houses respectively. What percentage of the streets value do 2 bedroom houses comprise in 2015?

A. 10%   B. 14%   C. 17%   D. 20%   E. 25%

END OF SECTION

Section D: Abstract Reasoning

For each question, decide whether each box fits best with Set A, Set B or with neither.

For each question, work through the boxes from left to right as you see them on the page. Make your decision and fill it into the answer sheet.

Answer as follows:
A = Set A
B = Set B
C = neither

Set 1:   Set A   Set B

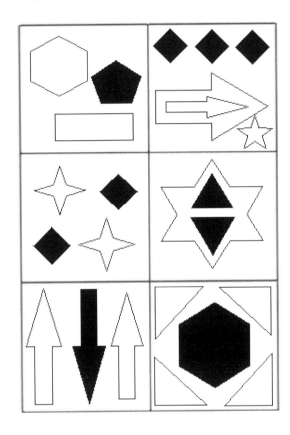 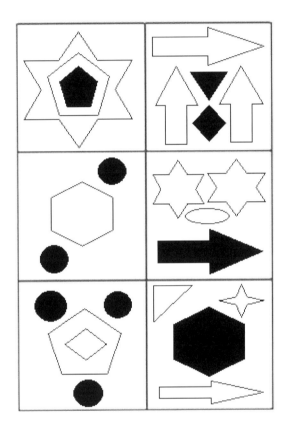

Questions 1-5:

 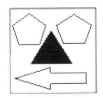  

Set 2: Set A Set B

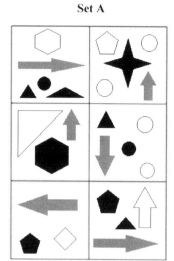 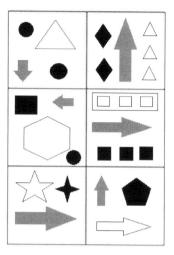

Questions 6-10:

Set 3: Set A Set B

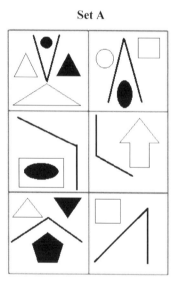 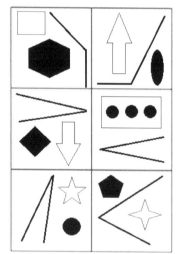

Questions 11-15:

    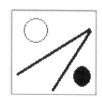

Set 4:   **Set A**   **Set B**

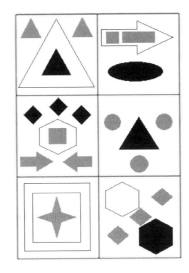 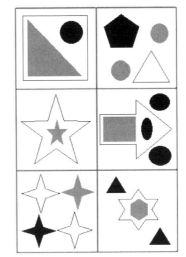

Questions 16-20:

Set 5:   **Set A**   **Set B**

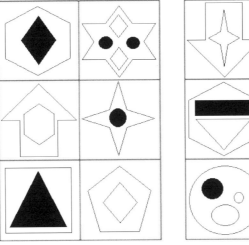 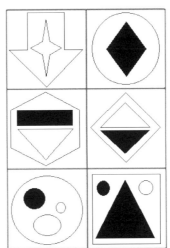

Questions 21-25:

  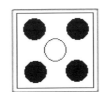  

Set 6

Which answer completes the series?

Question 26:

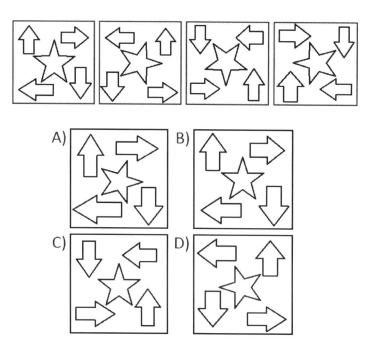

Question 27:

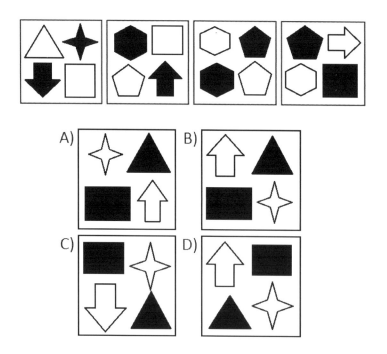

Question 28:

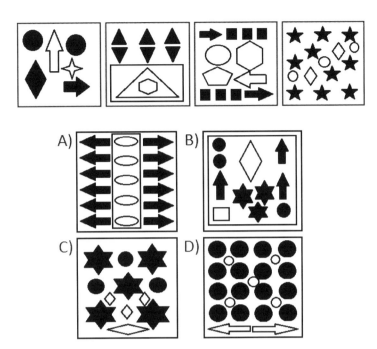

Question 29:

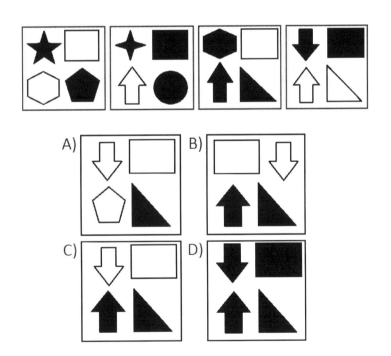

Question 30:

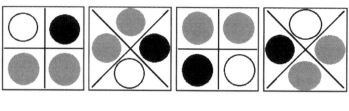

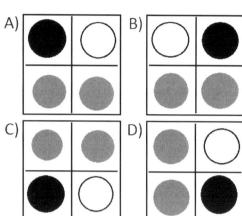

Set 7

Which answer completes the statement?

Question 31:

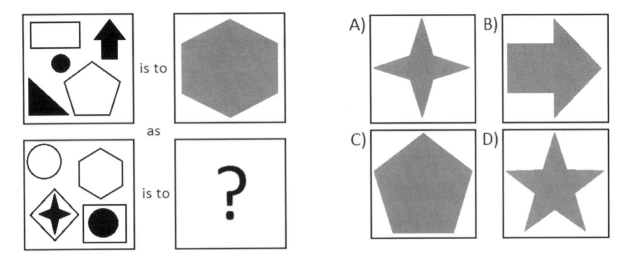

Question 32:

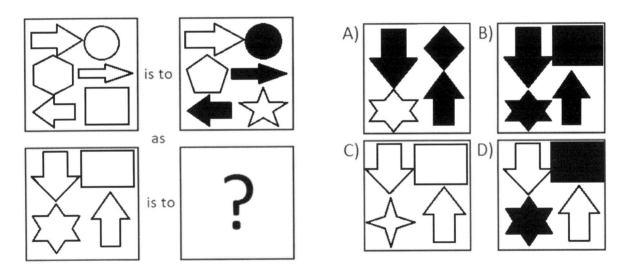

Question 33:

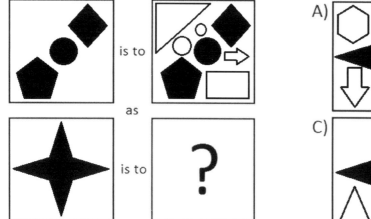 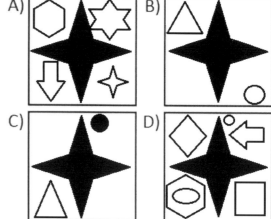

Question 34:

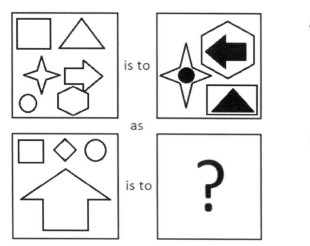 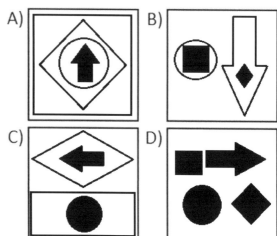

Question 35:

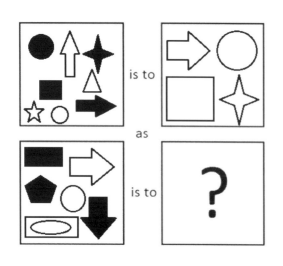 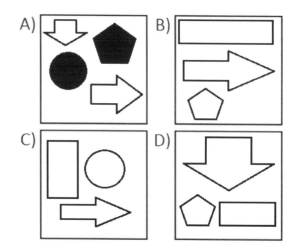

Set 8

Which of the four response options belongs to either set A or set B?

Set A   **Set B**

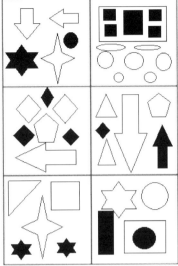

Question 36:

Set A?

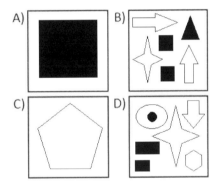

Question 37:

Set B?

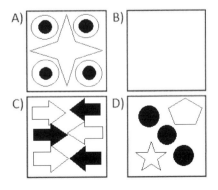

Question 38:

Set A?

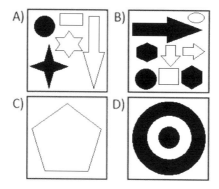

Question 39:

Set A?

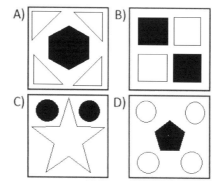

Question 40:

Set B?

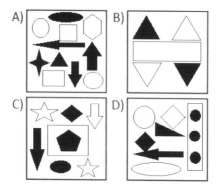

Set 9

Decide whether each box fits best with Set A, Set B or with neither.

Answer as follows:

A = Set A
B = Set B
C = neither

Set A   Set B

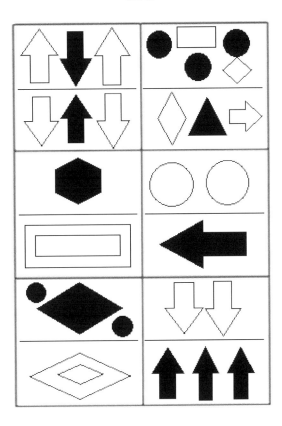 

Questions 41-45:

 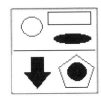 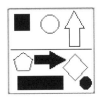 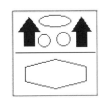 

Set 10   Set A   Set B

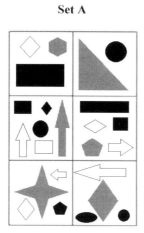 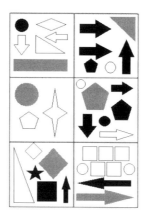

Questions 46-50:

Set 11   Set A   Set B

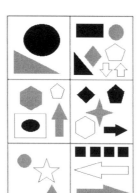 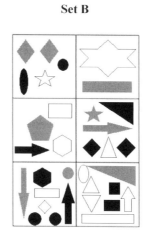

Questions 51-55:

END OF SECTION

Section E: Situational Judgement Test

Read each scenario. Each question refers to the scenario directly above. For each question, select one of these four options. Select whichever you feel is the option that best represents your view on each suggested action.

For questions 1 – 30, choose one of the following options:

A **A highly appropriate action**

B **Appropriate, but not ideal**

C **Inappropriate, but not awful**

D **A highly inappropriate action**

Scenario 1

Maya is a 2nd year medical student who has clinical placement once a week at the GP. She is asked by the GP to see a patient in clinic and take a full history. When she talks to the patient, she notices bruises down the patient's left arm. The patient is a young mother with children aged two and 4 months, and upon further questioning, Maya learns that the patient is going through domestic abuse at home at the hands of her husband. The patient is insistent that Maya tells no one of this information, as she fears that the situation could escalate at home.

How appropriate are the following responses to Maya dilemma?

1. Reassure the patient that she will tell no one of what is happening at home.
2. Reassure the patient that she will not say anything and then, once the patient has left the room, immediately phone the police and report what she has learnt.
3. Tell the patient that everything she says is confidential but if Maya is worried, she may discuss the situation with the GP and the GP may call the patient in again to talk to her.
4. Discuss the situation with the GP and leave it to her to take further action.
5. Discuss this with her parents and get their opinion on this issue.

Scenario 2

Abdul is a 3rd year medical student. He is on clinical placement for two days a week. He is starting his placement in the paediatric ward today. When he arrives at the hospital for 9am, the consultant he's supposed to see with regards to this placement isn't in and there is no one else on the ward he can refer to.

How appropriate are each of these responses to the situation?

6. Go home.
7. Ask a junior doctor or nurse on the ward, if there is anything of interest to see or any patients he can take a history from.
8. Join his friends on their placement, in a different ward.
9. Email his supervisor and ask them what to do.
10. Go around the ward and take some histories from patients as long as they are willing.

Scenario 3

Mike is a final year medical student working in the hospital. While in the canteen at lunch time, he hears some younger medical students he recognises from the ward round, discussing and joking about the case of a patient very loudly over their lunch. At that moment, a relative of the patient being discussed enters the canteen and overhears the students making a joke at the expense of the patient. They look very upset about what they've heard.

How appropriate are Mike's responses to this situation?

11. Tell the students that they are breaking confidentiality in front of everyone and then, immediately email their supervising consultant to report them.
12. Leave his lunch and seek out the senior consultant supervising them to report their behaviour.
13. Take the students aside and quietly tell them they are breaking confidentiality and ask them to apologise to the relative.
14. Take the relative aside and apologise on the student's behalf.
15. Do nothing.

Scenario 4

Alisha is a 4th year medical student on her way home after clinical placement. She is rushing as she is going on a date tonight and is already late. As she is walking home, she sees an elderly man collapsed and half-conscious on the side of the road. Everyone else is ignoring him and walking past him.

How appropriate are each of these responses to Alisha's situation?

16. Point someone in the direction of the elderly man and continue her journey.
17. Walk past him as well.
18. Stop and help him by performing first aid as best as possible and by calling an ambulance. Then call her date and inform them that she may be a little late.
19. Call an ambulance
20. Call a friend and ask them to call an ambulance.

Scenario 5

George is a 2nd year medical student. He is on the respiratory ward when the senior consultant he is shadowing is bleeped and has to leave him to go and see another patient. A nurse that is passing, mistakes him for a junior doctor and asks him to take some bloods from a patient. George does not know how to do this procedure, but the nurse has already left him to get the equipment.

How appropriate are these responses to the situation?

21. Do the procedure.
22. Tell the nurse that he is only a student when she comes back and that he hasn't been taught how to so this clinical skill and therefore does not want to carry out the procedure. It is not worth putting the patient at risk.
23. Ask an FY1 doctor in passing to do the procedure and slip out quietly.
24. Slip out quietly before the nurse comes back.
25. Tell the nurse he is just a student and offer to find someone else to do the procedure.

Scenario 6

Tracy is a first year medical student and it is her first day on the ward. An FY2 shows her around the ward and as he introduces one particular patient makes a racist joke. When Tracy looks uncomfortable, the FY2 tells her to 'lighten up' and that this patient doesn't understand English.

How appropriate are the responses to this situation?

26. Do nothing and laugh along.
27. Lodge a complaint with her senior as soon as possible.
28. Lodge a complaint with her senior the next week, when she is less busy.
29. Repeat the joke to her friends later as they sit in the canteen for lunch.
30. When she is out of earshot of the other patients on the ward, take the FY2 aside and explain to him how she feels about the joke and that she is going to take this further to her senior.

Scenario 7

Mark is a 3rd year medical student on a very busy ward. While he is looking at the drug charts in one of the patient's records, he realises that the patient has been prescribed a drug that they are allergic to by mistake. The chart hasn't been signed off yet.

How appropriate are these responses to Mark's situation?

31. Say nothing – everyone is busy.
32. Cross the medication off the list himself.
33. Alert the junior doctor he is with so that the chart may be changed immediately.
34. Wait until everyone becomes less busy, then alert them about the chart.
35. Remove the chart from the patient's records.
36. Tell the patient to advise the doctors when they try to give them the medication.

For questions 36 – 69, decide how important each statement is when deciding how to respond to the situation?

    A **Extremely important**
    B **Fairly important**
    C **Of minor importance**
    D **Of no importance whatsoever**

Scenario 8

Meena, a 5th year medical student is working on a project with her clinical partner, Tom. The project involves auditing patient records. Meena has been coming in early and staying late to look at the patient records, but she notices that Tom has been taking pictures of the records on his phone.

How important are the following factor's in Meena's situation?

37. The project is due in a week's time, and they still have a lot of records to get through.
38. Their supervisor could find out.
39. They have a shared mark.
40. Patient confidentiality could be broken if someone finds the pictures.
41. He could delete the pictures later.

Scenario 9

Maria is a 5th year medical student. She is taking an online course on digital professionalism and has learnt about the importance of using social media responsibly. When she goes into placement the next day, she hears some patients talking about how one of her colleagues has tagged the hospital page on a picture of a party she was at last night. They are not impressed. When she checks her own social media, she sees she has been tagged on the same post and it shows some of the other students present at the party, drinking and smoking.

How important are the following factors in Maria's situation?

42. The reputation of the hospital may be damaged because of this post.
43. Future employers may link her back to this post and this could impact her chances of getting a job.
44. Maria is not actually in the picture and so there is a slim chance anyone at the hospital will recognise her.
45. It's not her post, and therefore not her responsibility.
46. She will be leaving the hospital in a few weeks as her rotation will change.

Scenario 10

Muhammed has been invited to a neurology conference taking place in London, where he will have the chance to listen to some talks and interact with some consultants who are specialists in their field. He has an interest in neurology and is hoping to specialise in it in the future. However, the conference is on the same day as his meeting with the educational supervisor at the hospital to sign some forms, which he cannot miss as he may not have a chance to see his supervisor again before the deadline.

How important are the following factors in Muhammed's decision?

47. The conference will be taking place next year as well.
48. If he does not submit the forms in on time, then it may go down against his record as professionalism issues.
49. His educational supervisor has specifically made the time for Muhammed to come and see him.
50. Muhammed wanted to acquire some chances for work experience in a laboratory setting with one of the researchers in the summer.
51. The conference may not be in London next year.

Scenario 11

Melissa is a 2nd year medical student in the neonatal ward. While she is talking to a patient, she notices a nurse has dropped a piece of paper with patient information, as she walked past. The patient Melissa is with has just launched into an interesting story about her condition and Melissa feels it would be rude to stop her.

How important are the following factors in Melissa's decision?

52. The patient is in full flow with her story.
53. It would be rude to stop the patient.
54. Patient details are on full display.
55. Someone else could pick up the piece of paper.
56. It is a drug chart.
57. The patient is receiving palliative care.

Scenario 12

Jeremy is a 3rd year medical student at the GP on placement. His clinical partner, Tariq has been looking increasingly dishevelled every day. Jeremy has started to notice that some of the staff have been commenting about Tariq's appearance and some of the patients have been giving Tariq nasty looks.

How important are the following factors for Jeremy in deciding what to do?

58. Jeremy's friendship with Tariq.
59. Tariq's feelings.
60. Their daily contact with patients.
61. Tariq's reputation with the staff.
62. Issues with professionalism.

Scenario 13

Edgar is a 4th year medical student on surgical placement at the hospital. While he is observing an emergency surgery, he notices that one of the nurses does not use the correct sterile technique while putting on her gloves. He knows that the surgery must be done immediately in order to save the patient's life.

How important are the following factors in Edgar's decision?

63. The patient is losing a lot of blood very quickly.
64. Everyone else is gloved up and would have to wait for the nurse to leave the theatre and get some more gloves.
65. The patient could end up contracting an infection
66. They are on a tight time limit, as there are countless other patients waiting for surgery.
67. He is only a student.
68. Everything else is sterile.
69. The nurse doesn't actually touch the surgical site.

END OF PAPER

Mock Paper D

Section A: Verbal Reasoning

Passage 1

The Teeth of Children

Children have twenty temporary teeth, which begin making their appearance about the sixth or seventh month. The time varies in different children. This is the most dangerous and troublesome period of the child's existence, and every parent will do well to consult a reputable dentist. About the second or third year the temporary teeth are fully developed. They require the same care to preserve them as is exercised toward the permanent set.

About the sixth year, or soon after, four permanent molars, or double teeth, make their appearance. Some parents mistakenly suppose these belong to the first set. It is a serious error. They are permanent teeth, and if lost will be lost forever. No teeth that come after the sixth year are ever shed. Let every parent remember this.

At twelve years the second set is usually complete, with the exception of the wisdom teeth, which appear anywhere from the eighteenth to the twenty-fourth year. When the second set is coming in the beauty and character of the child's countenance is completed or forever spoiled. Everything depends upon proper care at this time to see that the teeth come with regularity and are not crowded together. The teeth cannot have too much room. When a little separated they are less liable to decay.

1. Children's teeth begin to appear:

A. Around June or July.
B. Typically before they have their first birthday.
C. At the same time in all children.
D. Before birth.

2. Which of the following statements are true according to the above passage:

A. Temporary teeth require less care than permanent teeth.
B. Temporary molars appear when the child is around six years of age.
C. All teeth that appear after six years of age are permanent.
D. Usually wisdom teeth are finished growing at twelve years of age.

3. Which of the following statements about permanent teeth are true:

A. Permanent teeth require the same level of care as temporary teeth.
B. Permanent teeth often appear when children are four years old.
C. Permanent teeth if lost are naturally regenerated by the body.
D. All teeth are permanent.

4. Which of the following statements is INCORRECT:

A. Wisdom teeth tend to appear when a person is a young adult.
B. You never naturally shed any teeth that appear after six years of age.
C. When teeth are a little separated, decay can easily develop in between.
D. Some parents mistake molars that appear when children are six as temporary teeth.

Passage 2

The Mysteries of Hypnotism

Animal magnetism is the nerve-force of all human and animal bodies, and is common to every person in a greater or less degree. It may be transmitted from one person to another. The transmitting force is the concentrated effort of will-power, which sends the magnetic current through the nerves of the operator to the different parts of the body of their subject. It may be transmitted through the eyes, or the application of the finger tips or the whole open hands, to different regions of the body of the subject, as well as to the mind. The effect of this force upon the subject will depend very much upon the health, mental capacity and general character of the operator. Its action in general should be soothing and quieting upon the nervous system; stimulating to the circulation of the blood, the brain and other vital organs of the body of the subject. It is the use and application of this power or force that constitutes hypnotism.

Magnetism is a quality that inheres in every human being, and it may be cultivated like any other physical or mental force of which men and women are constituted. From the intelligent operator using it to overcome disease, a patient experiences a soothing influence that causes a relaxation of the muscles, followed by a pleasant, drowsy feeling which soon terminates in refreshing sleep. On waking, the patient feels rested; all their troubles have vanished from consciousness and they are as if they had a new lease of life.

In the true hypnotic condition, when a patient voluntarily submits to the operator, any attempt to make suggestions against the interests of the patient can invariably be frustrated by the patient. Self-preservation is the first law of nature, and some of the best-known operators who have recorded their experiments assert that suggestions not in accord with the best interest of the patient could not be carried out. No one was ever induced to commit any crime under hypnosis, that could not have been induced to do the same thing much easier without hypnosis.

The hypnotic state is a condition of mind that extends from a comparatively wakeful state, with slight drowsiness, to complete somnambulism, no two subjects, as a rule, ever presenting the same characteristics. The operator, to be successful, must have control of their own mind, be in perfect health and have the ability to keep their mind concentrated upon the object they desires to accomplish with his subject.

5. Which of the following statements is correct:
A. Animal magnetism cannot be transmitted from one person to another.
B. Transmission of animal magnetism is limited to transmission through the eyes.
C. The effect of animal magnetism on a person is not at all related to the health of the operator.
D. Animal magnetism can be cultivated by men and women alike.

6. According to the passage:
A. The first law of nature is self-preservation.
B. No one has ever committed a crime whilst in a hypnotic state.
C. Upon waking from a hypnotic state the patient feels very tired and stressed.
D. All hypnotic operators are healthy people.

7. Which of the following statements is supported by the above passage:
A. The initial steps required to enter a hypnotic state involve relaxation of the patient's muscles.
B. The hypnotic state is often a turbulent state of mind and results in disturbed sleep.
C. Hypnotism is a skill that can only be performed by doctors.
D. Hypnotism causes raised blood pressure and increases risk of heart attack.

8. According to the above passage, refreshing sleep during hypnotism is preceded by:
A. A pleasant feeling that is accompanied with drowsiness.
B. Muscle ache and general soreness
C. Watching a metronome for extended periods of time.
D. Disturbed sleep.

Passage 3

Tea and Coffee

Tea is a nerve stimulant, pure and simple, acting like alcohol in this respect, without any value that the latter may possess as a retarder of waste. It has a special influence upon those nerve centres that supply will power, exalting their sensibility beyond normal activity, and may even produce hysterical symptoms, if carried far enough. Its principle active ingredient, theine, is an exceedingly powerful drug, chiefly employed by nerve specialists as a pain destroyer, possessing the singular quality of working toward the surface. That is to say when a dose is administered hypodermically for sciatica, for example, the narcotic influence proceeds outward from the point of injection, instead of inward toward the centres, as does that of morphia, atropia, etc. Tea is totally devoid of nutritive value, and the habit of drinking it to excess, which so many American women indulge in, particularly in the country, is to be deplored as a cause of our American nervousness. Coffee, on the contrary, is a nerve food. Like other concentrated foods of its class, it operates as a stimulant also, but upon a different set of nerves from tea. Taken strong in the morning, it often produces dizziness and that peculiar visual symptom of overstimulus which is called muscae volilantes--dancing flies. But this is an improper way to take it, and rightly used it is perhaps the most valuable liquid addition to the morning meal. Its active principle, caffeine, differs in all physiological respects from theine, while it is chemically very closely allied, and its limited consumption makes it impotent for harm.

9. Coffee:

A. Operates as a depressant.
B. If rightly used is the most important liquid ingested with breakfast.
C. Has an active principle named theine.
D. Acts on the same nerves that tea acts upon.

10. Which of these is NOT stated as an effect of tea:

A. Pain reduction.
B. American nervousness.
C. Nerve stimulation.
D. Hysteria

11. Which of the following statements does the above passage support?

A. Tea and coffee both are drunk habitually by American women.
B. Dizziness and peculiar vision symptoms are an effect of excess caffeine
C. Theine and caffeine are both chemically and physiologically closely allied
D. Theine is a weak drug.

12. Which of the following statements is NOT supported by the above passage:

A. Caffeine when taken in high dosage can cause flies to dance.
B. The active principles in tea and coffee are chemically similar.
C. Hypodermic administration of theine proceeds outward from the point of injection.
D. Tea does not have any nutritional value.

Passage 4

How to Care for a piano

The most important thing in the preservation of a piano is to avoid atmospheric changes and extremes and sudden changes of temperature. Where the summer condition of the atmosphere is damp all precautions possible should be taken to avoid an entirely dry condition in winter, such as that given by steam or furnace heat. In all cases should the air in the home contain moisture enough to permit a heavy frost on the windows in zero weather. The absence of frost under such conditions is positive proof of an entirely dry atmosphere, and this is a piano's most dangerous enemy, causing the sounding board to crack, shrinking up the bridges, and consequently putting the piano seriously out of tune, also causing an undue dryness in all the action parts and often a loosening of the glue joints, thus producing clicks and rattles. To obviate this difficulty is by no means an easy task and will require considerable attention. Permit all the fresh air possible during winter, being careful to keep the piano out of cold drafts, as this will cause a sudden contraction of the varnish and cause it to check or crack. Plants in the room are desirable and vessels of water of any kind will be of assistance. The most potent means of avoiding extreme dryness is .to place a single-loaf bread-pan half full of water in the lower part of the piano, taking out the lower panel and placing it on either side of the pedals inside. This should be refilled about once a month during artificial heat, care being taken to remove the vessel as soon as the heat is discontinued in the spring. In cases where stove heat is used these precautions are not necessary.

13. If the piano is exposed to humidity in summer, it must remain exposed to humidity during the following winter.

A. True.
B. False.
C. Can't tell.

14. Absence of frost indicates a dry atmosphere because:

A. Dryness makes frost melt.
B. Frost causes humidity.
C. Humidity during cold temperatures in winter causes frost.
D. The melting of frost causes dryness.

15. According to the above passage, avoiding dryness:

A. Is a straight forward process.
B. Requires plant life in the same room as the piano.
C. Is essential to maintaining a piano's tuning.
D. In the summer following a humid winter may cause clicks and rattles.

16. Which of the following is true according to the passage:

A. Out of tune keys produce clicks and rattles.
B. Cracking of the sounding board causes clicks.
C. Clicks and rattles are never produced by pianos in a humid environment.
D. Stiffness in action parts and glue joints is less likely to produce clicks and rattles

Passage 5

Theosophy

Much is said nowadays about theosophy, which is really but another name for mysticism. It is not a philosophy, for it will have nothing to do with philosophical methods; it might be called a religion, though it has never had a following large enough to make a very strong impression on the world's religious history. The name is from the Greek word theosophia--divine wisdom--and the object of theosophical study is professedly to understand the nature of divine things. It differs, however, from both philosophy and theology even when these have the same object of investigation. For, in seeking to learn the divine nature and attributes, philosophy employs the methods and principles of natural reasoning; theology uses these, adding to them certain principles derived from revelation.

Theosophy, on the other hand, professes to exclude all reasoning processes as imperfect, and to derive its knowledge from direct communication with God himself. It does not, therefore, accept the truths of recorded revelation as immutable, but as subject to modification by later and personal revelations. The theosophical idea has had followers from the earliest times. Since the Christian era we may class among theosophists such sects as Neo-Platonists, the Hesychasts of the Greek Church, the Mystics of mediaeval times, and, in later times, the disciples of Paracelsus, Thalhauser, Bohme, Swedenborg and others. Recently a small sect has arisen, which has taken the name of Theosophists. Its leader was an English gentleman who had become fascinated with the doctrine of Buddhism. Taking a few of his followers to India, they have been prosecuting their studies there, certain individuals attracting considerable attention by a claim to miraculous powers. It need hardly be said that the revelations they have claimed to receive have been, thus far, without element of benefit to the human race.

17. The idea of theosophy:

A. Is a combination of philosophy and theology.
B. Is based entirely upon scriptures.
C. Was born from the ancient Greeks.
D. Does not involve any natural reasoning.

18. Which of the following is NOT stated to be a sect of theosophy in the above passage:

A. Mystics from the mediaeval ages.
B. Hesychasts of the Roman Church.
C. Neo-Platonists.
D. Paracelsus' disciples.

19. The leader of the Theosophists:

A. Was a Buddhist monk.
B. Is a British lady that lives in India.
C. No longer practices Theosophy.
D. Was fascinated by Buddhist teachings.

20. According to the passage, the point of theosophy:

A. Is to learn about Buddhism and its teachings.
B. Is to learn how to use natural reasoning to understand religion.
C. Is to understand and learn more about God.
D. Is to spread teachings of God to the rest of the world.

Passage 6

Dreams

The Bible speaks of dreams as being sometimes prophetic, or suggestive of future events. This belief has prevailed in all ages and countries, and there are numerous modern examples, apparently authenticated, which would appear to favour this hypothesis. The interpretation of dreams was a part of the business of the soothsayers at the royal courts of Egypt, Babylon and other ancient nations.

Dreams and visions have attracted the attention of mankind of every age and nation. It has been claimed by all nations, both enlightened and heathen, that dreams are spiritual revelations to men; so much so, that their modes of worship have been founded upon the interpretation of dreams and visions. Why should we discard the interpretation of dreams while our mode of worship, faith and knowledge of Deity are founded upon the interpretation of the dreams and visions of the prophets and seers of old. Dreams vividly impressed upon the mind are sure to be followed by some event.

We read in the Holy Scripture the revelation of the Deity to His chosen people, through the prophet Joel: "And it shall come to pass, afterward, that I will pour out My Spirit on all flesh, and your sons and your daughters shall prophesy, your old men shall dream dreams, your young men shall see visions, and also upon the servants and the handmaids in those days will I pour out My Spirit." (Joel ii, 28.) Both sacred and profane history contain so many examples of the fulfilment of dreams that he who has no faith in them must be very sceptical indeed. Hippocrates says that when the body is asleep the soul is awake, and transports itself everywhere the body would be able to go; knows and sees all that the body could see or know were it awake; that it touches all that the body could touch. In a word, it performs all the actions that the body of a sleeping man could do were he awake. A dream, to have a significance, must occur to the sleeper while in healthy and tranquil sleep. Those dreams of which we have not a vivid conception, or clear remembrance, have no significance. Those of which we have a clear remembrance must have formed in the mind in the latter part of the night, for up to that time the faculties of the body have been employed in digesting the events of the day.

21. Dreams of significance only occur during a peaceful slumber.

A. True.
B. False.
C. Can't tell.

22. It is written in religious texts that dreams are divine interventions.

A. True.
B. False.
C. Can't tell.

23. What did Hippocrates NOT say about dreams:

A. That the body and soul are both alive and awake during the dreaming state.
B. That the soul sees all that the body would during the dreaming state.
C. Whilst dreaming, the soul can travel everywhere that the body would be able to in a waking state.
D. The soul can do everything the body can when we dream.

24. According to the passage, dreams:

A. Were never claimed to be spiritual by heathen nations.
B. Are only considered spiritual by religious people.
C. Have inspired some people in the way they worship deities.
D. That are vivid are less likely to be followed by an event suggested by the dream.

Passage 7

What Different Eyes Indicate

The long, almond-shaped eye with thick eyelids covering nearly half of the pupil, when taken in connection with the full brow, is indicative of genius, and is often found in artists, literary and scientific men. It is the eye of talent, or impressibility. The large, open, transparent eye, of whatever colour, is indicative of elegance, of taste, of refinement, of wit, of intelligence. Weakly marked eyebrows indicate a feeble constitution and a tendency to melancholia, Deep sunken eyes are selfish, while eyes in which the whole iris shows indicate erraticism, if not lunacy. Round eyes are indicative of innocence; strongly protuberant eyes of weakness of both mind and body. Eyes small and close together typify cunning, while those far apart and open indicate frankness. The normal distance between the eyes is the width of one eye; a distance greater or less than this intensifies the character supposed to be symbolized. Sharp angles, turning down at the corners of the eyes, are seen in persons of acute judgment and penetration. Well-opened steady eyes belong to the sincere; wide staring eyes to the impertinent.

25. Deep sunken eyes mean:

A. Lunacy.
B. Erraticism.
C. A lack of selflessness.
D. Melancholic depression.

26. If the distance between the eyes is twice the width of 1 eye means the person is more erratic.

A. True.
B. False.
C. Can't tell.

27. The passage states:

A. A person's entire personality is dependent on their eye shape.
B. A person with very small pupils are always lunatics.
C. That people with eyes that are far apart and open are more honest.
D. Elegant and intelligent people have thick eyelids.

28. Geniuses in the above passage are described as:

A. Having deep sunken eyes.
B. Always being scientists.
C. Artists or writers only.
D. Having long almond-shaped eyes.

Passage 8

Physical Exercise

The principal methods of developing the physique now prescribed by trainers are exercise with dumbbells, the bar bell and the chest weight. The rings and horizontal and parallel bars are also used, but not nearly to the extent that they formerly were. The movement has been all in the direction of the simplification of apparatus; in fact, one well-known teacher of the Boston Gymnasium when asked his opinion said: "Four bare walls and a floor, with a well-posted instructor, is all that is really required for a gymnasium."

Probably the most important as well as the simplest appliance for gymnasium work is the wooden dumbbell, which has displaced the ponderous iron bell of former days. Its weight is from three-quarters of a pound to a pound and a half, and with one in each hand a variety of motions can be gone through, which are of immense benefit in building up or toning down every muscle and all vital parts of the body.

The first object of an instructor in taking a beginner in hand is to increase the circulation. This is done by exercising the extremities, the first movement being one of the hands, after which come the wrists, then the arms, and next the head and feet. As the circulation is increased the necessity for a larger supply of oxygen, technically called "oxygen-hunger," is created, which is only satisfied by breathing exercises, which develop the lungs. After the circulation is in a satisfactory condition, the dumbbell instructor turns his attention to exercising the great muscles of the body, beginning with those of the back, strengthening which holds the body erect, thus increasing the chest capacity, invigorating the digestive organs, and, in fact, all the vital functions. By the use of very light weights an equal and symmetrical development of all parts of the body is obtained, and then there are no sudden demands on the heart and lungs.

29. The most important appliance for gymnasium work is the wooden dumbbell.

A. True.
B. False.
C. Can't tell.

30. Which of the following is NOT true:

A. The heaviest dumbbells are a pound and a half.
B. Increasing circulation is the first objective an instructor has when teaching a beginner.
C. Increasing circulation involves exercising all the limbs as well as the abdomen.
D. Use of light weights results in symmetrical muscle development.

31. Oxygen-hunger is:

A. Something that causes increased circulation.
B. Resolved by breathing pure oxygen.
C. A consequence of lunge development.
D. Resolved through breathing exercises.

32. The muscles of the back help maintain posture:

A. True.
B. False.
C. Can't tell.

Passage 9

Facts about Sponges

Sponges belong to the animal kingdom, and the principal varieties used commercially are obtained off the coasts of Florida and the West Indies; the higher grades are from the Mediterranean Sea, and are numerous in variety.

A sponge in its natural state is a different-looking object from what we see in commerce, resembling somewhat the appearance of the jelly fish, or a mass of liver, the entire surface being covered with a thin, slimy skin, usually of a dark colour, and perforated to correspond with the apertures of the canals commonly called "holes of the sponge." The sponge of commerce is, in reality, only the skeleton of a sponge. The composition of this skeleton varies in the different kinds of sponges, but in the commercial grades it consists of interwoven horny fibres, among and supporting which are epiculae of silicious matter in greater or less numbers, and having a variety of forms. The fibres consist of a network of fibriles, whose softness and elasticity determine the commercial quality of a given sponge. The horny framework is perforated externally by very minute pores, and by a less number of larger openings. These are parts of an interesting double canal system, an external and an internal, or a centripetal and a centrifugal. At the smaller openings on the sponge's surface channels begin, which lead into dilated spaces. In these, in turn, channels arise, which eventually terminate in the large openings. Through these channels or canals
definite currents are constantly maintained, which are essential to the life of the sponge. The currents enter through the small apertures and emerge through the large ones.

33. Commercial sponges look exactly like what you would find in the wild.

A. True.
B. False.
C. Can't tell.

34. According to the passage, sponges:

A. Are a type of jellyfish.
B. Are typically light coloured.
C. Can be found in many different varieties in the Mediterranean Sea.
D. Typically found of the coast of Mexico.

35. Definite currents:

A. Are critical to sponge life.
B. Enter through small apertures and exit through them as well.
C. Are the name given to waves that carry sponges onto the shore.
D. Fluctuate depending on the tide.

36. According to the above passage, the skeletal organisation of a sponge is uniform in all sponges.

A. True.
B. False.
C. Can't tell.

Passage 10

The Claims of Osteopathy

Strictly construing the claims of osteopathic doctors, it is an anti-medicine system of practice for the cure of every disease to which the human body is liable.

Dr. Andrew T. Still, who claims to have made the discoveries that led to the establishment of the school of Osteopathy, asserts that all diseases and lesions are the result of the luxation, dislocation, or breakage of some bone or bones; this, however, is not now maintained to any great extent by his followers. Osteopathists, though, do generally claim that all diseases arise from some maladjustment of the bones of the human body, and that treatment, therefore, must be to secure the normal adjustment of the bones and ligaments that form the skeleton. They claim that a dislocation is not always necessarily the result of external violence; it may be caused by the ulceration of bones, the elongation of ligaments, or excessive muscular action.

The constriction of an important artery or vein, which may be caused by a very slightly displaced bone, an indurated muscle, or other organ, may produce an excess of blood in one part of the body, thereby causing a deficiency in some other part. A dislocated member will generally show alteration in the form of the joint and axis of the limb; loss of power and proper motion; increased length or shortening of the limb; prominence at one point and depression at another; greatly impaired circulation, and pain due to the obstruction of nerve force in the parts involved.

37. Osteopathy is the study of bones.

A. True.
B. False.
C. Can't tell.

38. Dr Andrew T. Still:

A. Established the school of Osteopathy.
B. Lied in order to gain fame.
C. Imparted teachings to his followers which are not believed by all to be true today.
D. Was the first Osteopathist.

39. The passage states that the constriction of an artery:

A. Can cause a heart attack.
B. Is only fatal if the artery is important.
C. Always causes an excess of blood at the site of constriction.
D. Can be caused by a dislocated bone impinging upon it.

40. Most osteopathists believe:

A. That all medical issues occur due to breakage of bones.
B. The incorrect positioning of bones is involved to an extent in all diseases.
C. That if all bones are in their correct position then all ligaments must be okay as well.
D. The skeleton is the most important component of the human body.

Passage 11

Hints on Bathing

There has been a great deal written about bathing. The surface of the skin is punctured with millions of little holes called pores. The duty of these pores is to carry the waste matter off. For instance, perspiration. Now, if these pores are blocked up they are of no use, and the body has to find some other way to get rid of its impurities. Then the liver has more than it can do. Then we take a liver pill when we ought to clean out the pores instead. The housewife is very particular to keep her sieves in good order; after she has strained a substance through them they are washed out carefully with water, because water is the best thing known. That is the reason water is used to bathe in. But the skin is a little different from a sieve, because it is willing to help along the process itself. All it needs is a little encouragement and it will accomplish wonders. What the skin wants is rubbing. If you should quietly sit down in a tub of water and as quietly get up and dry off without rubbing, your skin wouldn't be much benefited. The water would make it a little soft, especially if it was warm. But rubbing is the great thing. Stand where the sunlight strikes a part of your body, then take a dry brush and rub it, and you will notice that countless little flakes of cuticle fly off. Every time one of these flakes is removed from the skin your body breathes a sigh of relief. An eminent German authority contends that too much bathing is a bad thing. There is much truth in this. Soap and water are good things to soften up the skin, but rubbing is what the skin wants. Every morning or every evening, or when it is most convenient, wash the body all over with water and a little ammonia, or anything which tends to make the water soft; then rub dry with a towel, and after that go over the body from top to toe with a dry brush. Try this for two or three weeks, and your skin will be like velvet.

41. Rubbing of the skin is essential to softening it.

A. True.
B. False.
C. Can't tell.

42. According to the passage, pores:

A. Are large holes found in the skin.
B. Are responsible for exuding waste material from the body.
C. Have the singular function of perspiring.
D. Are the only method employed by the body to exude waste.

43. Bathing without rubbing:

A. Does not clean you at all.
B. Is good for cleaning the skin if hot water is used.
C. Is how to make skin soft.
D. Causes no real harm or benefit to the skin.

44. Ammonia is used when bathing because:

A. It kills bacteria.
B. It dislodges dead skin from your pores.
C. There is no other alternative.
D. It softens the water.

END OF SECTION

Section B: Decision Making

1. Tennis balls come in packs of 5, 8 and 18. Which of the following denotes an exact number of balls that cannot be reached purchasing these packs.

   A. 188.   B. 44.   C. 13   D. 21.   E. 27.

2. "It is best for teenagers to start driving when they are 17.". Which statement gives the best supporting reason for this statement?

   A.   Teenagers are immature and driving a car before 17 is unsafe.
   B.   It is illegal to drive without a provisional license which can only be obtained at 17 years of age or older.
   C.   Most teenagers cannot afford to buy a car until they are 17.
   D.   At 17 years old most teenagers have the cognitive abilities to successfully drive a vehicle.

3. Professor Moriarty is only able to give lectures on physics. All lectures on philosophy are given by physics professors. Identity is a topic in philosophy. Some maths lectures are given by physics professors. Which statement below is TRUE.

   A. Professor Moriarty can give lectures on philosophy.
   B. Professor Moriarty can lecture on Identity.
   C. All professors that give philosophy lectures can also lecture on physics.
   D. All philosophers know a lot about physics.
   E. Professor Moriarty gives all of the maths lectures.

4. There are 4 different coloured cars parked in a parking lot. The yellow, red and blue cars are parked adjacent to each other. The blue car has green coloured cotton seats, the yellow car has fluffy pink cotton seats and the red car has shiny black leather seats. One of the cars with non-leather seats has white tires. A car next to the red car has orange tires. Either a car with pink or black seats has silver tires. The yellow car is opposite a car with gold tires. The red car is not next to the blue car. What colour are the yellow car's tires?

   A. Gold.   C. Silver.   E. Cannot tell.
   B. Orange.   D. White.

5. All engineers are intelligent. Some engineers are funny. Jamal is intelligent and Derek is funny. Choose a correct statement.

   A. An engineer can be funny as well as intelligent.
   B. Derek is both intelligent and funny.
   C. All funny people are intelligent as well
   D. Jamal is an engineer.
   E. None of the above.

6. Prakash, Ryan and Harvey are 100-metre sprinters. In order to qualify for district finals, they need to be able to set a time of less than 12 seconds. Prakash and Harvey are faster sprinters than Ryan. Ryan's best time is 12.2 seconds. Which statement must be correct?

A. Prakash and Harvey qualify.
B. All 3 sprinters qualify.
C. None of the three qualify.
D. Ryan doesn't qualify
E. None of the above.

7. Jennifer is drinking a cup of coffee which contains 95% water and 5% is dissolved coffee granules. The weight of the coffee in the cup is 100g. The cup weighs 10g. She leaves the coffee in the sun for an hour and then returns to it an hour later. Some of the water has evaporated in the sun and now the water content of the cup of coffee is 75%. What is the weight of the cup of coffee now?

A. 75g.   B. 110g.   C. 30g.   D. 20g.   E. 50g.

8. Julia, Polly and Fred own sweet shops. James and Fred own bakeries. Harrison recently became the owner of a grocery store. All the women own pharmacies, except for Holly who owns a sports shop. James, Polly and Harrison own shoe shops. Who owns the most shops?

A. Polly.   B. Fred.   C. James.   D. Julia.   E. Harrison.

9. Brian weighs less than Archith. Clarissa weighs less than Mike who weighs less than Archith. Archith and Christy weigh less than Billy. Who weighs the most?

A. Billy.   B. Christy.   C. Archith.   D. Mike.   E. Clarissa.

10. The odds of Sanjay losing his football match on a hot day is 0.1 and on a cold day is 0.3. The probability that Sanjay goes for ice cream on any day is 0.5. Sanjay has a match every single day of the week. Last week, it was cold for the last 3 days of the week. Sanjay thinks that last week he was more likely to lose his match in the cold and not eat ice cream after than he was to lose his match in the heat and eat ice cream after. Is he correct?

A. Yes.
B. No because the chance of winning the match when it is cold is greater than winning the match when it is hot.
C. No because he eats ice cream most days.
D. No because the chance of losing his match on a hot day is almost certain and the chance he loses his game on a cold day is less than half.

11. All athletes play sports. All footballers are athletes. Football and rugby are sports. Drogba is an athlete. Which statement is true?

A. All athletes play football or rugby
B. Drogba can play rugby.
C. All athletes play football as well as another sport.
D. Drogba plays football.
E. None of the above.

12. Homer has flipped 2 regular 10 pence coins 17 times. 11 times he has flipped the result has been 2 heads and 6 times the result has been 2 tails. Homer flips the 2 coins again and one lands as heads. Which statement is most likely to be correct?

A. The other coin also will land as heads.
B. The other coin will land as tails.
C. Homer is flipping both coins the same way.
D. Flipping heads is just as likely as flipping tails.
E. None of the above.

13. Alaric, David and Alex compete in a 500m cycling race. All of them cycle at a constant speed throughout the race. Alaric beats David by 50m and David beats Alex by 50m. How many metres does Alaric beat Alex by?

A. 100m.   B. 95m.   C. 90m.   D. 80m.   E. 150m.

Question 14:
Corrine has an exam today. She says that she knows she will cry later this evening. What assumption has Corrine made?

A. That the exam will have a lot of maths questions.
B. That she will be caught cheating during the exam by invigilators.
C. That the exam is going to cause Corrine stress and sadness.
D. That she will have to resit this exam multiple times.

15. "If the public elections have decided that Jed Arksey is to be the new prime minister, then the public should pay for his inauguration ceremony." Which of the following statements would weaken the argument?

A. The cost of the ceremony is too much to be funded locally.
B. Jed Arksey has the support of most of the MPs.
C. Many people did not vote for Jed Arksey.
D. Many politicians think that inauguration ceremonies are outdated.

16. Whilst discussing their family history, Isabella points to a name on their family tree and says to her son that "her sister's mother is the only daughter of my grandmother." How is the name that Isabella points to related to her?

A. Niece
B. Sister
C. Cousin
D. Aunt
E. Mother

17. In Oxford, a survey is done on a primary school. 5 children like all types of cake. 3 children like only chocolate and banana, and 13 like carrot. However, 2 like banana cake only. Only one child likes all flavours of cake but chocolate. Which Venn diagram represents this information?

A.

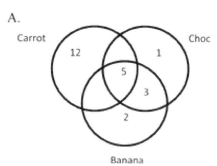

C.

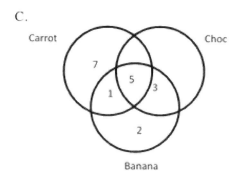

B.

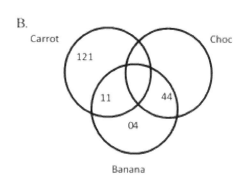

D.

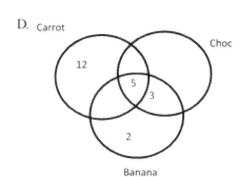

18. Harold and Fletcher play a game of 11. They have to take turns counting numbers up to 11. In a turn, they can say 1, 2 or 3 numbers. The person who says 11 wins the game. If Harold starts the game, how many numbers does he have to say to guarantee a win.

A. 1
B. 2
C. 3

D. It is impossible for Harold to win if he starts the game.

19. In a game of dance off, everyone puts their names in a hat. Each player takes turns to draw a name. If they pick their own name, it is placed back and they draw another. If the last person to pick draws their own name, everyone starts again. Imogen, Lucy and Norden play and pick in alphabetical order. Which statement is true?

A. Imogen has a one in three chance of picking and dancing against Norden.
B. Lucy is more likely to pick Imogen than picking Norden.
C. Imogen has an equal chance of dancing against Norden or Lucy.
D. Norden should reverse the order is he wants to increase his chances of dancing against Imogen.

20. Barry is sat to the right of Andrew. Charlie is sat next to Barry and Andrew. David is in front of Charlie. Eric is sat in front of Andrew. Where is David with respect to Eric?

A. David is behind Eric.
B. Eric is behind David.

C. Eric is to the right of David.

D. David is to the left of Eric.
E. Eric is to the left of David.

21. "The growing population is a growing problem, therefore there should be a tax on having more than one child." Which of the following statements would weaken the argument?

A. Having more than one child is already a high financial cost due to schooling costs.
B. People from low income backgrounds are more likely to have more than one child.
C. Cost has no effect on how many children people have.
D. People who only want 1 child will be unfairly benefited.

22. Three brothers, Peter, Max and Edwin are arguing over how many times they have been to the swimming pool this year. Peter says "We have been at least 5 times", Max says "No we have been less than 5 times", and Edwin says "We have been at least once". If only one of the brothers is telling the truth, how many times have they been to the swimming pool this year?

A. 3 B. 1 C. 0 D. 4 E. 6

23. "Living in cities is injurious to health." Which statement provides the best evidence for this conclusion?

A. People in cities exercise less than people in rural areas.
B. Fast food is cheaper in cities.
C. Cities have harmful levels of carbon monoxide in the air.
D. Increased traffic in cities means increased pollution.
E. Healthy food is more accessible in cities.

24. Will has many shoes in his wardrobe. He has 1 blue, 2 black, 4 orange and 10 red shoes. He is trying to find a pair of shoes but his house has had a power cut so he cannot see anything. How many shoes does Will have to pick to make sure he has 3 matching pairs?

A. 4. B. 10. C. 12. D. 5. E. 9.

25. Felix is taller than Bob. Bob's friend Gerard is taller than Felix. They all are basketball players. Felix is better at basketball than Gerard. Felix's brother Chris is the worst at basketball. Which statement is true?

A. Bob is better at basketball than Chris.
B. Gerard is better at basketball than Bob.
C. Bob is the shortest.
D. Chris is the tallest.
E. Felix is the worst at basketball.

26. 4 mice are placed in a square shaped maze. They each pick a random direction and walk along the edge of the square. Are the mice less likely to collide than not?

A. No because the likelihood of the mice colliding or not colliding is equal.
B. Yes as the probability they collide is 12.5%.
C. No because they are more likely to collide than they are to not collide.
D. Yes as mice do not like to be near each other.

27. "Lizards are not mammals. Some lizards are poisonous. All mammals produce offspring." Which of the following statements fits best with the given facts?

A. No lizards produce offspring.
B. Some lizards are not poisonous.
C. Some mammals are lizards.
D. Some lizards do not produce offspring.
E. Some lizards produce offspring.

28. Jacob played a game of football. He says every goal he scored was from a free kick and he scored every free kick his captain let him take. Which statement is true?

A. No goals were scored in the match that weren't from free kicks.
B. Jacob scored every goal in the match.
C. Jacob didn't have any other opportunities to score goals apart from free kicks.
D. Free kick goals were the only goals Jacob scored in the match.
E. Jacob didn't see any goals during the match that weren't from free kicks.

29. A survey was conducted in a city to learn about preferences for sport. 80 people said they liked football, 50 people said they liked rugby and 40 people said they liked cricket. 30 people said they liked all 3 sports. 0 people said they liked only football and rugby. 0 people said they only liked football and cricket. 10 people said they liked only rugby and cricket. How many people took the survey?

A. 75   B. 68   C. 110   D. 150   E. 100

END OF SECTION

Section C: Quantitative Reasoning

Data Set 1

Jake decided to take a trip up to Manchester from London. The journey by train took 2 hours 8 minutes over a distance of approximately 210 miles.

1. What was the average speed the train was travelling from London to Manchester?

    A. 104.7 mph
    B. 98.4 mph

    C. 89.6 mph
    D. 95.9 mph

    E. 100.2 mph

2. If there are 1.6 kilometres in a mile, how many kilometres did Jake travel over his journey?

    A. 346 km
    B. 340 km

    C. 336 km
    D. 353 km

    E. 338 km

3. Jake's sister Martha decided to drive to Manchester from London instead. It took her 3 hours 32 minutes (over approximately the same distance). How much slower was Martha's journey than Jake's, as a percentage to the nearest whole number?

    A. 54%   B. 74%   C. 47%   D. 59%   E. 66%

Data Set 2

Sam is a farmer and owns a field to grow his crops. The characteristics of the field are as follows:

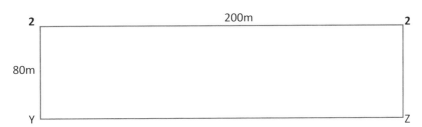

4. What is the area of the field?

    A. 8000m^2
    B. 16000m^2

    C. 10000m^2
    D. 18000m^2

    E. 16000m

5. Sam's wife, Helen, jogs along the edge of the field every Saturday morning to exercise. How long does it take her to jog around the field once? Assume that she jogs at a pace of 6km/hr.

    A. 5min 27sec
    B. 4min 54sec

    C. 5min 43sec
    D. 5min 36sec

    E. 5min 16sec

6. Sam uses the field to grow cabbages in rows parallel to the WX edge of the field. The first and last rows grow on the WX and YZ edges of the field, respectively. There are 50cm between each row. How many rows of cabbages are there on Sam's field?

    A. 150
    B. 142

    C. 161
    D. 156

    E. 148

Data Set 3

The table shows the population of Town X, divided up into age groups.

Age Group	Population	Age Group	Population
0 - 5	6900	46 - 50	9050
6 - 10	6750	51 - 55	8400
11 - 15	8200	56 - 60	7850
16 - 20	7150	61 - 65	8350
21 - 25	6350	66 - 70	7500
26 - 30	6400	71 - 75	6450
31 - 35	8750	76 - 80	6000
36 - 40	7750	81 - 85	5700
41 - 45	9950	86 and over	3650

7. How many people are older than 65?

   A. 25900   C. 29300   E. 23900
   B. 30650   D. 28600

8. If everyone in town X retires at 65 years, what percentage of the population is within 14 years of retiring?

   A. 18.76%   C. 22.87%   E. 16.68%
   B. 15.42%   D. 19.53%

9. Approximately what number of the population is of working age, assuming this starts at 16?

   A. 85000   C. 82000   E. 80000
   B. 79000   D. 87000

10. Approximately what fraction of the population is under 30 years?

   A. 1/4   B. 1/5   C. 1/3   D. 1/6   E. ½

Data Set 4

The pie chart below shows the popularity of different sports offered by a leisure centre.

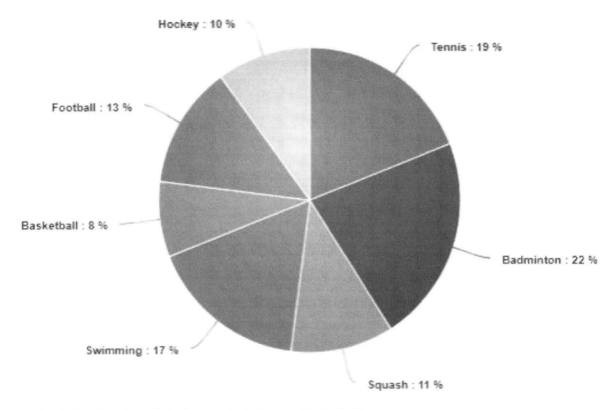

11. What is the ratio of popularity between badminton and basketball?

    A. 13:6   B. 11:5   C. 15:8   D. 11:4   E. 9:3

12. By how much is football less popular than tennis, as a fraction?

    A. 6/19   B. 7/18   C. 5/16   D. 7/19   E. 2/15

13. On average, approximately 570 people play a sport at the leisure centre each week. How many people play a racquet sport each week?

    A. 283   B. 301   C. 294   D. 308   E. 295

Data Set 5

The graph below shows company D's sales of hot chocolate over a year.

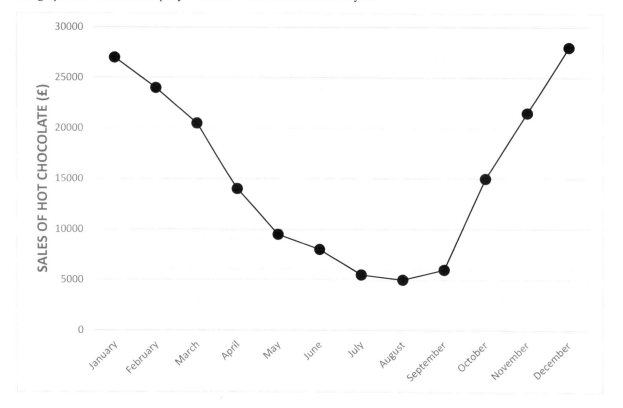

14. By how much do the sales increase from August to December?

    A. 21000   B. 21500   C. 22000   D. 22500   E. 23000

15. In October company D falls short of their expected sales by 12%, how much were they expecting to make to the nearest whole number?

    A. £16500   C. £18110   E. £15075
    B. £17045   D. £17000

16. If each hot chocolate costs £2.50, how many hot chocolates did the company make in January?

    A. 11400   B. 10800   C. 13200   D. 9900   E. 10000

Data Set 6

A company produced two variations of mango juice. They meet with two focus groups to decide which variation is preferred.

	Focus Group A	Focus Group B
Mango Juice 1	70%	55%
Mango Juice 2	25%	30%

17. What fraction of focus group B did not have a preference?

A. 3/20   B. 1/8   C. 7/40   D. 1/5   E. 7/25

18. If there are 108 people in focus group A and 110 in focus group B, how many people preferred mango juice 2 in total?

A. 55   B. 62   C. 58   D. 60   E. 63

Data Set 7

The following diagram presents student's percentages scored in the exam 'Methods in Mathematics'.

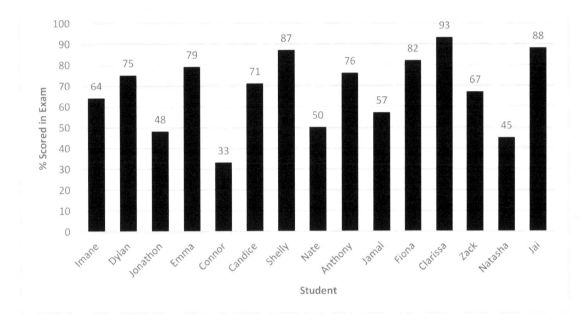

19. What is the average percentage obtained by the students?

A. 60.7%   C. 59.8%   E. 68.2%
B. 67.7%   D. 70.1%

20. If the school exam board decides that only up to 15% of students in the class should receive the maximum grade, what is the minimum threshold mark value for this?

A. 79%   C. 93%   E. 88%
B. 82%   D. 87%

21. If all students attended an extra class and the marks went up by 2% for all students, how many students would score higher than the 2 lowest scores combined?
A. 3   B. 5   C. 4   D. 1   E. 6

Data Set 8

The graph below shows the fluctuations of Company R's share price in 2007.

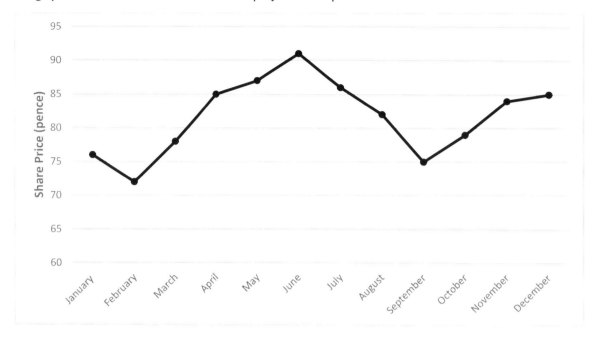

22. What fraction of the April share price is the September price?

A. 15/16   B. 14/21   C. 13/18   D. 15/17   E. 12/15

23. If the company's share price increased by 13% in January 2008, what would be its value?

A. 95.10 p   B. 92.04 p   C. 96.05 p   D. 100.00 p   E. 91.80 p

Data Set 9

The table right presents the Body Mass Index (BMI) for a group of people.

- ➤ BMI = Weight in kg / (Height in m)^2
- ➤ Normal Range: 18.5-24.9
- ➤ Overweight: 25.0-29.9
- ➤ Obese: 30.0 or more

Person	Weight (kg)	Height (m)
A	72	1.70
B	49	1.55
C	54	1.40
D	68	1.80
E	57	1.60

24. Which person has the lowest BMI?

A. A   B. B   C. C   D. D   E. E

25. Person F has a BMI of 17.6 but would like to be within the normal range. If person F currently weighs 45kg, how much weight would they need to gain?

A. 4kg   B. 7.5kg   C. 2.5kg   D. 1kg   E. 5kg

26. Person G is 1.7m and has a BMI of 28.0. Approximately what fraction of their weight would they have to lose to be within the normal BMI range?

A. 1/8   B. 1/10   C. 1/15   D. 1/6   E. 1/12

Data Set 10

The next set of questions concerns the cost of chicken and chips at a fast food restaurant.

27. If 3 portions of chicken and 4 portions of chips cost a total of £20.95, and 4 portions of chicken and 5 portions of chips cost £27.10, what equations would be used to determine the price of each item?

   A. $3x + 4y = 20.95$; $4x + 5y = 27.10$
   B. $6c + 8g = 2095$; $2c + y = 2600$
   C. $h + t = £6.10$; $2h + 3t = £14$
   D. $f = £4.20$; $e = £2.65$
   E. $2x + 6y = 20.95$; $5x + 4y = 27.10$

28. If the price of chicken increases by 10% and the price of chips increase by 5%, what would be the cost of a portion of chicken and chips, to the nearest penny?

   A. £7.54   B. £8.88   C. £6.00   D. £10.02   E. £6.65

29. The restaurant also accepts payment in other currencies. If a tourist purchases 2 portions of chicken and 5 portions of chips in dollars ($) at the original prices, how much would they pay? Assume that £1 = $1.39.

   A. $29.55   B. $32.46   C. $25.98   D. $27.52   E. $30.72

Data Set 11

A weightlifting team were required to check the weights of their entrants before a competition.

Team Member	Alex	Jay	Thomas	Kyran	James
Weight	78kg	13 stone 7lb	11 stone 11lb	82kg	97kg

1kg = 2.2 lb

1stone = 14 lb

30. How much does Kyran weigh in lb?

   A. 176.3 lb   B. 184.0 lb   C. 180.4 lb   D. 192.5 lb   E. 169.8 lb

31. How much does Alex weigh in stones and lb?

   A. 10 stone 4 lb   C. 12 stone 4 lb   E. 13 stone 2 lb
   B. 12 stone 0 lb   D. 11 stone 9 lb

32. What is the total weight of the boys in kg?

   A. 398.4kg   B. 417.9kg   C. 476.3kg   D. 502.8kg   E. 454.7kg

33. Approximately how much lighter is Alex than James, as a fraction?

   A. 1/5   B. 1/8   C. 1/4   D. 1/6   E. 1/10

Data Set 12

The graph shows the production of weed killer by a company in 10,000 litres between the years 1998-2005.

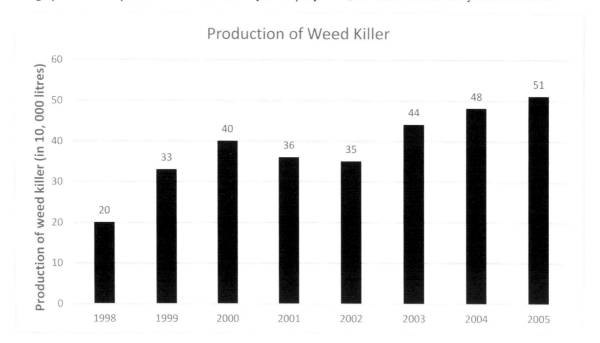

34. What was the percentage increase in the production of weed killer from 1998 to 1999?

A. 54%   B. 72%   C. 65%   D. 49%   E. 73%

35. In how many of the years was the production of weed killer more than the average production in the period shown above?

A. 1   B. 2   C. 3   D. 4   E. 5

36. In 2006 the company shut down one of the factories producing the weed killer, reducing the amount produced to 350,000 litres for that year. What was the decrease in production of weed killer from 2005 to 2006 as a decimal?

A. 0.431   B. 0.517   C. 0.298   D. 0.314   E. 0.277

END OF SECTION

Section D: Abstract Reasoning

For each question, decide whether each box fits best with Set A, Set B or with neither.

For each question, work through the boxes from left to right as you see them on the page. Make your decision and fill it into the answer sheet.

Answer as follows:
A = Set A
B = Set B
C = neither

Set 1:

Set A

Set B

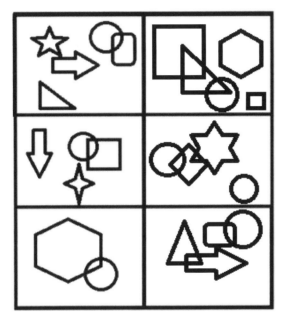
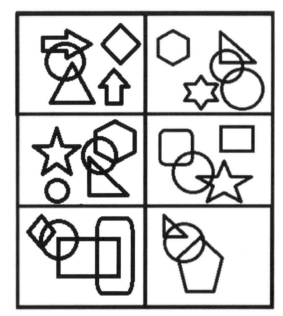

Questions 1-5

Set 2:   **Set A**   **Set B**

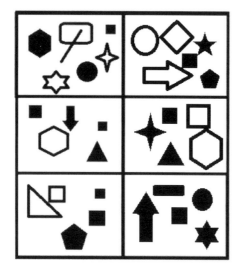

Question 6-10

Set 3:   **Set A**   **Set B**

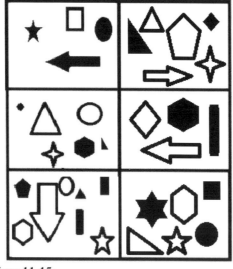
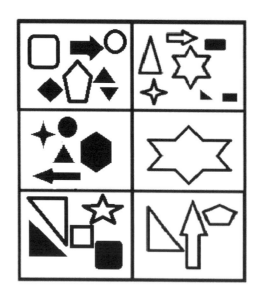

Questions 11-15:

Set 4:   Set A   Set B

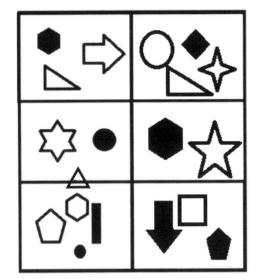 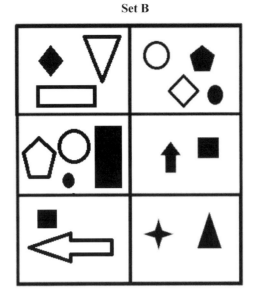

Question 16-20

Set 5:   Set A   Set B

 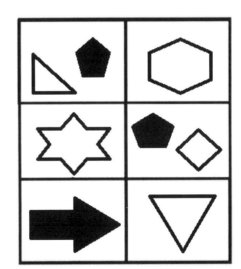

Questions 21-25

For the next set of questions, decide which figure completes the series.

Set 6:

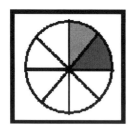 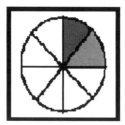 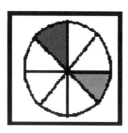 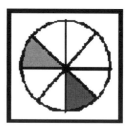

Question 26

| A | B | C | D |

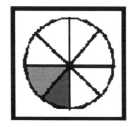 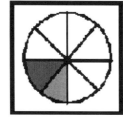 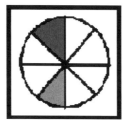 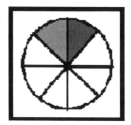

Set 7:

Question 27

| A | B | C | D |

For the next set of questions, decide which figure completes the statement

Set 8:

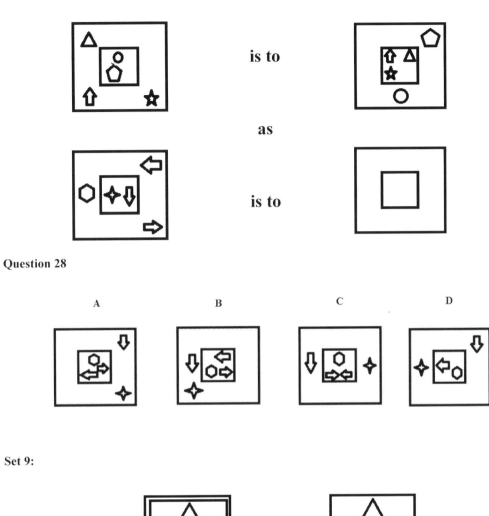

Question 28

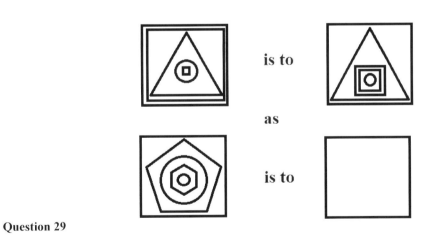

Set 9:

Question 29

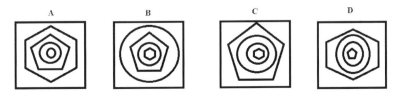

For the next set of questions, decide which of the options fits with the set stated.

Set 10:

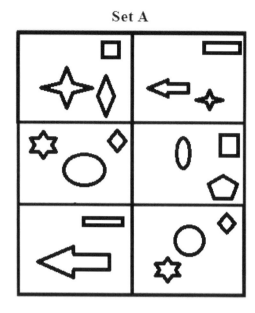 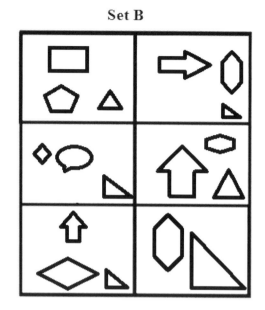

Set A   Set B

Question 30

Which of the following belongs in Set A?

A    B 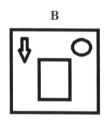   C 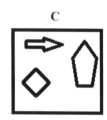   D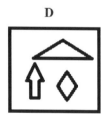

Question 31

Which of the following belongs in Set B?

A    B 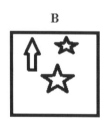   C 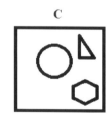   D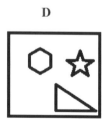

For the next set of questions, decide whether each box fits best with Set A, Set B or with neither.

Set 11:

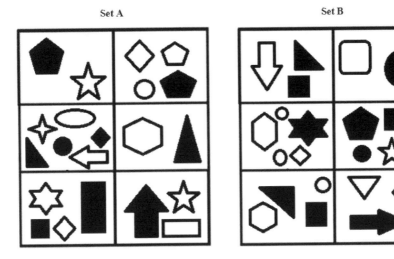

Questions 32-36

Set 12:

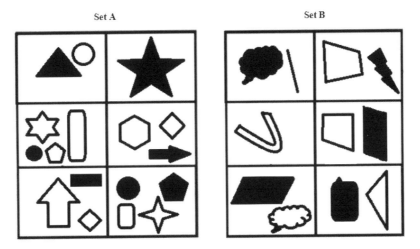

Questions 37-41

Set 13:

Set A   Set B

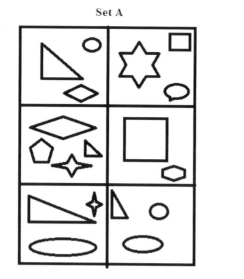 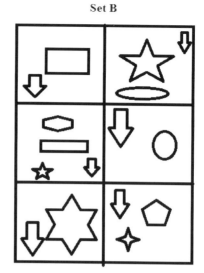

Questions 42-46

Set 14:

Set A   Set B

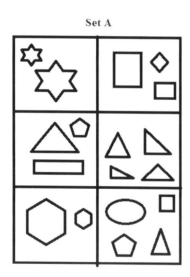 

Questions 47 – 51

For the next set of questions, decide which figure completes the series.

Set 15:

Question 52

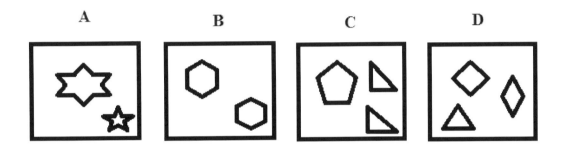

Set 16:

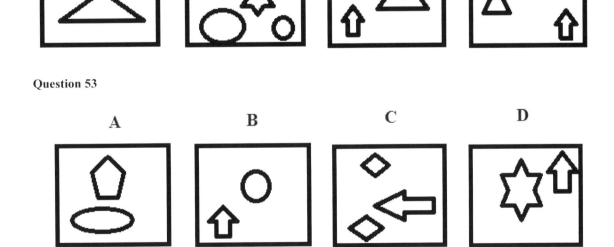

Question 53

For the next set of questions, decide which of the options fits with the set stated.

Set 17:

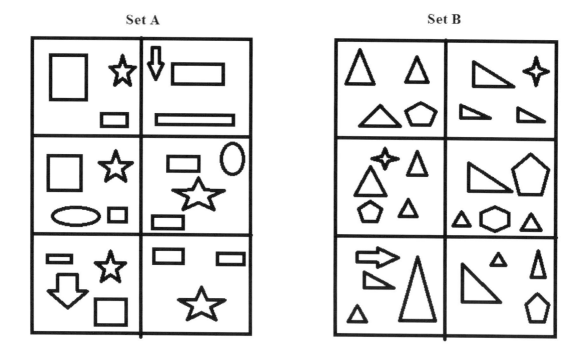

Question 54

Which of the following belongs in Set A?

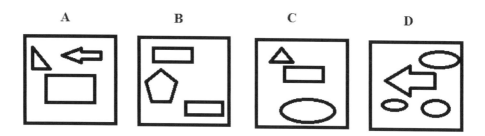

Question 55

Which of the following belongs in Set B?

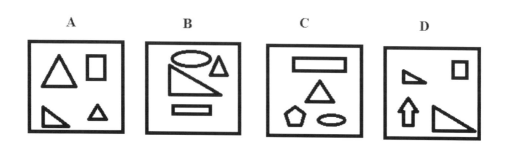

END OF SECTION

Section E: Situational Judgement Test

Read each scenario. Each question refers to the scenario directly above. For each question, select one of these four options. Select whichever you feel is the option that best represents your view on each suggested action.

For questions 1 – 30, choose one of the following options:

A	A highly appropriate action
B	Appropriate, but not ideal
C	Inappropriate, but not awful
D	A highly inappropriate action

Scenario 1

Afolarin is a first-year medical student. He plays for the university lacrosse team which has 3 training sessions a week, 1 match a week and 1 social a week. Having finished his first term, Afolarin knows he is considerably behind on work and is struggling to keep up due to his lacrosse commitments. Afolarin loves lacrosse and it is beneficial for his mental wellbeing. Afolarin has end of year exams coming and he is worried he will fail if he continues to play lacrosse.

How appropriate are the following actions to Afolarin's problem?

1. Study harder in the day time and try and fit more time to study into the day when he isn't playing lacrosse.
2. Stop playing lacrosse altogether for a month before exams.
3. Continue to play lacrosse but avoid going to the social so he has 1 more evening a week to study for his exams.
4. Speak to his coach about his concerns and try and create a new training schedule so that he has enough time to study for his exams.
5. Quit playing lacrosse as his medical degree is more important to him than his love of lacrosse.

Scenario 2

Jamal is a second-year medical student who has recently undertaken an additional project. This project will add value to Jamal and help develop his experimental technique. The project however is time consuming and the final submission for the project is 2 weeks before his 2nd year medical school exams. Jamal is keen to progress as a scientist however he does not want to jeopardise his medical exams.

How appropriate are the following responses by Jamal?

6. Quit the project and focus on his exams.
7. Continue with the project and fit preparation for his medical exams into the 2 weeks after the project submission deadline.
8. Allocate equal working hours to the project and to studying for his exams from now on.
9. Ask his peers who are also undertaking the project how they are planning to manage their time.
10. Speak to the project administrators and see if he can have an extension on the project deadline due to his upcoming exams.

Scenario 3

Jacob is a second-year medical student. He has been looking forward to the ski trip for a long time because he has booked a nice chalet with an amazing view and is staying with a group of his best friends from his course. He took out a small loan to be able to pay for this trip, and has spent a lot of time planning it. Unfortunately, the end-of-year exam results have just come out, and he has only passed 4 out of 5 exams. He only failed by 4%, but must re-sit this final exam in order to continue with his studies. The re-sit is scheduled to be only three days after his return from the ski trip.

How appropriate are the following responses to Jacob's dilemma?

11. Go on his trip and try and fit all the revision into the three days before the exam – he only failed by 4% anyway.
12. Cancel the entire trip and let his friends go without him.
13. Ask his friends if they can all reschedule for the following year.
14. Take revision with him on the ski trip – it'll be an active holiday but maybe he can study in his free time.
15. Go for part of the trip but return early to ensure sufficient time for revision.

Scenario 4

Ronit and Shloke are first year medical school students doing an experiment in biochemistry lab. They are using corrosive acids in an experiment. The professor has clearly instructed that for health and safety reasons everyone must wear lab coats and safety glasses. Shloke notices that Ronit isn't wearing his safety glasses and when prompted Ronit simply shrugs and dismisses it.

How appropriate are the following actions from Shloke?

16. Immediately alert the professor that Ronit isn't following protocol.
17. Do nothing as no harm will probably befall Ronit during the practical.
18. Tell all the surrounding classmates that Ronit isn't following protocol.
19. Explain to Ronit why following protocol is important and encourage him to wear his safety glasses.
20. Give Ronit his own safety glasses.

Scenario 5

Juan is a 4th year medical student who has just started his clinical training. During his induction he was told to just observe for the first 2 weeks of term and to not actively participate in any procedures. The following Friday is extremely busy in A&E and Juan is shadowing a nurse. She is applying pressure to a wound when she receives an urgent phone call. The nurse asks Juan to apply pressure to the wound just for a minute whilst she attends to the phone call.

How appropriate are each of the following responses by Juan in this situation?

21. Act like he didn't hear what the nurse said and go to the toilet.
22. Put on sterile gloves and immediately apply pressure to the wound.
23. Tell the nurse that he isn't allowed to and politely refuse.
24. Explain to the nurse what he was told at the induction and offer to go find someone else to cover the nurse whilst she attends to her phone call.
25. Tell the nurse okay and then ask the patient to apply pressure to their own wound.

Scenario 6

Nicolas and Jessica are first-year medical students. They are in their first anatomy class where they are observing dissected cadavers. As soon as the demonstrator removes the sheet covering the body Nicolas immediately starts to feel dizzy and becomes extremely pale but doesn't say anything. Jessica notices that Nicolas looks unwell.

How appropriate are the following actions by Jessica in this situation?

26. Say nothing as Nicolas isn't saying anything probably because he doesn't want attention.
27. Ask Nicolas if he is feeling okay and get him a glass of water.
28. Tell the demonstrator that Nicolas is unwell and needs to leave the room.
29. Open the window so Nicolas gets some fresh air.
30. Ask the demonstrator to re-cover the body and see if Nicolas feels better.

For questions 31 – 69, decide how important each statement is when deciding how to respond to the situation?

A   **Extremely important**
B   **Fairly important**
C   **Of minor importance**
D   **Of no importance whatsoever**

Scenario 7

Alvin is a 2nd year medical student. He has been offered the opportunity to interview a prominent scientist in an area of medicine he is interested in pursuing. The interview is in 2 days' time however Alvin has an essay due in for the day after the interview.

How important are the following factors for Alvin's decision?

31. The likelihood that this interview will benefit his career.
32. How quickly Alvin is able to write the essay.
33. Whether Alvin is going to have any other opportunities to interview this scientist.
34. How much this essay is going to help him in his end of year exams.
35. Alvin's reputation with the professor who will mark his essay.

Scenario 8

Damion and Phillipa are 5th year clinical medical students. They are currently collaborating on a statistical project which involves analysing patient data. Damion is one day speaking to his friend who tells him that Phillipa was explaining their project. Damion realises that Phillipa has accidentally been revealing patient data to some of her peers.

How important are the following factors for Damion in deciding what to do?

36. Completing the project before the deadline.
37. The breach of patient confidentiality by Phillipa.
38. The grade that Damion receives for the project is the same that Phillipa gets.
39. Clinical staff may find out about the breach in patient confidentiality.
40. The patient data revealed is largely harmless and will probably not be shared by Phillipa's peers.
41. The project may be invalidated.

Scenario 9

Carlos is 4th year medical student who has just started clinical school. In a clinical workshop he was lectured to about how important it is to look professional in the hospital. During his first rotation Carlos notices that the consultant he is shadowing frequently looks dishevelled and untidy when in the ward.

How important are the following factors for Carlos in deciding on what to do?

42. His rapport with the consultant.
43. His consultant will determine his final grade.
44. The consultant's appearance may reflect badly on Carlos in front of the patients.
45. The workshop on looking professional was optional to attend.
46. The consultant is extremely good at his job regardless of his appearance.

Scenario 10

Arran is a 6th year medical student. He has been invited to sit in on a rare surgical procedure that many medical students do not get to see. This is a fantastic opportunity to get experience that will stand him in good stead for the future. A day before the surgery Arran becomes ill with tonsillitis. Arran feels he is well enough to be able to attend the surgery however his illness could put the patient in the surgery at risk.

How important are the following factors for Arran to consider in deciding on what to do?

47. This opportunity will probably not arise again.
48. He has had tonsillitis before and he knows he can attend the surgery and remain focussed.
49. His illness could put the patient in the surgery at risk.
50. If he doesn't attend the surgery then the doctors may not offer similar opportunities in the future.
51. He knows that some of his friends have gone in to the hospital when they have been ill.

Scenario 11

Abraham is a 2nd year medical student. He has optional chemical pharmacology seminars on Wednesday afternoons. Abraham wants to attend these seminars so he can show his depth of knowledge in the end of year exams. Abraham is also a keen football player and football practice for the university team clashes with his chemical pharmacology seminars.

How important are the following factors for Abraham in deciding on what to do?

52. Abraham has been playing really well for his team recently and is in contention for the captaincy next year.
53. The chemical pharmacology lecturer is the examiner for the end of year exams.
54. The chemical pharmacology lecturer is a huge football fan and often comes to watch the university team play.
55. The information learned in the optional seminars will probably only improve his marks in the end of year exams a little.
56. The football team also has training sessions on other days of the week.

Scenario 12

Cameron and Brandon are 5th year medical students. One day when they are both in the clinic Cameron notices that Brandon looks extremely tired. Cameron and Brandon have suturing training later that day which involves patient contact. Cameron knows that Brandon went drinking last night and didn't get to sleep until quite late.

How important are the following factors to Cameron's situation?

57. Cameron knows that Brandon is skilled at suturing and will probably be alright.

58. Brandon's tiredness could compromise patient safety.

59. Brandon would be embarrassed if Cameron confronted him about being in hospital in an unsuitable state.

60. Brandon may look tired but maybe he feels completely fine.

61. Brandon has previously been in trouble with the medical faculty.

62. Cameron may cause problems in his friendship with Brandon if he confronts him.

Scenario 13

Jason is a 3rd year medical student currently working on a project which involves helping a research group with parts of their experiments. The research team lead offers Jason the opportunity to do some extra work during August in his summer holidays and in return he will get his name on the paper if it is published. Jason is very keen to get a publication in his name however he was planning to go on cricket tour with his teammates in August.

How important are the following factors for Jason in deciding what he should do?

63. Jason has been looking forward to this tour for a long time.

64. There is no guarantee that the paper will get published.

65. A publication in his name will help him in his future career.

66. Cricket tour occurs annually.

67. Jason's friends may be annoyed if he cancels on cricket tour.

68. Refusing to work in summer may create a bad image of Jason in the eyes of the research team lead.

69. Jason does not want to spend his summer cooped up in a lab.

END OF PAPER

Mock Paper E

Section A: Verbal Reasoning

Passage 1

"WAL, of all the dinners that ever a white man sot down to, this yere is the beat!"

The speaker was Godfrey Evans—a tall, raw-boned man, dressed in a tattered, brown jean suit. He was barefooted, his toil-hardened hands and weather-beaten face were sadly soiled and begrimed, and his hair and whiskers looked as though they had never been made acquainted with a comb. As he spoke he drew an empty nail-keg from its corner, placed a board over the top of it, and seating himself, ran his eye over the slender stock of viands his wife had just placed on the table.

The man's appearance was in strict keeping with his surroundings. The cabin in which he lived and everything it contained told of the most abject poverty. The building, which was made of rough, unhewn logs, could boast of but one room and a loft, to which access was gained by a ladder fastened against the wall. It had no floor and no windows, all the light being admitted through a dilapidated door, which every gust of wind threatened to shake from its hinges, and the warmth being supplied by an immense fire-place with a stick chimney, which occupied nearly the whole of one end of the cabin. There were no chairs to be seen—the places of these useful articles being supplied by empty nail-kegs and blocks of wood; and neither were there any beds—a miserable "shake-down" in one corner being the best in this line that the cabin could afford. Everything looked as if it were about to fall to pieces. Even the rough board table on which the dinner was placed would have tumbled over, had it not been propped up against the wall.

1. Which of the following statements is true regarding the above passage?

A. Godfrey Evans is a rich man
B. Godfrey Evans is unmarried
C. Godfrey Evans is clean-shaven
D. Godfrey Evans uses nail-kegs as chairs

2. Which of the following can be inferred from the above passage?

A. Godfrey Evans is an educated ma
B. Godfrey Evans is of Caucasian descent
C. Godfrey Evans lives in an urban area
D. Godfrey Evans has two children

3. Based on the above passage, what does Godfrey most likely do for a living?

A. Lawyer
B. Politician
C. Unemployed
D. Farmer

4. Which year is the above passage most likely to be set in?

A. 3007
B. 2018
C. 1996
D. 1878

Passage 2

William Allen, the eminent chemist, was born in London. His father was a silk manufacturer in Spitalfields, and a member of the Society of Friends. Having at an early period shown a predilection for chemical and other pursuits connected with medicine, William was placed in the establishment of Mr. Joseph Gurney Bevan in Plough Court, Lombard Street, where he acquired a practical knowledge of chemistry. He eventually succeeded to the business, which he carried on in connection with Mr. Luke Howard, and obtained great reputation as a pharmaceutical chemist. About the year 1804, Mr. Allen was appointed lecturer on chemistry and experimental philosophy at Guy's Hospital, at which institution he continued to be engaged more or less until the year 1827. He was also connected with the Royal Institution of Great Britain, and was concerned in some of the most exact experiments of the day, together with Davy, Babington, Marcet, Luke Howard, and Dalton. In conjunction with his friend Mr. Pepys, Allen entered upon his well known chemical investigations, which established the proportion of carbon in carbonic acid, and proved the identity of the diamond with charcoal; these discoveries are recorded in the 'Philosophical Transactions' of the Royal Society, of which he became a member in 1807. The 'Transactions' for 1829 also contain a paper by him, based on elaborate experiments and calculations, concerning the changes produced by respiration on atmospheric air and other gases. Mr. Allen was mainly instrumental in establishing the Pharmaceutical Society, of which he was president at the time of his death. Besides his public labours as a practical chemist, he pursued with much delight, in his hours of relaxation, the study of astronomy, and was one of the original members of the Royal Astronomical Society. In connection with this science, he published, in 1815, a small work entitled 'A Companion to the Transit Instrument.'

5. William Allen was a:

A. Scientist
B. Gardener
C. Farmer
D. Lawyer

6. Which of the following statements is true regarding the above passage?

A. William Allen always worked by himself
B. William Allen worked in a hospital
C. William Allen was only a member of one society
D. William Allen always worked by himself

7. Which of the following statements is FALSE regarding the above passage?

A. William Allen was the president of a Royal Society
B. William Allen made publications about astronomy
C. William Allen discovered carbonic acid
D. William Allen's father was a silk manufacturer

8. Based on the above passage, which of the following statements can be inferred?

A. William Allen made a significant contribution to science
B. William Allen was an only child
C. William Allen remained unmarried until his death
D. William Allen died of tuberculosis

Passage 3

Before earth and sea and heaven were created, all things wore one aspect, to which we give the name of Chaos—a confused and shapeless mass, nothing but dead weight, in which, however, slumbered the seeds of things. Earth, sea, and air were all mixed up together; so the earth was not solid, the sea was not fluid, and the air was not transparent. God and Nature at last interposed, and put an end to this discord, separating earth from sea, and heaven from both. The fiery part, being the lightest, sprang up, and formed the skies; the air was next in weight and place. The earth, being heavier, sank below; and the water took the lowest place, and buoyed up the earth.
Here some god—it is not known which—gave his good offices in arranging and disposing the earth. He appointed rivers and bays their places, raised mountains, scooped out valleys, distributed woods, fountains, fertile fields, and stony plains. The air being cleared, the stars began to appear, fishes took possession of the sea, birds of the air, and four-footed beasts of the land.

But a nobler animal was wanted, and Man was made. It is not known whether the creator made him of divine materials, or whether in the earth, so lately separated from heaven, there lurked still some heavenly seeds. Prometheus took some of this earth, and kneading it up with water, made man in the image of the gods. He gave him an upright stature, so that while all other animals turn their faces downward, and look to the earth, he raises his to heaven, and gazes on the stars.

9. Chaos is a god

A. True
B. False
C. Can't Tell

10. Which is the lightest part of these options?

A. The skies
B. Earth
C. Water
D. Woods

11. Humans are made in the image of God

A. True
B. False
C. Can't Tell

12. The Earth was created by:

A. God
B. Nature
C. God and Nature
D. Prometheus

Passage 4

MERLIN

Merlin was the son of no mortal father, but of an Incubus, one of a class of beings not absolutely wicked, but far from good, who inhabit the regions of the air. Merlin's mother was a virtuous young woman, who, on the birth of her son, entrusted him to a priest, who hurried him to the baptismal fount, and so saved him from sharing the lot of his father, though he retained many marks of his unearthly origin.

At this time Vortigern reigned in Britain. He was a usurper, who had caused the death of his sovereign, Moines, and driven the two brothers of the late king, whose names were Uther and Pendragon, into banishment. Vortigern, who lived in constant fear of the return of the rightful heirs of the kingdom, began to erect a strong tower for defence. The edifice, when brought by the workmen to a certain height, three times fell to the ground, without any apparent cause. The king consulted his astrologers on this wonderful event, and learned from them that it would be necessary to bathe the corner-stone of the foundation with the blood of a child born without a mortal father.

In search of such an infant, Vortigern sent his messengers all over the kingdom, and they by accident discovered Merlin, whose lineage seemed to point him out as the individual wanted.

13. Merlin has no human ancestry
A. True
B. False
C. Can't Tell

14. Based on the above passage, which of the following statements is true?
A. Merlin's father is immortal
B. Vortigern is of royal ancestry
C. There are no living heirs to the throne
D. Uther is the rightful king of Britain

15. Based on the above passage, which of the following statements is FALSE?
A. Vortigern is a superstitious king
B. Merlin's life is in danger
C. Merlin is a Christian
D. Merlin has no special abilities

16. Which correctly describes Vortigern's personality?
A. Benevolent
B. Loving
C. Just
D. Fearful

Passage 5

My Ancestors

I was born near Thorntown, Indiana, August 21, 1834.

My father, James P. Mills, third child of James Mills 2nd and Marian Mills, was born in York, Pennsylvania, August 22, 1808. His father, James Mills 2nd, was born October 1, 1770, and died December 3, 1808.

My father's mother died in 1816, leaving him an orphan at the age of eight. He lived with his Aunt Margery Mills Hayes for about two years, when he was "bound out" as an apprentice to a tanner by the name of Greenwalt, at Harrisburg, Pennsylvania. Here he was to serve until twenty-one, when he was to receive one hundred dollars and a suit of clothes. All the knowledge that he had of books was derived from night school, Greenwalt not permitting him to attend during the day. His apprenticeship was so hard he ran away when twenty, forfeiting the hundred dollars and the clothes.

His only patrimony was from his grandfather, James Mills I, who, as father told me, sent for him on his deathbed and, patting him on the head, said: "I want Jimmy to have fifty pounds."

After running away, my father went to Geneva, New York, and served as a journeyman until twenty-two. With his inheritance of $250, he and his brother Frank started West in a Dearborn wagon, crossing the Alleghenies. He travelled to Crawfordsville, Indiana, and here, about 1830, entered eighty acres of the farm on which I was born. The land was covered with walnut, oak and ash, many of the trees being one hundred feet high and three or four feet in diameter. Felling and burning the trees, he built his house with his own hands, neighbours aiding in raising the walls.

My father had little knowledge of his ancestors, other than that they were Quakers, but, by correspondence with officials of counties where his ancestors lived, I have learned that the first of his family came over with William Penn and settled in Philadelphia.

17. The author is:
A. American
B. Chinese
C. English
D. Indian

18. Which of the following statements is true regarding the above passage?
A. The author was born after the first world war
B. James P. Mills was not an educated man
C. The author's father, grandfather and great-grandfather had the same forename
D. The author's father inherited a large sum of money from his parents

19. James P. Mills was an only child
A. True
B. False
C. Can't Tell

20. The author is female
A. True
B. False
C. Can't Tell

Passage 6

FRANK AND BEN.

"Is your mother at home, Frank?" asked a soft voice.

Frank Hunter was stretched on the lawn in a careless posture, but looked up quickly as the question fell upon his ear. A man of middle height and middle age was looking at him from the other side of the gate. Frank rose from his grassy couch and answered coldly: "Yes, sir; I believe so. I will go in and see."

"Oh, don't trouble yourself, my young friend," said Mr. Craven, opening the gate and advancing toward the door with a brisk step. "I will ring the bell; I want to see your mother on a little business."

"Seems to me he has a good deal of business with mother," Frank said to himself. "There's something about the man I don't like, though he always treats me well enough. Perhaps it's his looks."

"How are you, Frank?"

Frank looked around, and saw his particular friend, Ben Cameron, just entering the gate.

"Tip-top, Ben," he answered, cordially. "I'm glad you've come."

"I'm glad to hear it; I thought you might be engaged."

"Engaged? What do you mean, Ben?" asked Frank, with a puzzled expression.

"Engaged in entertaining your future step-father," said Ben, laughing.

"My future step-father!" returned Frank, quickly; "you are speaking in riddles, Ben."

"Oh! well, if I must speak out, I saw Mr. Craven ahead of me."

"Mr. Craven! Well, what if you did?"

"Why, Frank, you must know the cause of his attentions to your mother."

"Ben," said Frank, his face flushing with anger, "you are my friend, but I don't want even you to hint at such a thing as that."

"Have I displeased you, Frank?"

"No, no; I won't think of it anymore."

"I am afraid, Frank, you will have to think of it more," said his companion, gravely.

"You surely don't mean, Ben, that you have the least idea that my mother would marry such a man as that?" exclaimed Frank, pronouncing the last words contemptuously.

21. Frank Hunter is an only child

A. True

B. False

C. Can't Tell

22. Based on the above information, which of the following statements can be inferred?

A. Frank dislikes the idea of his mother marrying

B. Frank is a university student

C. Mr. Craven is an unpleasant character

D. Frank dislikes Ben

23. Based on the passage, which season is it most likely to be?

A. Summer

B. Spring

C. Winter

D. Autumn

24. Based on the above information, which of the following statements is most likely to be true?

A. Frank is an only child

B. Frank and Ben are brothers

C. Frank's father is absent

D. Frank's mother is a schoolteacher

Passage 7

Cynthia, Countess of Hampshire, was sitting in an extraordinarily elaborate dressing-gown one innocent morning in June, alternately opening letters and eating spoonfuls of sour milk prepared according to the prescription of Professor Metchnikoff. Every day it made her feel younger and stronger and more irresponsible (which is the root of all joy to natures of a serious disposition), and since (when a fortnight before she began this abominable treatment) she felt very young already, she was now almost afraid that she would start again on measles, croup, hoops, whooping-cough, peppermints, and other childish ailments and passions. But since this treatment not only induced youth, but was discouraging to all microbes but its own, she hoped as regards ailments that she would continue to feel younger and younger without suffering the penalties of childhood.

The sour milk was finished long before her letters were all opened, for there was no one in London who had a larger and more festive post than she. Indeed, it was no wonder that everybody of sense (and most people of none) wanted her to eat their dinners and stay in their houses, for her volcanic enjoyment of life made the dullest of social functions a high orgy, and since nothing is nearly so infectious as enjoyment, it followed that she was much in request.

Even in her fiftieth year she retained with her youthful zest for life much of the extreme plainness of her girlhood, but time was gradually lightening the heaviness of feature that had once formed so remarkable an ugliness, and in a few years more, no doubt, she would become as nice looking as everybody else of her age.

25. Based on the above passage, which of the following statements is true?
A. Cynthia is a child
B. Cynthia lives in London
C. Professor Metchnikoff is Cynthia's father
D. Cynthia eats sour milk for breakfast

26. Cynthia is 60 years old
A. True
B. False
C. Can't Tell

27. Which of the following statements correctly summarises the passage?
A. Cynthia is a countess attempting to appear more youthful
B. Cynthia is a countess looking for love
C. Professor Metchnikoff is a Psychology professor performing an experiment on Cynthia
D. The benefits and side effects of regular sour milk ingestion

28. With regards to the passage above, which of the following statements is FALSE?
A. Cynthia receives a large amount of post
B. Cynthia is a rich lady
C. Cynthia has been drinking sour milk for the past 2 weeks
D. Cynthia was good-looking in youth

Passage 8

The old writers tell how Long Island was once the happy hunting ground of wolves and Indians, the playing place of deer and wild turkeys; and how the seals, the turtles, grampuses and pelicans loved its long, quiet beaches. Seals and whales are still occasional visitors, and its coasts are rich in lore of wrecks, of pirates and of buried treasure.

A hundred years ago it could boast of hamlets only less remote from civilisation than are to-day the villages of that other "Long Island"—the group of the Outer Hebrides—which, for an equal distance, extends along the Scottish coast from Butt of Lewis to Barra Head. The desultory stage then occupied a week on the double journey between Brooklyn and Sag Harbour. Beyond the latter, Montauk Point thrusts its lighthouse some fifteen miles out into the Atlantic breakers. Here the last Indians of the island lingered on their reservation, and here the whalers watched for the spouting of their prey in the offing.

A ridge of hills runs along the island near the northern shore, rising here and there into heights of three or four hundred feet which command the long gradual slope of woods and meadows to the south, with the distant sea beyond them; to the north, across the narrow Sound, rises the blue coast line of Connecticut.

It is on the slopes below the highest of these points of wide vision that the Whitman homestead lies, one of the pleasant farms of a land which has always been mainly agricultural. Large areas of the island are poor and barren, covered still with scrub and "kill-calf" or picturesque pine forest, as in the Indian days. But the land here is productive.

29. Long Island is in Scotland
A.   True
B.   False
C.   Can't Tell

30. Which of the following statements correctly summarises the above passage?
A.   The passage describes the history of Long Island
B.   The passage describes the history of Indians
C.   The passage provides an account of the Scottish Coast
D.   The passage describes a love story in New York

31. Which of the following statements is true regarding the above passage?
A.   Montauk Point is in Scotland
B.   Long Island is in Connecticut
C.   There are no more Indians on Long Island
D.   The Outer Hebrides are in Scotland

32. Which of the following animals have never been seen on Long Island?
A.   Wolves
B.   Seals
C.   Whales
D.   Tigers

Passage 9

In the year 2126, England enjoyed peace and tranquillity under the absolute dominion of a female sovereign. Numerous changes had taken place for some centuries in the political state of the country, and several forms of government had been successively adopted and destroyed, till, as is generally the case after violent revolutions, they all settled down into an absolute monarchy. In the meantime, the religion of the country had been mutable as its government; and in the end, by adopting Catholicism, it seemed to have arrived at nearly the same result: despotism in the state, indeed, naturally produces despotism in religion; the implicit faith and passive obedience required in the one case, being the best of all possible preparatives for the absolute submission of both mind and body necessary in the other.

In former times, England had been blessed with a mixed government and a tolerant religion, under which the people had enjoyed as much freedom as they perhaps ever can do, consistently with their prosperity and happiness. It is not in the nature of the human mind, however, to be contented: we must always either hope or fear; and things at a distance appear so much more beautiful than they do when we approach them, that we always fancy what we have not, infinitely superior to anything we have; and neglect enjoyments within our reach, to pursue others, which, like ignes fatui, elude our grasp at the very moment when we hope we have attained them.

33. The passage is set in the future
A. True
B. False
C. Can't Tell

34. The overall theme of the passage is:
A. War
B. The future
C. Human nature
D. Feminism

35. Which of the following statements is true regarding the above passage?
A. The passage is based in the USA
B. The main religion in England is Islam
C. There is no government in England in 2126
D. There is a single female leader in England in 2126

36. The author is expressing admiration for human compassion
A. True
B. False
C. Can't Tell

Passage 10

That evening at eight o'clock we met at the old Edinburgh Hotel (now no longer in existence), and after dinner he told me his very remarkable tale.

"Some years ago," he said, "I was staying in a small coast town in Fife, not very far from St Andrews. I was painting some quaint houses and things of the sort that tickled my fancy at the time, and I was very much amused and excited by some of the bogie tales told me by the fisher folk. One story particularly interested me."

"And what was that?" I asked.

"Well, it was about a strange, dwarfish, old man, who, they swore, was constantly wandering about among the rocks at nightfall; a queer, uncanny creature, they said, who was 'aye beckoning to them,' and who was never seen or known in the daylight. I heard so much at various times and from various people about this old man that I resolved to look for him and see what his game really was. I went down to the beach times without number, but saw nothing worse than myself, and I was almost giving the job up as hopeless, when one night 'I struck oil,' as the Yankees would say."

"Good," I said, "let me hear."

"It was after dusk," he proceeded, "very rough and windy, but with a feeble moon peeping out at times between the racing clouds. I was alone on the beach. Next moment I was not alone."

"Not alone," I remarked. "Who was there?"

"Certainly not alone," said Ashton. "About three yards from me stood a quaint, short, shrivelled, old creature. At that time the comic opera of 'Pinafore' was new to the stage-loving world, and this strange being resembled the character of 'Dick Deadeye' in that piece. But this old man was much uglier and more repulsive. He wore a tattered monk's robe, had a fringe of black hair, heavy black eyebrows, very protruding teeth, and a pale, pointed, unshaven chin. Moreover, he possessed only one eye, which was large and telescopic looking."

"What a horrid brute," I said.

37. The ghost in the passage had 1 eye
A. True
B. False
C. Can't Tell

38. Which of the following statements regarding the above passage is true?
A. The main protagonist is called Ashton
B. The passage is set in the morning
C. The ghost is that of an old King
D. The protagonist is not fond of the ghost

39. If you had to describe the ghost in one word, what would it be?
A. Terrifying
B. Worried
C. Happy
D. Pitiful

40. Ashton is interested in ghosts
A. True
B. False
C. Can't Tell

Passage 11

It was a lovely twilight evening at Lytton Springs, India. These famous springs were very high up in the Araville hills; Mandavee was the nearest city, situated on a small island in the Arabian sea. The great red sun was slowly sinking as the bells were ringing the Angelus from an ancient Hindu temple. The sacred chimes pealed forth melodiously, the sweet sounds echoing forth the harmony of those bells. Inside of this ancient temple sweet incense was burning on a beautiful golden altar. A dark, handsome prince and his family were praying around this sacred altar. Here they would often see beautiful visions of angels and their loved ones who had died in this same faith years ago. This faith was a strange, mysterious, mythical religion, handed down from the ancient Indians. It was a mixture of Catholicism and Hinduism. The Prince and his family were highly educated and great musicians; they were all great Psychics, and often spent hours in this old temple praying. They lived in constant communion with their saints, who constantly watched over them and protected them. At the other side of this altar a strange veiled princess was silently praying. After sunset they all left the temple with bowed heads. They went to their summer homes in the hills. Sita, the Prince's only daughter, felt sorry for the lonely stranger and invited her to their lovely home in the mountains.

41. Which country is the passage based in?

A. Sri Lanka

B. China

C. Japan

D. India

42. The religion that the Princes practice is Hinduism

A. True

B. False

C. Can't tell

43. Which of the following statements regarding the above passage is true?

A. There are only royal characters in the passage

B. The Princes had special abilities

C. The Princes live in Mandavee

D. Sita is the Prince's wife

44. Sita is the veiled princess in the passage

A. True

B. False

C. Can't Tell

END OF SECTION

Section B: Decision Making

1. In 2007 AD, Halley's Comet and Comet Encke were observed in the same calendar year. Halley's Comet is observed on average once every 73 years; Comet Encke is observed on average once every 104 years. Based on this, estimate the calendar year in which both Halley's Comet and Comet Encke are next observed in the same year.

A. 9559 AD   B. 2114 AD   C. 5643 AD   D. 3562 AD   E. 1757 AD

2. "One has to be at least 18 to vote in the UK". Which statement gives the best supporting reason for this statement?

A. At 18, citizens have the right level of maturity to vote sensibly
B. One must legally be able to buy alcohol before they can vote
C. Too many people would vote in the UK otherwise
D. Citizens do not understand policies below the age of 18

3. Adam, Ben, Caitlyn, Joe, James, Simon and David are sitting around a circle facing the centre. Joe is sitting between Adam and David. Simon is second to the right of David and James is second to the right of Simon. Caitlyn is not an immediate neighbour of David. Which of the below statements is true?

A. Caitlyn is to the left of Simon
B. Simon is to the left of Adam
C. David is to the right of Simon
D. Joe is between Simon and Caitlyn
E. Caitlyn is directly opposite to David

4. In a certain Code '8 2 9' means 'how are you,' '9 5 8' means 'you are good' and '1 5 8 7 3' means 'I good and you bad'. Based on this, what is the code for 'you'?

A. 1   B. 5   C. 8   D. 7

5. 'Smoking should not be condemned because smokers pay for their healthcare through the tax on cigarettes'. Which option is the best argument against the above statement?

A. Smokers cause detrimental effects on people around them
B. Smoking should be condemned because it is bad for your health
C. That argument doesn't work if the healthcare system is private
D. Smoking makes you more likely to drink, which also causes further health problems.

6. Jason, Peter, John and Alan are four brothers. Jason is older than Peter and John. Alan is younger than John. Peter is older than John. Which of the following correctly represents the four brothers in age order (from youngest to oldest)?

A. Peter, Alan, John, Jason
B. Jason, Peter, John, Alan
C. Alan, John, Peter, Jason
D. Jason, John, Peter, Alan

7. If 'TOILET' is related to 'WDJEBW' and 'SOUNDS' is related to 'YDHILY', which of the following options is related to 'SOILED'?

A. QUDHSJQ   C. YDJEBL
B. FGHDJR   D. AJSHDG

8. 'Surgeon's mortality rates should not be available to the public'. Which of the following statements provides the best argument against this statement?

A. Surgeons are more willing to do surgery if their mortality rates are published
B. Publishing mortality rates promotes a more transparent healthcare system and embeds public trust in healthcare professionals
C. Mortality rates will increase if mortality rates are published
D. Surgeons are more likely to not leave the country if mortality rates are published

9. In a school there are 40 more girls than there are boys. The boys make up a percentage of 40% of the school. What is the number of students in the school?

A. 150   B. 200   C. 300   D. 500

10. David is 4 years older than Anna. Anna and May are in the same school year. May's younger brother Isaac is best friends with David's brother Mike.

Which of the following conclusions is true based on the above information?
A. Anna is older than Isaac
B. David is younger than Isaac
C. David and Anna are related
D. Isaac is older than Mike
E. May is older than David

11. During the school day, there are 6 lessons timetabled. Maths is never second or fifth. English always follows Science. French is not fourth and there is a lesson between French and History. There is always a break before Geography.

Which of the following statements is true?
A. Science is the second lesson of the day
B. Geography is the second lesson of the day
C. English is the third lesson of the day
D. History is the first lesson of the day
E. Maths is the fourth lesson of the day

12. 'Euthanasia should be legalised to allow people a dignified death'. Which of the following options is the best argument for this statement?

A. Elderly people may be pressured into euthanasia by their relatives
B. If euthanasia is legalised, then why not legalise murder?
C. There is no need for euthanasia; palliative patients can be kept perfectly comfortable as it is
D. Legalisation is pointless; those who want to be euthanised go abroad to do it anyway

13. If one day on Earth corresponds to 0.4 days on Mars, and there are 365 days in a year on Earth, how frequent would the Olympics be held if they were being held on Mars?

A. 10 years   B. 12 years   C. 32 years   D. 48 years

14. There has recently been an initiative established in a developing country for schoolteachers to teach the illiterate members of society after school hours. Which of the options provides the best argument as to why the lessons are being offered after hours?

A. So that the illiterate cannot interact with schoolchildren
B. So that the schooling does not interfere with any jobs that the illiterate members of society during the day
C. So that the illiterate members of society can understand how difficult getting an education is
D. So that the streets are less crowded at night time

15. Jay either walks to school or takes the bus, depending on whether it is raining or not. If he walks to school, there is a 60% chance that he is late. If he takes the bus to school, there is a 20% chance that he is late. Given that one morning, the probability that it will rain is 70%, what is the probability that Jay will be late to school?

A. 32%   B. 67%   C. 23%   D. 44%

16. If the number 273546 is rearranged such that the digits within the number are rearranged in ascending order (from left to right), how many digits would not change position within the number?

A. 0   B. 1   C. 2   D. 3

17. In a bank, there are 150 employees. Of these, there are 55 English employees, 58 married employees and 55 employees which PhDs. There are 19 employees who are both English and have a PhD, there are 21 employees who are both married and have a PhD, and there are 26 employees who are both English and married. Based on this information, how many employees are neither English, nor married, nor have a PhD?

A. 26   B. 38   C. 22   D. 41

18. All buses are motor vehicles. Motor vehicles include trucks. Some trucks are cars. Which of the following statements is true?

A. All cars are motor vehicles
B. All trucks are buses
C. Some cars are buses
D. All trucks are motor vehicles
E. Some buses are not motor vehicles

19. All sharks are fish. Fish, dolphins and whales are aquatic animals. Dolphins are mammals. Whales are not fish. Which of the following statements is true based on this information?

A. Whales are mammals
B. Whales and dolphins are related
C. Sharks are aquatic animals
D. Dolphins are fish
E. Sharks are mammals

20. Alex is going on a hike. He starts off in the morning walking towards the sun rise. He walks 5km, turns 90 degrees clockwise, walks 2km in that direction, then turns 180 degrees, walks 3km in that direction before turning 90 degrees anticlockwise to walk the final 4km home. Which direction is he walking for the last 4km?

A. East   B. North   C. South   D. West

21. 'Possession of a gun should be a federal crime'. Which of the following statements provides the strongest argument in favour of this statement?

A. There would be reduced murders if possession of guns is made illegal
B. People can use knives rather than guns to protect themselves
C. Making possession illegal will reduce black market purchasing of guns
D. People will be more afraid to use guns in public if they know it is a federal crime to possess one

22. 70% of the Earth's surface is covered by water. There are 5 oceans in the world. The largest of these is the Pacific Ocean. A body of water which is partially enclosed by land is called a sea. The largest sea in the world is the Arabian Sea. Which of the following statements is true?

A. The Pacific Ocean covers 70% of the Earth's surface
B. Seas cover 70% of the Earth's surface
C. The Pacific Ocean is larger than the Arabian Sea
D. Most of the water on Earth's surface is in the oceans
E. The Arabian Sea is partially enclosed by land

23. In a medical school, 56% of the 300 students are boys. Of these, 18% are Chinese. If 25% of the girls are Chinese, how many female medical students are Chinese?

A. 33   B. 41   C. 23   D. 27

24. 'People who are obese should pay for their own healthcare'. Which of the following options provides the best argument for the above statement?

A. It is often the poorest of society who have the worst diet and therefore are most predisposed to becoming obese
B. Obesity is not an expensive morbidity
C. Obese people contribute more to the national health system than thin people
D. Obesity is a problem of genetics, it is unfair for people to have to pay for what they have inherited

25. Susan is the mother of Johnathan, Sally and Christina. Christina is married to Karl, whose brother, Sam, is married to Leanne. Sam and Leanne have 3 children-David, Nat and Liam. Which of the following statements are true based on this information?

A. Sam and Christina are related
B. Nat and Susan are related
C. Karl and David are related

D. Johnathan and Liam are not related
E. Sally and Sam are related

26. To get to school, Joanne takes the school bus every morning. If she misses this, then she can take the public bus to school. The school bus arrives at 08:15, which if she misses will come again at 08:37. The public bus comes every 17 minutes, starting at 06:56. The school bus takes 24 minutes to get to her school; the public bus takes 18 minutes. If she arrives at the bus stop at 08:25, which bus must she catch to get to school first?

A. The 08:37 school bus
B. The 08:26 public bus

C. The 08: 38 public bus
D. The 08: 31 public bus

27. Jason, Karen, Liam, Mason, Neil, Obie, Pari and Richard are sitting around a circular table facing the centre. Each of them was born in a different year-1992, 1996, 1997, 2000, 2001, 2004, 2005 and 2010, but not necessarily in the same order. Mason is sitting second to the right of Karen. Liam is sitting third to the right of Jason. Only the one born in 2004 is sitting exactly between Jason and Karen. Neil, who is the eldest is not an immediate neighbour of Jason and Mason. Richard is older than only Mason. Richard is sitting second to the left of Pari. Pari is not an immediate neighbour of Neil. Jason is younger than Liam and Obie. Karen was born before Obie but she is not second eldest. Based on this information, which of the following statements is true and which are false?

A. Neil is opposite Karen
B. Liam was born in 2000
C. Mason was born in 2005

D. Neil is between Richard and Jason
E. Obie was born in 2010

28. 'Zoos should be made illegal'. Which of the following options provides the best argument against this statement?
A. Human entertainment is more important than animal welfare
B. Children wouldn't learn about animals without zoos
C. Zoos provide an important method by which critically endangered species can be protected in conservation until their numbers recover
D. Many people would be out of a job without zoos

29. All humans are apes. The apes include the related families of chimpanzees, gorillas and orangutans. All apes are mammals. Most mammals produce live offspring. Based on this information, which of the following statements are true and which are false?

A. All humans are mammals
B. All apes are humans
C. Chimpanzees and humans are unrelated
D. All mammals produce live offspring
E. All mammals which produce live offspring are apes

END OF SECTION

Section C: Quantitative Reasoning

Data Set 1

The following graph describes the number of people visiting a store at different times of the day. Study the graph, then answer the following six questions.

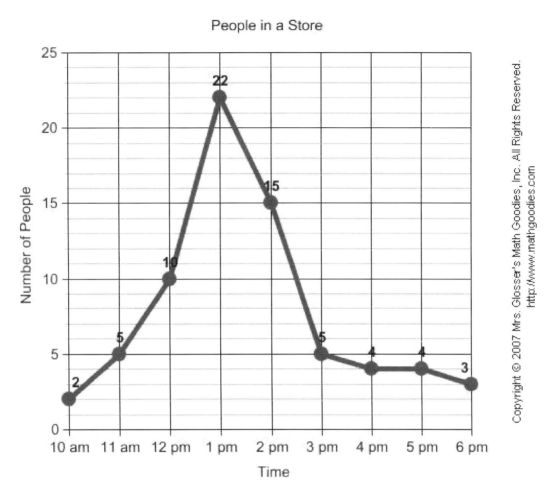

1. What is the mode number of people visiting the store over the day?

A. 22   B. 10   C. 2   D. 5   E. 4.5

2. If one was to visit the shop, what would be the best time to visit?

A. 12pm   B. 4pm   C. 6pm   D. 10am   E. 5pm

3. What is the range?

A. 15   B. 17   C. 20   D. 12   E. 9

4. If each customer spends on average £5 in the shop, what is the total income for the shop (to the nearest pound)?

A. £200   B. £176   C. £250   D. £350   E. £421

Data Set 2

The following table shows the currency exchange rates for different currencies.

	GBP	CAD	EUR	HKD	JPY	CHF	USD
GBP		0.5364	0.7470	0.0860	0.0057	0.7564	0.6667
CAD	1.8595		1.3908	0.1601	0.0105	1.4064	1.2413
EUR	1.3368	0.7178		0.1156	0.0076	1.0115	0.8922
HKD	11.6112	6.2364	8.6461		0.0658	8.7820	7.7513
JPY	176.4800	94.7300	131.9100	15.1900		133.5500	117.7500
CHF	1.3171	0.7061	0.9836	0.1132	0.0075		0.8776
USD	1.4980	0.8046	1.1204	0.1290	0.0085	1.1330	

5. Sally is planning to travel to Canada for a holiday, and wishes to convert £500 into Canadian Dollars (CAD). How many Canadian dollars can she get?

A. $836.43   B. $929.75   C. $736.44   D. $827.39   E. $283.33

6. After her trip, Sally had 150 CAD left over, which she converted back into Great British Pounds (GBP). How many pounds would she get back?

A. £80.46   B. £80.67   C. £92.18   D. £67.44   E. £81.19

7. How much money is she losing in this exchange?

A. £1.09   B. £0.89   C. £0.21   D. £1.58   E. £0.71

8. Shoko is planning to send money to her family in Japan from the USA. One travel agency is offers her a rate of 118.6300. If she is planning to send USD 700, how much more Japanese Yen (JPY) does she get from using this travel agency as opposed to the one above?

A. JPY 82425   B. JPY 83041   C. JPY 616   D. JPY 5.236   E. JPY 782

Data Set 3

The following graph shows how the Earth's surface temperature has varied with time.

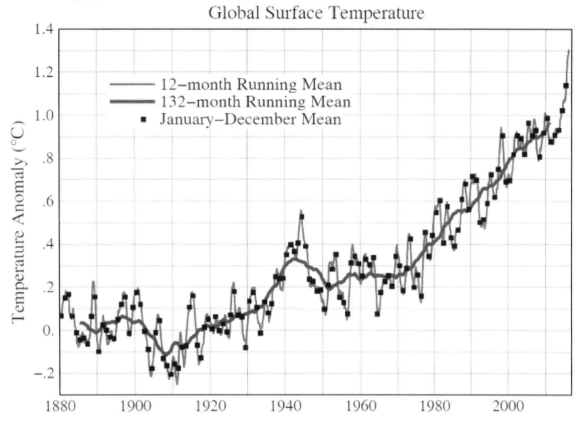

9. If the surface temperature is 22.6°C today, what was the temperature in 1940?

A. 20.4°C   B. 21.6°C   C. 22.0°C   D. 19.8°C   E. 20.7°C

10. What should the normal surface temperature be?

A. 21.3°C   B. 22.1°C   C. 20.8°C   D. 21.1°C   E. 20.2°C

11. During which period was the rate of temperature change the greatest?

A. 1880-1890   C. 1920-1930   E. 2010 onwards
B. 1900-1910   D. 1980-1990

12. What is the rate of temperature change between 1980 and 2000?

A. 0.12°C per year   C. 0.06°C per year   E. 0.25°C per year
B. 0.2°C per year   D. 0.09°C per year

Data Set 4

The following table shows how a group of students performed in their end of year examination. The number in the table gives the percentage that each student achieved in each exam

The Numbers in the Brackets give the Maximum Marks in Each Subject.

Student	Subject (Max. Marks)					
	Maths	Chemistry	Physics	Geography	History	Computer Science
	(150)	(130)	(120)	(100)	(60)	(40)
Ayush	90	50	90	60	70	80
Aman	100	80	80	40	80	70
Sajal	90	60	70	70	90	70
Rohit	80	65	80	80	60	60
Muskan	80	65	85	95	50	90
Tanvi	70	75	65	85	40	60
Tarun	65	35	50	77	80	80

13. Which student performed the best overall?

A. Muskan   B. Tanvi   C. Tarun   D. Sajal   E. Ayush

14. What mark did Rohit score in Computer Science?

A. 27   B. 25   C. 24   D. 31   E. 18

15. Which exam had the highest average score?

A. Maths   C. Physics   E. Computer Science
B. Chemistry   D. Geography

16. What mark did Sajal get in Maths?

A. 128   B. 122   C. 141   D. 135   E. 109

Data Set 5

The following graph shows how much rainfall Atlanta has been getting over the past 20 years.

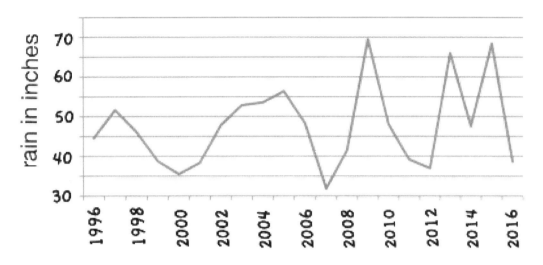

17. How much rain fell in 2010?

A. 55 inches
B. 40 inches

C. 45 inches
D. 30 inches

E. 70 inches

18. In which year did Atlanta receive the maximum rainfall?

A. 2009

B. 2003

C. 1996

D. 2016

E. 2001

19. What was the difference in rainfall between 2007 and 2009?

A. 28 inches
B. 34 inches

C. 38 inches
D. 51 inches

E. 12 inches

20. What is the average rainfall between 2012 and 2016?

A. 52 inches
B. 35 inches

C. 40 inches
D. 65 inches

E. 22 inches

Data Set 6

The following train timetable shows the times that trains reach certain stations.

	1st	2nd	3rd	4th	5th	6th
Depot	07:30	07:45	08:00	08:15	08:30	08:45
Green St	07:40	07:55	**08:10**	08:25	08:40	08:55
High St	07:45	08:00	08:15	**08:30**	08:45	?
Central Park	**07:48**	08:03	08:18	**08:33**	08:48	09:03
Railway Station	07:53	08:08	08:23	08:38	?	09:08
Shopping Centre	08:00	08:15	08:30	08:45	09:00	09:15
Brown St	08:06	08:21	08:36	08:51	09:06	09:21
Church St	08:08	08:23	08:38	08:53	09:08	09:23
St Georges School	08:15	08:30	**08:45**	09:00		
Library	08:20	08:35	08:50	09:05	**09:20**	09:35
Hospital	08:25	08:40	08:55	09:10	09:25	09:40
Friary Walk	08:33	08:48	09:03	09:18	09:33	09:48
St Marys School	**08:42**	08:57	09:12			
Forest Rd	08:48	09:03	**09:18**	09:33	09:48	10:03
Swimming Pool	09:00	?	**09:30**	09:45	**10:00**	10:15

21. If there are 5km between the depot and Green St, what speed is the train travelling in km/h?

A. 40km/h   B. 30km/h   C. 37km/h   D. 43km/h   E. 15km/h

22. If this is the average speed of the train, what is the distance between Brown St and Church St?

A. 1km   B. 2.5km   C. 3km   D. 4.2km   E. 7km

23. If Jamie misses her train from Friary Walk at 08:33 by 7 minutes, how long must he wait until the next train?

A. 9 minutes   B. 7 minutes   C. 12 minutes   D. 8 minutes   E. 6 minutes

24. Dani must walk 8 minutes to get to Central Park station from his house. He takes the train to Forest Road, from which he must walk another 12 minutes to get to his work. If he must be at work by 9am, when does he have to set out from his house?

A. 08:03   B. 07:15   C. 07:56   D. 07:28   E. 07:40

Data Set 7

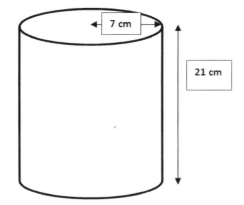

25. What is the surface area of the above cylinder to the nearest centimetre?

A. 949cm²   B. 1232cm²   C. 814cm²   D. 795cm²   E. 892cm²

26. What is the volume of the above cylinder to the nearest centimetre?

A. 3233cm³   B. 2647cm³   C. 4536cm³   D. 2222cm³   E. 5647cm³

27. What is the circumference of the cross-section to the nearest centimetre?

A. 37cm   B. 28cm   C. 44cm   D. 50cm   E. 19cm

28. If the radius increases by a factor of 2, what does the volume increase by?

A. 2   B. 4   C. 8   D. 16   E. 32

Data Set 8

Hayley buys a new car for £9000. Every year that Hayley owns the car, its value depreciates by 15%.

29. What will the value of the car be in 3 years (to the nearest pound)?

A. £5527   B. £7250   C. £6125   D. £4480   E. £8400

30. After how many years will the value of the car be £3394 (to the nearest pound)?

A. 4 years   B. 3 years   C. 5 years   D. 6 years   E. 7 years

31. As part of the cross-country race in school, the children must run 4 laps of the school field, which has a perimeter of 2.4km. If the runner in the lead is running at 12km/h, and the runner at the back is running at 8km/h, what time will the runner in the lead lap the runner in the back?

A. 20 minutes   B. 25 minutes   C. 36 minutes   D. 24 minutes   E. 17 minutes

32. James has a bag of 78 blue counters and David has a bag of 32 red counters. They must divide the counters into boxes with equal numbers of counters in each box. If there is no remainder, find the largest number of counters that can be put into a box.

A. 2   B. 4   C. 6   D. 8   E. 12

Data Set 9

The pie chart below represents a survey conducted on 300 people asking what their favourite book genre is.

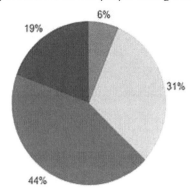

33. If the largest proportion represents romantic novels, how many people indicated this preference?

A. 44   B. 88   C. 132   D. 126   E. 182

34. If the smallest proportion represents horror, what angle does this slice of the pie make (to the nearest degree)?

A. 22°   B. 18°   C. 26°   D. 32°   E. 45°

Data Set 10

The following table shows the tax brackets for a country

Tax Rate	Single	Married (joint)/ Widow(er)	Married (separate)	Head of household
10%	$0 - $9,325	$0 - $18,650	$0 - $9,325	$0 - $13,350
15%	$9,326 - $37,950	$18,651 - $75,900	$9,326 - $37,950	$13,351 - $50,801
25%	$37,951 - $91,900	$75,901 - $153,100	$37,951 - $76,550	$50,801 - $131,200
28%	$91,901 - $191,650	$153,101 - $233,350	$76,551 - $116,675	$131,201 - $212,500
33%	$191,651 - $416,700	$233,351 - $416,700	$116,676 - $208,350	$212,501 - $416,700
35%	$416,701 - $418,400	$416,701 - $470,700	$208,351 - $235,350	$416,701 - $444,550
39.6%	$418,401 +	$470,701 +	$235,351 +	$445,551 +

35. For a single lawyer earning $176,000 a year, how much of that is deducted by tax?

A. $41461   B. $26354   C. $78263   D. $25362   E. $29283

36. If a doctor earning $200,000 gets married, what is his change in tax rate?

A. 5% increase   B. 7% increase   C. 5% reduction   D. 8% reduction   E. 3% increase

END OF SECTION

Section D: Abstract Reasoning

For each question, decide whether each box fits best with Set A, Set B or with neither.

For each question, work through the boxes from left to right as you see them on the page. Make your decision and fill it into the answer sheet.

Answer as follows:
A = Set A
B = Set B
C = neither

Set 1:   Set A   Set B

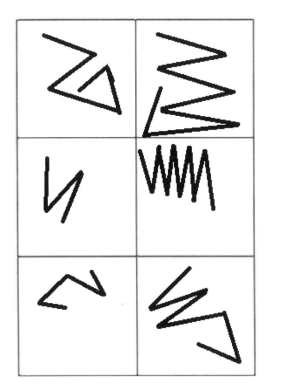 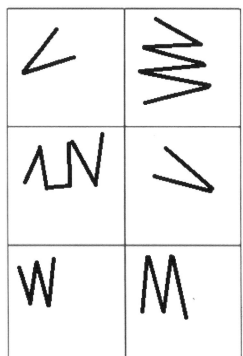

Questions 1-5:

Set 2:   Set A   Set B

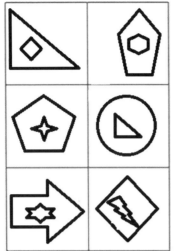 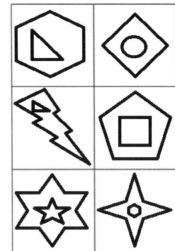

Questions 6-10:

Set 3:   Set A   Set B

 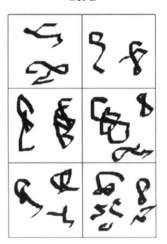

Questions 11-15:

Set 4:   Set A   Set B

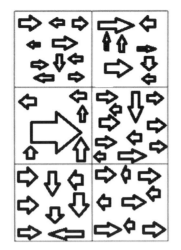 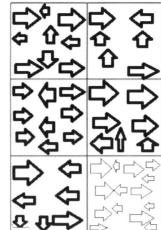

Questions 16-20:

Set 5:   Set A   Set B

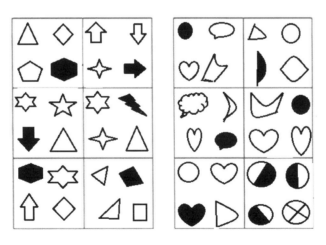

Questions 21-25:

Set 6:   Set A   Set B

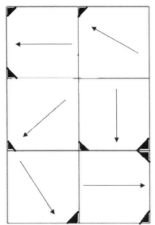 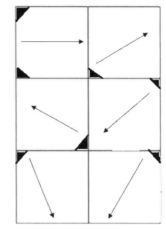

Questions 26-30:

Set 7:   Set A   Set B

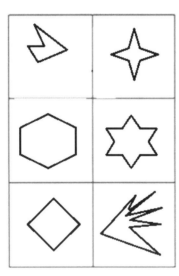 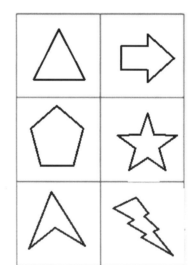

Questions 31-35:

Set 8:   Set A   Set B

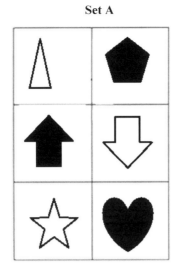 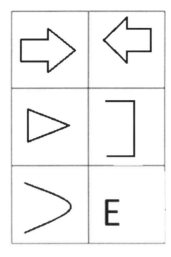

Questions 36-40:

Set 9:   Set A   Set B

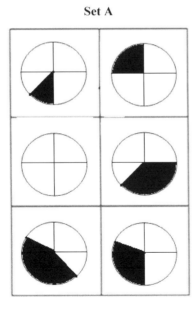 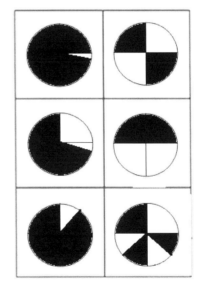

Questions 41-45:

Set 10:

Questions 46: Which figure completes the series?

| A. | B. | C. | D. |

Set 11:

Questions 47: Which figure completes the series?

| A. | B. | C. | D. |

Set 12:

E is to B

as

A is to

Question 48: Which figure completes the statement?

| A. | B. | C. | D. |

G   B   U   D

Set 13:

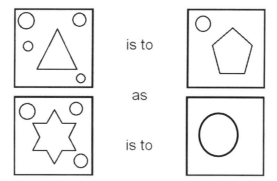

Question 49: Which figure completes the statement?

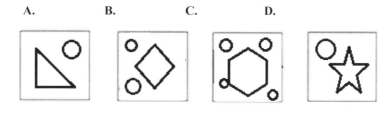

Set 14:

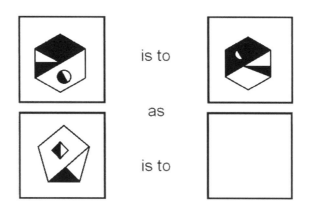

Question 50: Which figure completes the statement?

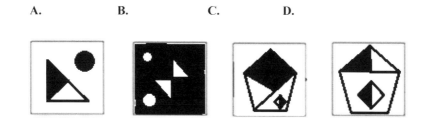

Set 15:

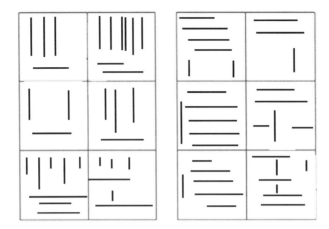

Question 51: Which of the following belongs in Set A?

A.   **B.**   **C.**   **D.**

Question 52: Which of the following belongs in Set A?

A.   **B.**   **C.**   **D.**

Question 53: Which of the following belongs in Set A?

A.   **B.**   **C.**   **D.**

Question 54: Which of the following belongs in Set B?

A. B. C. D.

Question 55: Which of the following belongs in Set B?

A. B. C. D.

END OF SECTION

Section E: Situational Judgement Test

Read each scenario. Each question refers to the scenario directly above. For each question, select one of these four options. Select whichever you feel is the option that best represents your view on each suggested action.

For questions 1 – 30, choose one of the following options:
  A   A highly appropriate action
  B   Appropriate, but not ideal
  C   Inappropriate, but not awful
  D   A highly inappropriate action

Scenario 1

James is a fourth-year medical student. In preparation for his end of year exams, James has been asking students in the year above for questions that they had in their end of year exams, and is planning to create a question bank using these responses. Once the question bank is completed, James plans to distribute the bank to his fellow students. You are James's good friend and are aware of his plans.

How appropriate are your responses to this situation?

1.   Report James to the medical school, as this is cheating
2.   Don't say anything to anyone, so that you can also benefit from the question bank
3.   Talk to James about what he is doing, and convey that what he is doing is technically cheating
4.   Inform James's parents of his plans, so that they can intervene
5.   Publicly expose James on social media, and warn your colleagues that anyone using James's question bank will be reported for cheating.

Scenario 2

Rashid is a clinical medical student. As part of his academic year, He has a 6-week research placement in which he can pursue any academic interest of his choice. During this placement, his family has arranged for him to attend a relative's wedding in Sri Lanka. This will involve a two-week trip. The trip has already been booked. Rashid wants to go, as he plays an important role in the wedding, but he is worried that if he asks permission from the clinical school then they might refuse.

How appropriate are Rashid's responses to this situation?

6.   Go on the holiday without telling the clinical school
7.   Ask the clinical school for permission before travelling
8.   Sweet-talk his supervisor so he is more willing to pass Rashid's placement before he travels
9.   Tell his parents that they should cancel his ticket
10. Tell the clinical school and go on the holiday regardless of their response

Scenario 3

Jamie is a sixth form student. As part of his A-level Chemistry, he must write a practical paper, which is worth 20% of his grade, in the classroom. However, as the paper is conducted in the classroom, different schools take the test at different times. As a result, one of Jamie's friends at another college already has done the exam, and offers Jamie the questions the week before his own exam, so that he can prepare answers beforehand.

How appropriate are Jamie's responses to this situation?

11. Take the questions-he is not cheating, his friend is
12. Take only enough questions such that he can achieve an A grade in the practical paper
13. Refuse the questions
14. Report the friend for cheating to the exam board
15. Take the questions and distribute them to all his classmates so that he is not at an unfair advantage

Scenario 4

You are a student on placement in a rural practice. Two of the patients you have been asked to see are married, and have arranged separate appointments with you to discuss the results of their STI screen. They wanted to do this to make sure both partners had no infections before they started family planning. You meet the husband first, who is clear of infections and you discuss how to optimise chances of getting pregnant with him. Later, you meet the wife, whose blood test reveals that she is HIV positive. She tells you that she does not want you telling the husband.

How appropriate are your responses to this situation?

16. Tell the wife that you have a responsibility to the husband as his doctor to tell him if he is at risk of developing HIV
17. Try to convince the wife to speak to him before they try family planning
18. Call up the husband after the wife has left
19. Call the husband into the consultation room with the wife
20. Make up an excuse for the wife as to why she has HIV so that she can tell the husband that

Scenario 5

You are a student on placement on a surgical ward. You are printing off the lists in preparation for the ward round, when your consultant walks in looking very untidy. It soon becomes obvious that he is drunk, and there is an obvious smell of alcohol on his breath.

How appropriate are your responses to this situation?

21. Consult your registrar
22. Pretend that you don't notice
23. Report the consultant to the hospital board
24. Take the consultant into a private room, make him drink plenty of water and tell him to go home
25. Tell all the patients that the consultant is too drunk to do the ward round this morning

Scenario 6

You are a fifth-year medical student approaching your finals. One of your good friends, Neil, is also preparing for his finals, but you know that every weekend Neil stays awake most of the night taking illicit drugs. It has not affected his performance at medical school so far but you've noticed that his drug use has been getting worse and worse in the run-up to finals. You are worried that Neil might go off the rails before finals and end up failing.

How appropriate are your responses to this situation?

26. Inform the university pastoral care about Neil's situation
27. Inform the medical school that Neil is taking illicit drugs
28. Try and speak to Neil, and make sure that he is ok and voice your concerns with regards to the drug use
29. Sneak into Neil's room when he is in hospital and steal his drugs
30. Inform the police that Neil is holding drugs

For questions 31 – 69, decide how important each statement is when deciding how to respond to the situation?

A   Extremely important
B   Fairly important
C   Of minor importance
D   Of no importance whatsoever

Scenario 7

Janice is a fourth-year medical student. During the ward round, the consultant on the ward offers her the opportunity to assist in a liver transplant. However, the same time that the surgery is scheduled to begin, Janice has compulsory teaching which is part of her course requirements.

How important are the following factors for Janice in deciding what to do?

31. The teaching is likely to come up in her end of year exams
32. She is unlikely to get this sort of opportunity during the rest of her time in medical school
33. Janice is very interested in transplant surgery and has been waiting for an opportunity such as this
34. The teaching is small group and it will be very noticeable is Janice is not present
35. Assisting in a transplant surgery will add to her portfolio, strengthening her application for transplant surgeon in the future.

Scenario 8

Johnathan is a second-year medical student. This year, Johnathan has been selected for the university cricket team, which involves a rigorous training schedule during the last two months of the calendar year. However, the second-year final exams are also at the end of the academic year, so most of his classmates are gearing up to sit those exams.

How important are the following factors for Johnathan in deciding what to do?

36. Second year exams determine whether Johnathan can begin his clinical training or not
37. There are prizes given for excellent achievement in the second-year final exams
38. Johnathan's father has spent a lot of money in coaching Johnathan in preparation for university cricket
39. Johnathan is only just starting to integrate into the team
40. Johnathan has a bet with his friends on who will do the best at the end of the year

Scenario 9

Hayley, a final year medical student, has been offered the opportunity to present a case report at a national conference by a consultant on the ward she is working on. Unfortunately, she has a meeting with her director of studies about elective planning the day she is flying.

How important are the following factors for Hayley in deciding what to do?

41. Her director of studies is a prestigious professor in the university, who has a very tight schedule
42. She cannot go on her elective without university funding, which must be approved by her director of studies
43. The conference is in Sydney, a city Hayley has always hoped to visit
44. The consultant is only taking one student with him
45. Hayley is guaranteed a publication if she agrees to presenting

Scenario 10

Latisha and Tim are clinical partners. Tim finds out that over the Christmas holidays, Latisha's boyfriend broke up with her. Since returning, Latisha has been arriving late to the ward round, and often leaves in a fit of tears during the middle of consultations. The doctors on the ward say nothing at the time, but Tim hears them whispering about Latisha quietly once the ward round has finished. Tim has spoken to Latisha before but she insisted that she was fine.

How important are the following factors for Tim in deciding what to do?

46. Latisha's reputation with the doctors
47. The consultant on the ward round determines whether they pass or fail that placement
48. The appearance to patients
49. Latisha may never speak to Tim again if he insists on the matter
50. Tim's friendship with Latisha's ex-boyfriend

Scenario 11

Javad is a final year medical student. One day during lunch in the hospital canteen, he overhears his friend laughing loudly about a patient he saw on the ward who is in a vulnerable position, and loudly revealing details of the patient for all to hear. Later, an emergency meeting is called with the clinical dean, who reveals that a breach of confidentiality has been reported, and it is known that it was a medical student.

How important are the following factors for Javad in deciding what to do?

51. Javad and his friend have been friends since kindergarten
52. The clinical school are sure to exclude Javad's friend if they find out it was him
53. The hospital would be heavily fined if the news got out that there was a confidentiality breach
54. This will be the first time Javad's friend has been in trouble with the clinical school
55. Javad knows that his friend has been really struggling in the run-up to finals.

Scenario 12

Ross and Alex are final year medical students currently on their Emergency Medicine rotation. One night when Ross is working, Alex comes in heavily inebriated. He takes Ross aside, and explains to him that he is working at 6am tomorrow. Alex then begs Ross to give him a bed in the Emergency Department and hook him up to an IV fluid drip, so that he can wake up sober and ready for work the next day.

How important are the following factors for Ross in deciding what to do?

56. Alex and Ross are the only students on Emergency Medicine now
57. There is a shortage of beds in the Emergency Department
58. The consultant will be doing a ward round at 5am, for which Ross will be a part of
59. IV fluids are a prescription which require inpatient admittance
60. Alex has stated that 'he would do the same' for Ross
61. Ross is going on holiday with Alex for two weeks in June

Scenario 13

Surya and Leah are two first year medical students doing their first clinical placement. When they ask the FY2 on the ward who would be a good patient to examine, the FY2 recommends a patient in one of the side rooms of the ward. He also says that the patient is MRSA positive, but that if they're quick enough then there should be nothing to worry about.

How important are the following factors for the two of them in deciding what to do?

62. MRSA requires strict isolation
63. The nurses have told them that they should not go in unless necessary
64. The junior doctor knows more about the ward than they do
65. They need to present their examination findings to the junior doctor when they're done
66. The patient is clearly unwell
67. An abdominal examination on such a patient would provide invaluable learning experience
68. It is not essential for them to examine the patient
69. Surya and Leah are going to a party later

END OF PAPER

Mock Paper F

Section A: Verbal Reasoning

Passage 1

The English Sparrow

The English Sparrow was first introduced into the United States at Brooklyn, New York, in the years 1851 and '52. The trees in our parks were at that time infested with a canker-worm, which wrought them great injury, and to rid the trees of these worms was the mission of the English Sparrow.

In his native country this bird, though of a seed-eating family (Finch), was a great insect eater. The few which were brought over performed, at first, the duty required of them; they devoured the worms and stayed near the cities. With the change of climate, however, came a change in their taste for insects. They made their home in the country as well as the cities, and became seed and vegetable eaters, devouring the young buds on vines and trees, grass-seed, oats, rye, and other grains.

Their services in insect-killing are still not to be despised. A single pair of these Sparrows, under observation an entire day, were seen to convey to their young no less than forty grubs an hour, an average exceeding three thousand in the course of a week. Moreover, even in the autumn he does not confine himself to grain, but feeds on various seeds, such as the dandelion, the sow-thistle, and the groundsel; all of which plants are classed as weeds. It has been known, also, to chase and devour the common white butterfly, whose caterpillars make havoc among the garden plants.

The good he may accomplish in this direction, however, is nullified to the lovers of the beautiful, by the war he constantly wages upon our song birds, destroying their young, and substituting his unattractive looks and inharmonious chirps for their beautiful plumage and soul-inspiring songs.

1. Which of the following statements about the English Sparrows is true:

A. The English Sparrows kills song birds
B. The English Sparrow was first discovered in the mid 19th century
C. The English Sparrow sings soul-inspiring songs
D. The English Sparrow eats canker-worms all year round

2. Which of the following statements are INCORRECT:

A. English Sparrows inhabit both the cities and countryside
B. English Sparrows accomplish more good than bad
C. English Sparrows eat seeds as well as insects
D. English Sparrows are unattractive

3. According to the above passage:

A. English Sparrows prefer insects to grains
B. Song birds are more harmonious than English Sparrows
C. English Sparrows are omnivores
D. English Sparrows eat caterpillars

4. A single pair of sparrows were observed to:

A. Convey over 40 grubs per hour to their young
B. Convey at least 40 grubs per hour to their young
C. Convey over 3000 grubs to their young per week
D. Convey over 12,000 grubs to their young per month

Passage 2

Greek Gods

In appearance, the gods were supposed to resemble mortals, whom, however, they far surpassed in beauty, grandeur, and strength; they were also more commanding in stature, height being considered by the Greeks an attribute of beauty in man or woman. They resembled human beings in their feelings and habits, intermarrying and having children, and requiring daily nourishment to recruit their strength, and refreshing sleep to restore their energies. Their blood, a bright ethereal fluid called Ichor, never engendered disease, and, when shed, had the power of producing new life.

The Greeks believed that the mental qualifications of their gods were of a much higher order than those of men, but nevertheless, as we shall see, they were not considered to be exempt from human passions, and we frequently behold them actuated by revenge, deceit, and jealousy. They, however, always punish the evil-doer, and visit with dire calamities any impious mortal who dares to neglect their worship or despise their rites. We often hear of them visiting mankind and partaking of their hospitality, and not infrequently both gods and goddesses become attached to mortals, with whom they unite themselves, the offspring of these unions being called heroes or demi-gods, who were usually renowned for their great strength and courage. But although there were so many points of resemblance between gods and men, there remained the one great characteristic distinction, that is to say, that the gods enjoyed immortality. Still, they were not invulnerable, and we often hear of them being wounded, and suffering in consequence such exquisite torture that they have earnestly prayed to be deprived of their privilege of immortality.

5. According to the above passage:

A. Gods always enjoy immortality.
B. Gods infrequently have children with mortals.
C. Gods do not need food or sleep.
D. Gods are more intelligent than humans.

6. Which of the following are not differences between gods and humans?

A. Height
B. IQ
C. Mortality
D. Emotions

7. According to the passage, which of the following statements is true of demi-gods?

A. Demi-gods are more courageous than gods.
B. Hercules is a demi-god.
C. Heroes and demi-gods can be used interchangeably to mean the same thing.
D. Demi-gods live longer than humans.

8. Which of the following statements are INCORRECT?

A. The biggest difference between gods and humans is the fact that gods are immortal.
B. Gods are taller than humans.
C. On the whole, Greeks find shorter women more attractive.
D. Gods cannot transmit illnesses between themselves.

Passage 3

The Effort of Digestion

Digestion is a huge, unappreciated task, unappreciated because few of us are aware of its happening in the same way we are aware of making efforts to use our voluntary muscles when working or exercising. Digestion begins in the mouth with thorough chewing. If you don't think chewing is effort, try making coleslaw in your own mouth. Chew up at least half a big head of cabbage and three big carrots that have not been shredded. Grind each bit until it liquefies and has been thoroughly mixed with saliva. I guarantee that if you even finish the chore your jaw will be tired, and you will have lost all desire to eat anything else, especially if it requires chewing. Making the saliva you just used while chewing the cabbage is by itself, a huge and unappreciated chemical effort.

Once in the stomach, chewed food has to be churned in order to mix it with hydrochloric acid, pepsin, and other digestive enzymes. Manufacturing these enzymes is also considerable work! Churning is even harder work than chewing but normally, people are unaware of its happening. While the stomach is churning (like a washing machine) a large portion of the blood supply is redirected from the muscles in the extremities to the stomach and intestines to aid in this process. Anyone who has tried to go for a run, or take part in any other strenuous physical activity immediately after a large meal feels like a slug and wonders why they just can't make their legs move the way they usually do. So, to assist the body while it is digesting, it is wise to take a siesta as los Latinos do instead of expecting the blood to be two places at once like los norteamericanos.

After the stomach is through churning, the partially digested food is moved into the small intestine where it is mixed with more pancreatin secreted by the pancreas, and with bile from the gall bladder. Pancreatin further solubilises proteins. Bile aids in the digestion of fatty foods. Manufacturing bile and pancreatic enzymes is also a lot of effort. Only after the carbohydrates (starches and sugars), proteins and fats have been broken down into simpler water-soluble food units such as simple sugars, amino acids and fatty acids, can the body pass these nutrients into the blood thorough the little projections in the small intestines called villi.

9. Why is digestion unappreciated?

A. Because we use voluntary muscles like when working or exercising.
B. Because we aren't conscious of the effort it takes our body to digest food.
C. Because it requires the whole digestive system.
D. Because we need food as energy for our bodies.

10. How many aspects of digestion in the passage are described as effort or hard work?

A. 2   B. 3   C. 4   D. 5

11. According to the passage, which of the following statements about digestion is true:

A. It requires blood to be diverted from muscles in the arms and legs to the stomach
B. It is as much effort as chewing half a head of cabbage and three big carrots
C. It requires amylase
D. It partially occurs in the gall bladder.

12. Which of the following statements about pancreatin is INCORRECT?

A. It is secreted into the small intestine
B. It helps with the formation of amino acids
C. It requires a lot of effort to be made
D. It only digests proteins.

Passage 4

Queen Victoria: Early years

When she was five years old the Princess Victoria began to have lessons, chiefly with a governess, Miss von Lehzen—"my dearly beloved angelic Lehzen," as she called her. These two remained devotedly attached to one another until the latter's death in 1870. The young Princess was especially fond of music and drawing, and it was clear that if she had been able to devote more time to study she would in later years have excelled in both subjects.

Her education was such as to fit her for her future position of Queen of England. The Princess did not, however, know that she was likely at any future time to be Queen. She read much, chiefly books dealing with history, and these were often chosen for her by her uncle, the King of the Belgians.

The family life was regular and simple. Lessons, a walk or drive, very few and simple pleasures made up her day. Breakfast was at half-past eight, luncheon at half-past one, and dinner at seven. Tea was allowed only in later years as a great treat.

The Queen herself said: "I was brought up very simply—never had a room to myself till I was nearly grown up—always slept in my mother's room till I came to the throne."

13. According to the passage:

A.   Princess Victoria had lessons with her governess for 5 years.
B.   Princess Victoria excelled in both music and drawing.
C.   Princess Victoria died in 1870.
D.   Princess Victoria was fond of her governess.

14. Which of the following was NOT part of Princess Victoria's day-to-day activities as a young child?

A.   Breakfast, lunch and dinner at set times.
B.   Lessons with her governess.
C.   Tea
D.   Music

15. Which of the following statements is true of Princess Victoria?

A.   She was unaware that she was likely to become Queen.
B.   She sometimes slept in her mother's room before becoming Queen.
C.   She read about Belgian history.
D.   She enjoyed music more than reading.

16. Which is INCORRECT regarding the Princesses education:

A.   It was tailored to her future position as Queen
B.   It involved reading history books
C.   Her uncle was involved in her education
D.   She was solely taught by her governess

Passage 5

Outdoor exercise in pregnancy

Outdoor exercise is indispensable to good health. It benefits not only the muscles, but the whole body. By this means the action of the heart is strengthened, and consequently all the tissues receive a rich supply of oxygen. Exercise also promotes the digestion and the assimilation of the food. It stimulates the sweat glands to become more active; and, for that matter, the other excretory organs as well. It invigorates the muscles, strengthens the nerves, and clears the brain. There is, indeed, no part of the human machine that does not run more smoothly if its owner exercises systematically in the open air; and during normal pregnancy there is no exception to this rule. Only in extremely rare cases—those, namely, in which extraordinary precautions must be taken to prevent miscarriage—will physicians prohibit outdoor recreation and, perhaps, every other kind of exertion. Under such circumstances the good effects that most persons secure from exercise should be sought from the use of massage.

The amount of exercise which the prospective mother should take cannot be stated precisely, but what can be definitely said is this— she should stop the moment she begins to feel tired. Fatigue is only one step short of exhaustion—and, since exhaustion must always be carefully guarded against, the safest rule will be to leave off exercising at a point where one still feels capable of doing more without becoming tired. Women who have laborious household duties to perform do not require as much exercise as those who lead sedentary lives; but they do require just as much fresh air, and should make it a rule to sit quietly out of doors two or three hours every day. It will be found, furthermore, that the limit of endurance is reached more quickly toward the end of pregnancy than at the beginning; a few patients will find it necessary to stop exercise altogether for a week or two before they are delivered.

17. A woman's exercise tolerance decreases throughout pregnancy.

A. True
B. False
C. Can't tell

18. Exhaustion can cause miscarriages.

A. True
B. False
C. Can't tell

19. Pregnant women should exercise until they go into labour.

A. True
B. False
C. Can't tell

20. Women who have household duties do not need to exercise.

A. True
B. False
C. Can't tell

Passage 6

Spaghetti or Macaroni with Butter and Cheese

This is the simplest form in which the spaghetti may be served, and it is generally reserved for the thickest pasta. The spaghetti are to be boiled until tender in salted water, taking care to remove them when tender, and not cooked until they lose form. They should not be put into the water until this is at a boiling point.

Take as much macaroni as will half fill the dish in which it is to be served. Break into pieces two and a half to three inches long if you so desire. The Italians leave them unbroken, but their skill in turning them around the fork and eating them is not the privilege of everybody. Put the macaroni into salted boiling water, and boil twelve to fifteen minutes, or until the macaroni is perfectly soft. Stir frequently to prevent the macaroni from adhering to the bottom. Turn it into a colander to drain; then put it into a pudding-dish with a generous quantity of butter and grated cheese. If more cheese is liked, it can be brought to the table so that the guests can help themselves to it.

The macaroni called "Mezzani" which is a name designating size, not quality, is the preferable kind for macaroni dishes made with butter and cheese.

21. Butter and cheese is the simplest accompaniment to pasta.

A. True
B. False
C. Can't tell

22. Only Italians can eat unbroken macaroni.

A. True
B. False
C. Can't tell

23. Which of the following is INCORRECT regarding the preparation of this dish?

A. If spaghetti is cooked for too long it will lose its form.
B. Spaghetti should be cooked for 12-15 minutes.
C. Stirring the pasta as it boils will prevent it from sticking to the bottom.
D. Italians eat macaroni unbroken.

24. According to the passage:

A. Pasta must be cooked in salted boiling water
B. Pasta should only be put into the water at boiling point.
C. Pasta can be drained in a colander.
D. Mezzani is the best kind of pasta for butter and cheese.

Passage 7

The Penguins of Antarctica

The penguins of the Antarctic regions very rightly have been termed the true inhabitants of that country. The species is of great antiquity, fossil remains of their ancestors having been found, which showed that they flourished as far back as the eocene epoch. To a degree far in advance of any other bird, the penguin has adapted itself to the sea as a means of livelihood, so that it rivals the very fishes. This proficiency in the water has been gained at the expense of its power of flight, but this is a matter of small moment, as it happens.

In few other regions could such an animal as the penguin rear its young, for when on land its short legs offer small advantage as a means of getting about, and as it cannot fly, it would become an easy prey to any of the carnivora which abound in other parts of the globe. Here, however, there are none of the bears and foxes which inhabit the North Polar regions, and once ashore the penguin is safe.

The reason for this state of things is that there is no food of any description to be had inland. Ages back, a different state of things existed: tropical forests abounded, and at one time, the seals ran about on shore like dogs. As conditions changed, these latter had to take to the sea for food, with the result that their four legs, in course of time, gave place to wide paddles or "flippers," as the penguins' wings have done, so that at length they became true inhabitants of the sea.

25. According to the above passage, which of the following statements is correct?

A. Penguins can fly
B. Penguins were the first land inhabitants of the Antarctic region
C. Seals benefit from the meal of penguin chicks
D. Penguin chicks cannot find food on land.

26. Penguins live in the North Polar regions

A. True
B. False
C. Can't tell

27. Penguins are safe when searching for food on land

A. True
B. False
C. Can't tell

28. Which of the following is incorrect?

A. Penguin are easy prey to foxes, bear and seals.
B. Penguins are highly adapted to the sea environment.
C. Penguins legs evolved over time into fins.
D. Seals could eat Penguin chicks.

Passage 8

Indian birds

In India winter is the time of year at which the larger birds of prey, both diurnal and nocturnal, rear up their broods. Throughout January the white-backed vultures are occupied in parental duties. The breeding season of these birds begins in October or November and ends in February or March. The nest, which is placed high up in a lofty tree, is a large platform composed of twigs which the birds themselves break off from the growing tree. Much amusement may be derived from watching the struggles of a white-backed vulture when severing a tough branch. Its wing-flapping and its tugging cause a great commotion in the tree. The boughs used by vultures for their nests are mostly covered with green leaves. These last wither soon after the branch has been plucked, so that, after the first few days of its existence, the nest looks like a great ball of dead leaves caught in a tree.

The nurseries of birds of prey can be described neither as picturesque nor as triumphs of architecture, but they have the great merit of being easy to see. January is the month in which to look for the eyries of Bonelli's eagles (Hieraetus fasciatus); not that the search is likely to be successful. The high cliffs of the Jumna and the Chambal in the Etawah district are the only places where the nests of this fine eagle have been recorded in the United Provinces. Mr. A. J. Currie has found the nest on two occasions in a mango tree in a tope at Lahore. In each case the eyrie was a flat platform of sticks about twice the size of a kite's nest. The ground beneath the eyrie was littered with fowls' feathers and pellets of skin, fur and bone. Most of these pellets contained squirrels' skulls; and Mr. Currie actually saw one of the parent birds fly to the nest with a squirrel in its talons.

29. Which of the following is incorrect?

A. Some birds of prey incorporate squirrels into their nests
B. Nests of Bonelli's eagles have been found in Jumna and Chambal
C. One is most likely to successfully find the nest of a Bonelli Eagle in January
D. The breeding season of the white-backed vulture begins in January

30. The Latin name for Bonelli's eagles is Hieraetus faciatus

A. True
B. False
C. Can't tell

31. The kite's nest is made of squirrels' skulls

A. True
B. False
C. Can't tell

32. Which is correct?

A. Bonelli's eagles build their nests on flat areas of land.
B. Kites have short and rounded tails
C. Winter is when the larger birds of prey in India breed
D. Only the highest cliffs of the Jumna and the Chambal will suit White-backed vultures

Passage 9

Argentinian history

The province of Buenos Aires, the largest in the country, has always been the most populated, and its lands have always commanded the highest prices, and these have risen tremendously, but not so much of late years in proportion as land in the northern provinces. During the years 1885, 1886, 1887, and 1888, there was a great boom in land. Foreigners were pouring in, bringing capital; great confidence was put by foreign capitalists in the country, several railways had run out new branches, new railways were built, new banks were opened, and a very large extent of land was opened up and cultivated, and put under wheat and linseed, harvests were good and money was flowing into the country. Then came a very bad year, 1889; the harvest was practically lost owing to the heavy and continuous rains which fell from December till July with hardly a clear day. This, together with a bad government and the revolution of 1890, created a great panic and a tremendous slump in all land, from which it took a long time to recover. Where people had bought camps and mortgaged them, which was the general thing to do in those days, the mortgagees foreclosed, and, when the camps were auctioned off, they did not fetch half what the properties had been bought for in the first instance, some four or five years previously. This, naturally, had a serious effect on the credit, soundness, and finances of the country, but really, the crisis was not felt until some three or four years after, and it was 1896 and 1897 which were very serious years for the country.

33. Which statement is incorrect?

A. There was a boom in land prices in the late 1880s.
B. Rain caused the land prices to fall.
C. 1889 was the worst year of the century.
D. It took a few years for the drop-in land prices to badly affect the city.

34. Which is the correct statement?

A. The government brought about the boom in land prices in the late 1880s.
B. The government foreclosed mortgages after 1890.
C. Foreign investment enables land prices to rise.
D. Many locals bought camps to try and save their land.

35. Buenos Aires is the smallest province in Argentina.

A. True
B. False
C. Can't tell

36. Monsoon season in Buenos Aires begins in December.

A. True
B. False
C. Can't tell

Passage 10

Mosquitoes

Mosquito eggs are laid in water or in places where water is apt to accumulate, otherwise they will not hatch. Some species lay their eggs in little masses that float on the surface of the water, looking like small particles of soot. Others lay their eggs singly, some floating about on the surface, others sinking to the bottom where they remain until the young issue. Some of the eggs may remain over winter, but usually those laid in the summer hatch in thirty-six to forty-eight hours or longer according to the temperature.

When the larvæ are ready to issue they burst open the lower end of the eggs and the young wrigglers escape into the water. The larvæ are fitted for aquatic life only, so mosquitoes cannot breed in moist or damp places unless there is at least a small amount of standing water there. A very little will do, but there must be enough to cover the larvæ or they perish. The head of the larvæ of most species is wide and flattened. The eyes are situated at the sides, and just in front of them is a pair of short antennæ which vary with the different species. The mouth-parts too vary greatly according to the feeding habits. Some mosquito larvæ are predaceous, feeding on the young of other species or on other insects. These of course have their mouth-parts fitted for seizing and holding their prey. Most of the wrigglers, however, feed on algæ, diatoms, Protozoa and other minute plant or animal forms which are swept into the mouth by curious little brush-like organs whose movements keep a stream of water flowing toward the mouth.

A few kinds feed habitually some distance below the surface, others on the bottom, while still others feed always at the surface. With one or two exceptions, the larvæ must all come to the surface to breathe. Most species have on the eighth abdominal segment a rather long breathing-tube the tip of which is thrust just above the surface of the water when they come up for air.

37. Mosquitoes require water to lay their eggs.

A. True
B. False
C. Can't tell

38. Eggs laid in Summer hatch thirty-six to forty-eight hours after eggs laid in Winter.

A. True
B. False
C. Can't tell

39. Select the incorrect statement.

A. Larvae are hatched wriggling.
B. Larvae must be wet when hatched.
C. The eyes of the larvae cannot see when first hatched.
D. Different groups of larvae feed at different levels underwater

40. Select the true statement.

A. All larvae feed on algae.
B. All larvae must come to the surface to breathe.
C. With a few exceptions all species of larvae have a breathing tube.
D. Larvae have a different circulatory system than mosquitoes.

Passage 11

Mites

The mites are closely related to the ticks, and although none of them has yet been shown to be responsible for the spread of any disease, their habits are such that it would be entirely possible for some to transmit certain diseases from one host to another, from animal to animal, from animal to man, or from man to man. A number of these mites produce certain serious diseases among various domestic animals and a few are responsible for certain diseases of men.

Face-mites. Living in the sweat-glands at the roots of hairs and in diseased follicles in the skin of man and some domestic animals are curious little parasites that look as much like worms as mites. Such diseased follicles become filled with fatty matter, the upper end becomes hard and black and in man are known as blackheads. If one of these blackheads is forced out and the fatty substance dissolved with ether the mites may be found in all stages of development. The young have six legs, the adult eight. The body is elongated and transversely wrinkled. In man they are usually found about the nose and chin and neck where they do no particular harm except to mar the appearance of the host and to indicate that his skin has not had the care it should have. Very recently certain investigators have found that the lepræ bacilli are often closely associated with these face mites and believe that they may possibly aid in the dissemination of leprosy. It is also thought that they may sometimes be the cause of cancer, but as yet these theories have not been proven by any conclusive experiment.

In dogs and cats these same or very similar parasites cause great suffering. In bad cases the hair falls out and the skin becomes scabby. Horses, cattle and sheep are also attacked. The disease caused by these mites on domestic animals is not usually considered curable except in its very early stages when salves or ointments may help some.

41. Select the correct statement:

A. Mites can spread diseases from one human to another.
B. Mites can spread diseases from themselves to ticks.
C. Salves prevent domestic animal infection.
D. Leprosy is caused by the bacteria lepræ bacilli

42. The main problem mites cause in humans are black heads.

A. True
B. False
C. Can't tell

43. Mites inhabit domestic animals for longer than they do humans.

A. True
B. False
C. Can't tell

44. Select the true statement.

A.   Mites affect humans in a worse way than they affect domestic animals.
B.   Black heads caused by mites are commonly found on extensor surfaces such as elbows.
C.   Face mites cause a similar facial appearance in humans and other animals.
D.   As the mites mature they develop 2 more legs.

END OF SECTION

Section B: Decision Making

1. In Manchester, a survey is done on a school. 6 children like cheese and onion crisps, 5 children like salt and vinegar crisps and 9 children like ready salted. 2 children like all three flavours, 1 child likes salt and vinegar and ready salted, 1 child likes salt and vinegar and cheese and onion, and 1 child likes cheese and onion and ready salted. All of the children like crisps. How many children took the survey?

    A. 25   B. 13   C. 20   D. 15   E. 37

2. Sam is shorter than Tom who is shorter than Alex. Tom is shorter than Joseph who is shorter than Lilly. Alex is taller than Lilly. Who is the shortest?

    A. Lilly   B. Joseph   C. Alex   D. Sam   E. Tom

3. All dancers are strong. Some dancers are pretty. Alexandra is strong, and Katie is pretty. Choose a correct statement.

    A. Alexandra is a dancer   C. A dancer can be strong and pretty
    B. Katie is not a dancer   D. A dancer can be strong and ugly

4. There is a chocolate cake. Nick takes 1/6 of the cake, then John takes 1/4 of the remainder. Alice and Jane attempt to share the rest of the cake but get full after they have eaten 2/3 of what was left. What proportion of the original cake is left?

    A. 1/12   B. 1/6   C. 3/24   D. 5/24

5. At a barbecue, 1/4 of the guests are vegetarian. Of the meat eaters, 1/2 want 2 pieces of chicken each, and 1/2 want 1 piece of chicken. 50 pieces of chicken were bought, which includes 5 spare pieces for anyone who changes their mind. How many guests are at the barbecue?

    A. 35   B. 40   C. 45   D. 50

6. A pig travels 5 km north then he turns to his right and walks 3 km. He then turns to his right and moves 5km forward. Now in which direction is he from his starting point.

    A. North   C. East   E. North
    B. South   D. West   West

7. What is the next number in the following sequence: 144, 196, 256,

    A. 346   B. 324   C. 289   D. 300

8. If horse A completes a job in 10 days and B completes the same work in 15 days - in how many days is the work completed if they work together?
    A. 2   B. 3   C. 4   D. 5   E. 6

9. The ratio of a:b is 2:3 – if the present age of a is 20 years, find the age of b after 5 years.

    A. 10   B. 15   C. 25   D. 30   E. 35

10. In a queue, Billy is 11th from the front of a queue and Amy is 22nd from the back. There are 4 people between Amy and Billy. If 3 people get money from the ATM then leave the queue, how many people are in the queue?

A. 31   B. 32   C. 34   D. 35   E. 38

11. Southampton is bigger than only Romsey. Oxford is bigger than Cambridge but not as big as London. Which is second biggest city?

A. Southampton   C. Oxford   E. London
B. Romsey   D. Cambridge

12. There are 5 giraffes, J is taller than D, but shorter than V and M. V is shorter than only R. If the height of the second tallest giraffe is 160cm and second shortest giraffe is 135cm, what is the possible height of M?

A. 130cm   C. 155cm   E. None of these
B. 162cm   D. Cannot be determined

13. All professional jobs require a university degree. Chemistry is a university degree. All lawyers are professionals. Which of the following statements is true based on this information?

A. All Chemistry teachers are professionals
B. All professionals require a chemistry degree
C. All lawyers have a degree
D. All lawyers have a chemistry degree
E. All universities offer law as a degree course

14. Read the following statements. Which of the options is correct regarding whether the conclusions drawn from the statements are true or not?
   Statements:
   No man is a lion.
   Joseph is a man.
   Conclusions:
   I. Joseph is not a lion.
   II. All men are not Joseph.

A. Conclusion I is TRUE and Conclusion II is TRUE
B. Conclusion I is TRUE and FALSE
C. Conclusion I is TRUE and Conclusion II is CAN'T TELL
D. Conclusion I is FALSE and Conclusion II is CAN'T TELL
E. Conclusion I is FALSE and Conclusion II is TRUE

15. The radius of a circle is 2/3 of the side of a square. The area of the square is 441 cm^2. Find the perimeter of the circle.

A. 27pi   C. 29pi   E. None of
B. 28pi   D. 30pi   these

16. The sum of 2 numbers is 3430. If 12% of one of these numbers is equal to 28% of the other number, find the largest number.

A. 1029   B. 1032   C. 1041   D. 2401   E. 2453

17. Billy is James's father. 4 years ago, Billy's age was 4 times that of James. After 6 years, the ages of the Billy and James are in the ratio of 5:2. How old is Billy?

A. 12   B. 13   C. 14   D. 15   E. 16

18. Pointing to a photograph, a man said, "I have no brother or sister but that man's father is my father's son." Who owns the photograph?

A. His own   B. His nephew's   C. His father's   D. His son's   E. His sister's

19. The average marks scored by 12 students is 73. If the scores of Bea, Bay and Boe are included, the average becomes 73.6. If Bea scored 68 marks and Boe scored 6 more than Bay, what was Bay's score?

A. 75   B. 76   C. 77   D. 78   E. 79

20. Calculate the ratio of curved surface area and total surface area of a cone whose diameter is 40m and height is 21m.

A. 15:18   B. 26:28   C. 10:11   D. 2:3   E. 24:29

21. In a family, there is the father, the mother, two sons and two daughters. All the ladies were invited to a dinner. Both sons went out to play. The father is at work. Who was at home?

A. Only the mother was at home   C. Only the sons were at home
B. All the females were at home   D. Nobody was at home

22. There are five books A, B, C, D and E placed on a table. If A is placed below E, C is placed above D, B is placed below A, and D is placed above E, then which of the following books touches the surface of the table?

A. A   B. B   C. C   D. D

23. A man covers a distance in 1hr 24min by covering 2/3 of the distance at 4 km/h and the rest at 5km/h. What distance does he cover?

A. 5km   B. 6km   C. 7km   D. 8km

24. Two people starting from the same place walk at a rate of 5kmph and 5.5kmph respectively. What time will it take for them to be 8.5km apart, if they walk in the same direction?

A. 15   B. 16   C. 14   D. 18   E. 17

25. If 34 men completed 2/5th of a work in 8 days working 9 hours a day, how many more men should be employed to finish the rest of the work in 6 days working 9 hours a day?

A. 68   B. 108   C. 34   D. 36   E. 62

26. Given these statements:
Robots are Machines.
Machines are not human.

Which of the following are correct?
i) Jack is a human. Therefore, Jack is not a machine.
ii) All machines are robots.

A.   True AND True
B.   True AND False
C.   True AND Can't Tell

D.   Can't Tell AND Can't Tell
E.   Can't Tell AND False

27. Artists are generally whimsical. Some of them are frustrated. Frustrated people are prone to be drug addicts. Based on these statements which of the following conclusions is true?

A.   All frustrated people are drug addicts
B.   Some artists may be drug addicts

C.   All drug addicts are artists
D.   All frustrated people are whimsical

28. Three ladies X, Y and Z marry three men A, B and C. X is married to A, Y is not married to an engineer, Z is not married to a doctor, C is not a doctor and A is a lawyer. Then which of the following statements is correct?

A.   Y is married to C who is an engineer
B.   Z is married to C who is a doctor

C.   X is married to a doctor
D.   None of these

29. 'Knowing the times tables is an important skill for a primary school child.' Which of the following statements provides the best argument against this statement?

A.   Vigorously learning the times tables in primary school provides an unnecessary level of stress
B.   Literacy is a more important skill than mathematics
C.   Knowing the times tables provides no benefit for children in later life
D.   There is no need for mental arithmetic when calculators are so readily available

END OF SECTION

Section C: Quantitative Reasoning

Data Set 1

John purchases a 360g box of chai latte powder. The label states the following:

	Per 100g (dry powder)	Per 18g serving (with 200g whole milk)
Energy (kJ)	1555	845
kCal	372	200
Protein (g)	8.9	8.0
Carbohydrate (g)	65.2	20.4
Fat (g)	8.4	9.6

INSTRUCTIONS

Put 3 heaped teaspoons (18g) into a mug, Add 200ml of hot milk and stir well. For a diet drink, use semi-skimmed or skimmed milk. Semi-skimmed milk contains 1.7g fat per 100ml. Skimmed milk contained 0.3g of fat per 100ml.

1.   How much fat is contained in 200ml whole milk (to the nearest 0.1g)?

A. 1.2g   B. 3.4g   C. 6.8g   D. 8.1g   E. 9.6g

2.   John is on a diet so decides to use a combination of semi-skimmed milk and water instead. He then adds 200ml of the mixed liquid to 18g of the chai powder. The total fat content of the resulting drink is 2.464g. Given that water does not contain any fat, what volume of semi skimmed milk did he use?

A. 56.0ml   B. 81.5ml   C. 144.9ml   D. 118.5ml   E. 170.4ml

3.   John uses up the whole box of chai latte powder, making hot drinks according to the instructions. How many pints of milk will he have used (1pint = 570ml)?

A. 0.35   B. 4.4   C. 7.0   D. 14.1   E. 20.0

4.   John buys semi-skimmed milk with a label that says its protein content is 3.3g per 100ml. What is the total amount of protein that he will have drunk once he has made chai latte with the entire box using standard proportions and semi-skimmed instead of whole milk (to the nearest gram)?

A. 66g   B. 160g   C. 164g   D. 244g   E. 272g

5.   The calories of food is measured in kCal and it calculated by adding the calorific values of fat, protein and carbohydrate. The calorific value of 1g of protein is 4kCal and of 1g carbohydrate is 4kCal. What is the calorific value of 1g fat to the nearest gram?

A. 4 kCal   B. 6 kCal   C. 9 kCal   D. 21 kCal   E. 44 kCal

Data Set 2
Parking at Great Ormond Street Hospital

Parking Fees

This car park uses three separate changing time zones

Time zone 1: Monday – Friday

9:00am-1:00pm: 20p for each period of 10 minutes

1:00pm-6:00pm: 30p for each period of 10 minutes

Time zone 2: Saturday, Sunday and Bank Holiday

9:00am-6:00pm: 15p for each period of 20 minutes

Time zone 3: 6:00pm-9:00am any day

£1 standard fee regardless of the duration

NOTICE

1 – Payment is made as you exit the car park

2 – Any period stated should be paid for in full e.g. if a driver stays in the car park for 15 minutes on a Monday afternoon they should pay for a full 20 minutes (i.e. 2 x 10-minute periods)

3 – If a car is parked in the car park over several charging zones then a calculation will be made separately for each zone, each being subject to its own minimum charge. For example, if a driver arrived at 12:59pm on a Monday and stayed 20 minutes, they would be charged 20p for the period between 12:59pm and 1:00pm and 60p for the period for 1:00pm to 1:19pm, a total of 80p.

6. A driver parks his car on Monday morning at 9:00am and leaves it in the car park until Sunday evening 6:00pm. How much will it cost him?

   A. £59.10   B. £65.10   C. £71.10   D. £77.10   E. £83.10

7. A driver enters the car park at 5:55pm on a Monday afternoon and leaves the car park at 6:15pm. How much will he be charged?

   A. 40p   B. 60p   C. £1   D. £1.15   E. £1.30

8. A driver arrives at the car park at 9am on a Monday and stay in the car park for 37 minutes. How much will he be charged?

   A. 20p   B. 50p   C. 74p   D. 80p   E. £1

9. A driver uses the car park all year round and contacts the council to obtain a weekly season ticket. The season ticket he requires is for commuters and only allows him to use the car park Monday to Friday between 8:30am and 6:00pm. The cost of the weekly season ticket is £54. How much money will be save in comparison to paying full price for that period?

   A. £5   B. £10   C. £15   D. £20   E. £25

10. A driver comes into the car park at 12:58pm and leaves 4 minutes later at 1:02pm on a Monday lunch time after having dropped his wife and son off at the hospital. How much will he be charged?

   A. 10p   B. 20p   C. 30p   D. 40p   E. 50p

11. A driver only has £4.90 in his pocket to pay for car park fees. He enters the car park on Friday afternoon at 5pm. Until when can he stay in the car park?

   A. Friday 9:00pm   B. Saturday 9:00am   C. Saturday 11:00am   D. Saturday 11:40am   E. Saturday 1:40pm

Data Set 3

The following table shows the gold price per 10 grams of cold in India from 1925-2011. The price is given in Indian Rupees.

Gold Prices in India from 1925 - 2011 (per 10 gm)

Year	Price	Year	Price	Year	Price	Year	Price	Year	Price
1925	18.75	1943	51.05	1961	119.35	1979	937	1997	4725
1926	18.43	1944	52.93	1962	119.75	1980	1330	1998	4045
1927	18.37	1945	62	1963	97	1981	1800	1999	4234
1928	18.37	1946	83.87	1964	63.25	1982	1645	2000	4400
1929	18.43	1947	88.62	1965	71.75	1983	1800	2001	4300
1930	18.05	1948	95.87	1966	83.75	1984	1970	2002	4990
1931	18.18	1949	94.17	1967	102.5	1985	2130	2003	5600
1932	23.06	1950	99.18	1968	162	1986	2140	2004	5850
1933	24.05	1951	98.05	1969	176	1987	2570	2005	7000
1934	28.81	1952	76.81	1970	184.5	1988	3130	2006	8400
1935	30.81	1953	73.06	1971	193	1989	3140	2007	10800
1936	29.81	1954	77.75	1972	202	1990	3200	2008	12500
1937	30.18	1955	79.18	1973	278.5	1991	3466	2009	14500
1938	19.93	1956	90.81	1974	506	1992	4334	2010	18500
1939	31.74	1957	90.62	1975	540	1993	4140	2011	26400
1940	36.04	1958	95.38	1976	432	1994	4598		
1941	37.43	1959	102.56	1977	486	1995	4680	jagoinvestor.com	
1942	33.05	1960	111.87	1978	685	1996	5160		

12. What is the price of 1kg of gold in 2010?

A. 1 750 000 Rupees
B. 1 850 000 Rupees
C. 18500 Rupees
D. 1850 Rupees
E. 18 500 000 Rupees

13. If there are 85 rupees in 1 Great British pound, what is the price of 1kg of gold in 2010 in Great British Pounds (to the nearest pound)?

A. £22547
B. £24536
C. £26534
D. £22981
E. £21765

14. How much gold would £25000 buy in 1940 (to the nearest kilogram)?

A. 625kg
B. 590kg
C. 716kg
D. 821kg
E. 678kg

15. In 1940 however, £1 was worth 110 Indian Rupees. Based on this, how much gold would £25000 buy in 1940 (to the nearest kilogram)?

A. 783kg
B. 779kg
C. 763kg
D. 634kg
E. 921kg

16. During school break, Aaron and Mason must arrange 46 blue pens and 38 red pens in equal quantities into pencil cases. If there is no remainder, what is the maximum number of pens that can go into each pencil case?

A. 2   B. 4   C. 6   D. 5   E. 7

Data Set 4

The following graph shows how the value (in US dollars) of the crypto-currency, Bitcoin, has varied over the past 8 years.

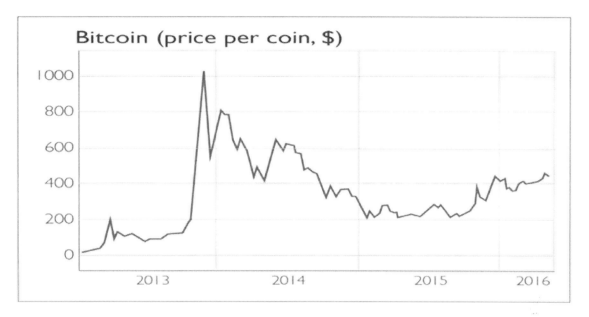

17. At what point did Bitcoin's price reach its peak?

A.   July 2013
B.   September 2013

C.   Decemeber 2013
D.   July 2015

E.   July 2016

18. What was the price of Bitcoin at this stage?

A. $1020   B. $1350   C. $1256   D. $1789   E. $898

19. What is the rate of increase of the value of Bitcoin between July 2013 and December 2013 (to the nearest dollar)?

A. $150 per month
B. $230 per month

C. $143 per month
D. $184 per month

E. $350 per month

20. Jamil is a young entrepreneur who believed that Bitcoin would rapidly increase its value at some point. As a result, he invested $1000 into Bitcoin shares in January 2013, when the price of a Bitcoin share was 50 cents. He sold the shares in March 2013.How much more money would he have made had he sold it in December 2013?

A. $364537
B. $1600000

C. $2736281
D. $271928

E. $17728

Data Set 5

The following cone has radius (r) of 19cm and perpendicular height (h) of 65cm.

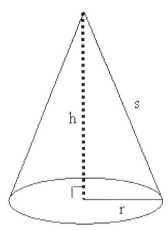

21. What is the volume of the above shape (to the nearest metre cubed)?

A. 1.79m³   B. 1.25m³   C. 1.40m³   D. 0.25m³   E. 2.34m³

22. Given that the value s=79cm, what is the surface area of the above shape (to the nearest metre squared)?

A. 0.79m²   B. 0.92m²   C. 0.28m²   D. 0.33m²   E. 0.58m²

23. If the angle made between h and s=34°, what is the angle between s and r?

A. 64°   B. 56°   C. 23°   D. 39°   E. 51°

Data Set 6

The following bar chart shows how many deaths are caused by several cancers.

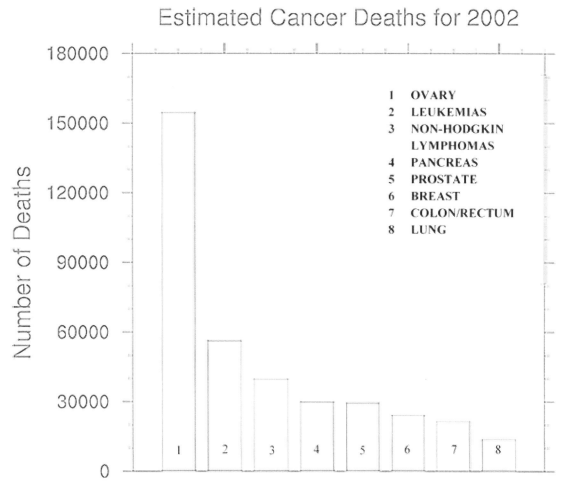

Estimated Cancer Deaths for 2002

1 OVARY
2 LEUKEMIAS
3 NON-HODGKIN
   LYMPHOMAS
4 PANCREAS
5 PROSTATE
6 BREAST
7 COLON/RECTUM
8 LUNG

24. What is the mode of the above data set?

A. Ovary cancer
B. Breast cancer
C. Colon/rectum cancer
D. Lung cancer
E. Pancreas cancer

25. What is the mean of the above data set?

A. 46250   B. 78350   C. 87290   D. 17350   E. 22350

26. What is the range of the above data set?

A. 120 000   B. 130 000   C. 150 000   D. 140 000   E. 200 000

Data Set 7

The following graph shows how the human population has changed since 1750.

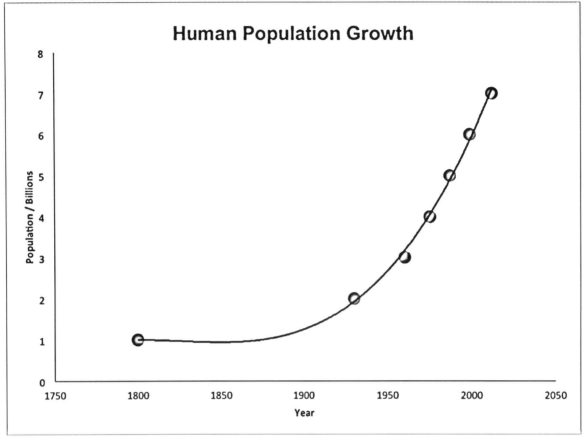

27. Estimate the human population in 1900

A.   1 billion
B.   1.5 billion

C.   1.2 billion

D.   800 million

E.   1.8 billion

28. What is the rate of population increase between 1960 and 2010?

A.   40 million per year
B.   50 million per year

C.   60 million per year
D.   70 million per year

E.   80 million per year

29. Assuming the same rate of increase, what will the population be in 2125?

A.   14.5 billion
B.   15.9 billion

C.   16.2 billion
D.   17.8 billion

E.   20.2 billion

Data Set 8

Ali is the CEO of an accounting firm. In 2016, the company had 3600 accounts. In 2017, their business saw a massive drop, falling to 2250 accounts.

30. What percentage drop does this represent?

A. 37.5%   B. 21.5%   C. 31.5%   D. 41.2%   E. 67.5%

31. If each client contributes a revenue of $150 000 to the firm, how much money did the firm lose between 2016 and 2017?

A. $390 million   C. $215 million   E. $512 million
B. $420 million   D. $202.5 million

32. If the firm must pay each employee between $30 000 and $100 000, what is the minimum number of employees that Ali must fire because of the revenue loss in 2017?

A. 5675   B. 7275   C. 1465   D. 265   E. 2025

Data Set 9

The following table is a bus timetable for Ludlow Town

701 Ludlow Town Service **Saturday** Schedule commences 27 January 2014

Service No	701	701	701	701	701	701	701	701	701	701	701	701
Ludlow, Tourist Info Centre	..	0740	0840	0940	1040	1140	1240	1340	1440	1540	1640	1740
Ludlow, Railway Station	..	0745	0845	0945	1045	1145	1245	1345	1445	1545	1645	1745
Ludlow, Henley Road	..	0749	0849	0949	1049	1149	1249	1349	1449	1549	1649	1749
Ludlow, Rocks Green Estate	0705	0751	0851	0951	1051	1151	1251	1351	1451	1551	1651	1751
Ludlow, Weyman Road, Potters Close Jct	0707	0753	0853	0953	1053	1153	1253	1353	1453	1553	1653	1753
Ludlow, Bringewood Road, opp Bringewood Close	0709	0755	0855	0955	1055	1155	1255	1355	1455	1555	1655	1755
Ludlow, Sandpits Road, Whitefriars Jct	0711	0759	0859	0959	1059	1159	1259	1359	1459	1559	1659	1759
Ludlow, Clee View	0712	0800	0900	1000	1100	1200	1300	1400	1500	1600	1700	1800
Ludlow, Sidney Road	0716	0804	0904	1004	1104	1204	1304	1404	1504	1604	1704	1804
Ludlow, Sheet Road, Kennet Bank Jct	0717	0805	0905	1005	1105	1205	1305	1405	1505	1605	1705	1805
Ludlow, Eco Park & Ride	0720	0810	0910	1010	1110	1210	1310	1410	1510	1610	1710	1810
Ludlow, Tollgate	0723	0814	0914	1014	1114	1214	1314	1414	1514	1614	1714	..
Ludlow, Parys Road, opp Vashon Close Jct	0725	0816	0916	1016	1116	1216	1316	1416	1516	1616	1716	..
Ludlow, Greenacres	0729	0820	0920	1020	1120	1220	1320	1420	1520	1620	1720	..
Ludlow, Steventon New Road, Churchill Close Jct	0732	0823	0923	1023	1123	1223	1323	1423	1523	1623	1723	..
Ludlow, Weeping Cross Lane, Friarfields Jct	0733	0824	0924	1024	1124	1224	1324	1424	1524	1624	1724	..
Ludlow, Upper Galdeford, Cooperative Store	0735	0826	0926	1026	1126	1226	1326	1426	1526	1626	1726	..
Ludlow, opp Railway Station	0737	0828	0928	1028	1128	1228	1328	1428	1528	1628	1728	..
Ludlow, Tourist Info Centre	0740	0831	0931	1031	1131	1231	1331	1431	1531	1631	1731	..

33. If the bus is driving at a speed of 40 miles per hour between Sidney Road and Tollgate, what is the distance between these two stops?

A. 9 miles   B. 10 miles   C. 6 miles   D. 4.66 miles   E. 3 miles

34. If the total distance between Henley Road and Greenacres is 34 miles, what is the average speed of the bus?

A. 50mph   B. 60mph   C. 36mph   D. 42.5mph   E. 68mph

35. Susan catches the 0733 bus from Weeping Cross Lane to the Railway station every day, in order for her to get to work by 8. One morning, Susan misses her bus by 2 minutes. If there is a 1.5 mile walk to the station from Weeping Cross Lane, how fast does she have to walk to make the 0752 train?

A. 4.8mph   B. 3.2mph   C. 5.3mph   D. 6.4mph   E. 8mph

36. Tom and Jack have a bag of skittles, which have 53 skittles in them. If there 12 blue skittles, 13 red skittles and 14 yellow skittles, what is the probability that they take out a purple skittle, given that skittles come in either blue, red, yellow or purple?

A. 0.284   B. 0.298   C. 0.217   D. 0.264   E. 0.156

END OF SECTION

Section D: Abstract Reasoning

For each question, decide whether each box fits best with Set A, Set B or with neither.

For each question, work through the boxes from left to right as you see them on the page. Make your decision and fill it into the answer sheet.

Answer as follows:
A = Set A
B = Set B
C = neither

Set 1	Set A	Set B

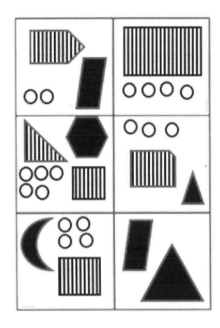 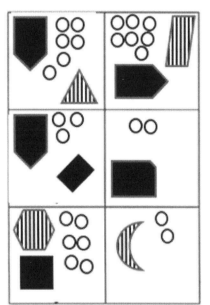

Questions 1-5:

Set 2

Question 6: Which figure completes the series?

        A.   B.   C.   D.

Set 3

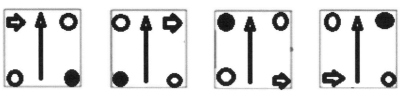

Question 7: Which figure completes the series?

        A.   B.   C.   D.

Set 4

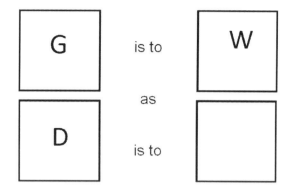

Question 8: Which figure completes the statement?

A.   B.   C.   D.

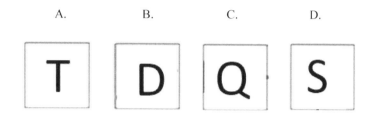

Set 5

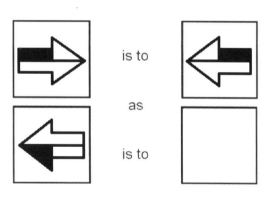

Question 9: Which figure completes the statement?

A.   B.   C.   D.

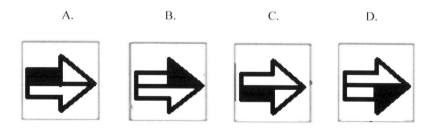

Set 6

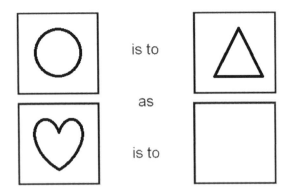

Question 10: Which figure completes the statement?

A.   B.   C.   D.

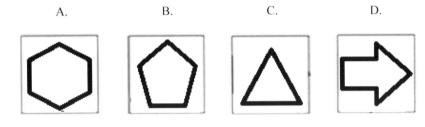

Set 7   Set A   Set B

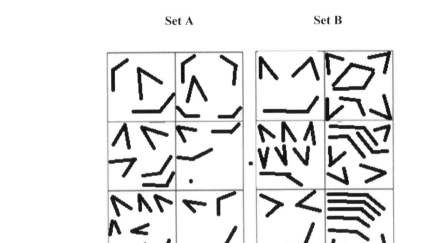

Question 11: Which of the following belongs in Set A?

A.   B.   C.   D.

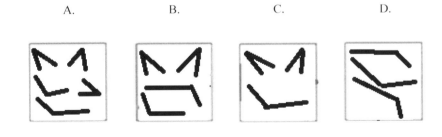

Question 12: Which of the following belongs in Set A?

 A.   B.   C.   D.

Question 13: Which of the following belongs in Set A?

A.   B.   C.   D.

Question 14: Which of the following belongs in Set B?

A.   B.   C.   D.

Question 15: Which of the following belongs in Set B?

A.   B.   C.   D.

Set 8   **Set A**   **Set B**

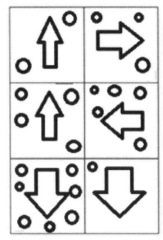 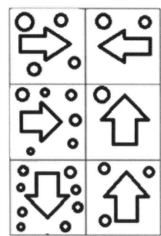

Questions 16-20:

Set 9   **Set A**   **Set B**

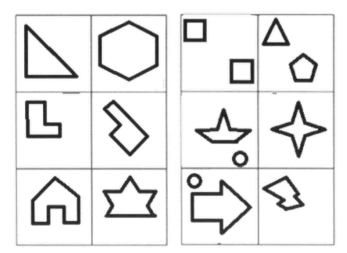

Questions 21-25:

Set 10   **Set A**   **Set B**

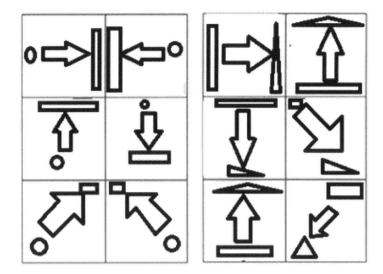

Questions 26-30:

Set 11   **Set A**   **Set B**

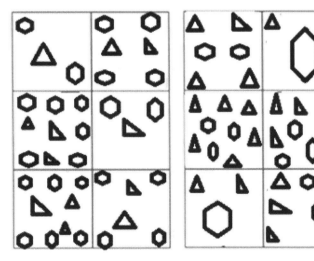

Questions 31-35:

Set 12   **Set A**   **Set B**

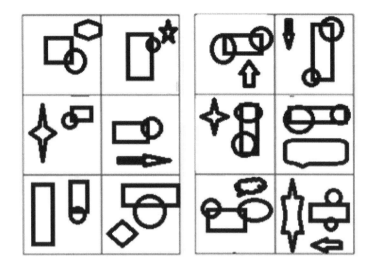

Questions 36-40:

Set 13   **Set A**   **Set B**

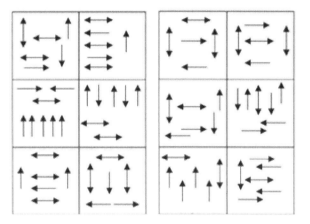

Question 41-45:

Set 14   **Set A**   **Set B**

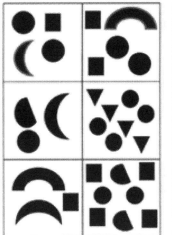 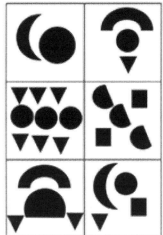

Question 46-50:

Set 15   **Set A**   **Set B**

Question 51-55:

END OF SECTION

Section E: Situational Judgement Test

Read each scenario. Each question refers to the scenario directly above. For each question, select one of these four options. Select whichever you feel is the option that best represents your view on each suggested action.

For questions 1 – 30, choose one of the following options:

A   A highly appropriate action
B   Appropriate, but not ideal
C   Inappropriate, but not awful
D   A highly inappropriate action

Scenario 1

John is a 4th year medical student doing a placement on a general medical ward. One morning he notices that his fellow medical student, Tom, has slightly slurred speech. He can also smell alcohol on Tom's breath.
How appropriate are the following responses to this situation?

1. Do nothing, it is not your responsibility.
2. Report him to the head of the medical school.
3. Speak to Tom and express your concerns, and that you think it is not appropriate for him to be on the ward round today. Let him open up to you about any problems he may be having, and encourage him to seek help if he needs it.
4. Discuss this occurrence with your fellow medical students and seek their advice on how to manage the situation.
5. Discuss this occurrence with your medical supervisor and seek their advice on how to manage the situation.

Scenario 2

Jacob is a 3rd year medical student, who has a compulsory communications skills session about 'breaking bad news' to patients next week. His mother was diagnosed with cancer a month ago and this feels very close to home. He is not keen on attending the session.
How appropriate are the following responses to this situation?

6. Don't show up even though it was compulsory, it is too difficult for him to cope with a session like this right now.
7. Show up and tolerate the session even though he doesn't feel comfortable with the idea of it.
8. Speak to the doctor running the session and inform them of his circumstances and why he will not be able to attend the session.
9. Tell his friend, who knows about his mother, to tell the doctor he cannot come for personal reasons.
10. Show up to the session and leave half way when he thinks that he can't cope with being there anymore.

Scenario 3

Lucy, a third year medical student, is captain of the University hockey team but is also very interested in paediatrics. A week before the final hockey match she realises that an interesting looking paediatrics conference is being held in the university on the same day in the afternoon. She is unsure what to do.
How appropriate are the following responses to this situation?

11. Inform the members of the hockey team that she can't attend the match and attend the conference instead.
12. Let the team members know about the conference, and see if there is a way for her to play in the majority of the match and be subbed in for the time when the conference begins.
13. Attend the match fully and sign up for the conference, and turn up to the conference mid-way through once the match is over.
14. Don't attend the conference, she is only a third year after all and there will be one next year when she is no longer captain of the team.
15. Pretend to be ill and attend the conference instead.

Scenario 4

Tara is a 4th year medical student on the consultant-led ward round on a medical ward. Yesterday on the ward round, the consultant shouted at her in front of a patient for turning up two minutes late. She tried to brush it off, but today she was shouted at again, for not writing in the notes quickly enough. Tara found both experiences very unpleasant and distressing.

How appropriate are the following responses to this situation?

16. Report the consultant to the head of the medical school.
17. Speak to her clinical supervisor about her concerns and how to proceed.
18. Speak to the consultant about how uncomfortable she felt in this situation.
19. Do nothing about it and continue the placement on the ward.
20. Stop attending the placement and pretend to be sick.

Scenario 5

Alice is a second year medical student. She failed one of her exams and has to retake them in the beginning of September. However, her sister's wedding is two days before her retake and as the maid of honour she will have a lot of wedding-related work to do in the weeks approaching the wedding.

How appropriate are the following responses to the situation?

21. Don't revise for the exams until two days before (when the wedding is over), she didn't fail by much and won't have time before her sister's wedding.
22. Hand over some of the maid of honour responsibilities to another family member, giving her more time to focus on her retake at the same time.
23. Attempt to take on the responsibilities of both being maid of honour (and all this entails) whilst also revising for the retake.
24. Ask her sister if she could move the date of her wedding.
25. Ask her sister if she would be happy for another family member/friend to be maid of honour.

Scenario 6

Luke is a final year medical student. On the way out of the hospital, in the lift, he overhears a student on the phone speaking loudly about a patient in the hospital, using the patient's name.

How appropriate are the following responses to the situation?

26. Report the student to the head of the medical school.
27. Ignore the student and pretend not to know them.
28. Speak to his supervisor about the most appropriate action to take.
29. Speak to the student and explain that he should not be talking about patients publically.
30. Tell the student that he should not be speaking about patients whilst still in the hospital.

For questions 31 – 69, decide how important each statement is when deciding how to respond to the situation?

A   **Extremely important**
B   **Fairly important**
C   **Of minor importance**
D   **Of no importance whatsoever**

Scenario 7

Eva is a final year medical student, who has her finals coming up in a month. She has a mock OSCE (clinical examination) this week, which is the only one before the exam. However she comes down with the flu, she does not feel too well but is determined not to miss this opportunity. The medical school states that you should not attend if you are unwell.
How important are the following factors for Eva in making her decision?

31. This mock OSCE is a good opportunity to practice on real patients before finals.
32. Eva having the flu could put the patients at risk.
33. There are no further mocks after this one.
34. The patients participating in the mock are not going to be extremely unwell.
35. The medical school has stated that if ill you should not take part.
36. Eva's finals are in a month, so if she allows herself to recover she will still have time to go on the wards and practice on patients for the exam.

Scenario 8

Lucy is a 5th year medical student. She was lucky to get first choice in the ballot for GP placement and so has chosen a GP surgery because it is a 10 minute walk away. Alice is near the bottom of the ballot and so had to choose a practice which is either 30 minutes driving or 1 hour 30 minutes by bus, and she cannot drive. Lucy can drive. Alice has asked to swap with Lucy in exchange for the top place in the ballot for psychiatry placements.
How important are the following factors for Lucy in making her decision?

37. The medical school would reimburse Lucy for petrol so she would not be losing out financially by driving to the further GP practice.
38. The half an hour drive Lucy would have to do is much more convenient than the 1.5 hour bus ride Alice would have to take.
39. She would be higher in the psychiatry placement in return.
40. Lucy would be compromising her morning by giving up the GP placement.
41. There are other people in nearby GP practices who can drive and swap places with Alice.
42. Lucy prefers GP as a speciality to psychiatry.

Scenario 9

Nick is a second year medical student. He struggles with bad insomnia whereby he is wide awake through most of the night but then falls asleep in the morning. He has missed many morning lectures because of this and is unsure of what to do.
How important are the following factors for Nick in making his decision?

43. Whether he learns better by going to lectures or by reading a textbook.
44. If he attends the lectures in the morning, he will either fall asleep during them or compromise his concentration for the teaching in the afternoon.
45. Whether the other students say the lectures are useful.
46. Whether the lectures are available online.
47. Whether someone is able to record the lectures for him.
48. He has a medical problem, so should not be afraid to seek help from the medical school

Scenario 10

Lorna is a 3rd year medical student on her paediatrics placement. She really enjoys paediatrics and really likes her consultant - and her consultant has offered to give her extra teaching sessions on Thursday evening. However, she is also a member of the hockey team and this is when they have their practice.

How important are the following factors for Lorna in making her decision?

49. The consultant is only free on a Thursday evening.
50. The teaching may not be useful for the exams.
51. Hockey practice is only on a Thursday evening.
52. Lorna is considering paediatrics as a speciality.
53. Lorna's team mates will be disappointed if she doesn't show up to practice.
54. Lorna does not carry out any other extracurricular activities.

Scenario 11

Chris has written an essay for assignment. When handed back their essays, he notices that Joe's essay is an exact replica of an essay they were both sent by a medical student in the year above. He is unsure as to what to do.

How important are the following factors for Chris in making his decision?

55. Chris and Joe are good friends.
56. Plagiarism is against University rules.
57. It is unfair to the medical student in the year above that Joe copied his essay word for word.
58. Joe's essay was very good.
59. This essay only counts for 5% of the grade this year.
60. Plagiarism is inevitable - has occurred in the past and will occur in the future.

Scenario 12

Chen is a final year student on his medical elective in Swaziland. He is assisting surgery on a girl who has a deeply infected hand. If her fingers are not amputated, the infection will spread and could make her very unwell. The surgeon asks Chen to assist him in amputating the girl's fingers with scissors, as this is the only equipment available. The surgeon wants Chen's assistance as it is a difficult task. Chen is not sure whether he is comfortable with this situation.

How important are the following factors for Chen in making his decision?

61. The girl will become very ill and possibly die if her fingers are not amputated.
62. Chen is likely to find the procedure psychologically traumatic.
63. The UK and Swaziland have very different resources (e.g. only scissors for amputation) to the UK, as well as policies regarding how much students are allowed to do.
64. Chen is the only person available to help the surgeon with this procedure.
65. In the UK, Chen would not be allowed to do anything of this calibre as a medical student.

Scenario 13

Jasmine is a third year medical student. She has completed a poster for the genetics project she has been working on. Her supervisor says that if she continues coming to the labs three times a week for the next 2 months, she will have enough data for a publication. However, her exams are also coming up in a couple of months.

How important are the following factors for Jasmine in making her decision?

66. Going to the lab three times a week could interfere with her upcoming exams.
67. A publication will be useful on Jasmine's CV and application for jobs.
68. A publication also benefits the lab – Jasmine does not want to let them down.
69. Jasmine also plays university level netball and hockey.

END OF PAPER

ANSWERS

Answer Key

PAPER A

VR	DM	QR	AR		SJT	
1. B	1. D	1. B	1. C	46. C	1. D	46. B
2. D	2. B	2. B	2. C	47. B	2. B	47. B
3. D	3. A	3. A	3. B	48. C	3. C	48. A
4. B	4. A	4. D	4. A	49. B	4. C	49. C
5. D	5. E	5. C	5. B	50. D	5. A	50. D
6. C	6. E	6. B	6. A	51. C	6. C	51. A
7. D	7. C	7. C	7. B	52. A	7. C	52. B
8. A	8. B	8. D	8. C	53. B	8. C	53. B
9. C	9. D	9. C	9. A	54. A	9. B	54. C
10. C	10. A	10. C	10. C	55. D	10. D	55. A
11. D	11. A	11. B	11. B		11. D	56. D
12. B	12. B	12. C	12. C		12. B	57. C
13. A	13. D	13. E	13. C		13. D	58. A
14. C	14. B	14. C	14. C		14. D	59. A
15. D	15. E	15. C	15. C		15. A	60. A
16. D	16. C	16. C	16. B		16. C	61. D
17. A	17. C	17. C	17. B		17. A	62. D
18. C	18. B	18. A	18. C		18. D	63. D
19. D	19. C	19. D	19. C		19. C	64. A
20. C	20. E	20. B	20. A		20. A	65. D
21. A	21. A	21. C	21. A		21. C	66. D
22. A	22. A	22. E	22. A		22. A	67. D
23. C	23. B	23. B	23. A		23. C	68. D
24. D	24. E	24. D	24. A		24. C	69. A
25. B/A	25. A	25. A	25. B		25. D	
26. C	26. D	26. D	26. C		26. B	
27. D	27. C	27. C	27. C		27. A	
28. C	28. A	28. E	28. A		28. B	
29. B	29. B	29. C	29. B		29. D	
30. A		30. A	30. A		30. C	
31. C		31. D	31. B		31. D	
32. A		32. E	32. N		32. A	
33. B		33. C	33. A		33. B	
34. A		34. D	34. A		34. B	
35. C		35. C	35. C		35. C	
36. C		36. D	36. B		36. C	
37. C			37. C		37. D	
38. C			38. C		38. B	
39. C			39. C		39. A	
40. C			40. A		40. B	
41. C			41. A		41. A	
42. D			42. B		42. B	
43. C			43. A		43. A	
44. B			44. C		44. A	
			45. B		45. A	

PAPER B

VR	DM	QR	AR		SJT	
1. A	1. C	1. B	1. A	46. A	1. A	46. B
2. B	2. E	2. B	2. B	47. B	2. C	47. C
3. D	3. B	3. A	3. C	48. A	3. B	48. C
4. A	4. B	4. B	4. C	49. C	4. D	49. B
5. A	5. B	5. D	5. B	50. B	5. C	50. A
6. B	6. A	6. B	6. B	51. D	6. A	51. C
7. D	7. B	7. C	7. C	52. B	7. D	52. B
8. D	8. C	8. A	8. B	53. C	8. A	53. B
9. C	9. D	9. C	9. A	54. B	9. D	54. B
10. D	10. E	10. A	10. N	55. A	10. D	55. A
11. C	11. C	11. B	11. C		11. A	56. D
12. C	12. A	12. B	12. B		12. C	57. B
13. C	13. A	13. C	13. C		13. D	58. A
14. C	14. D	14. C	14. A		14. C	59. A
15. A	15. D	15. A	15. B		15. A	60. B
16. D	16. B	16. B	16. B		16. D	61. A
17. B	17. D	17. B	17. B		17. C	62. A
18. D	18. B	18. A	18. B		18. A	63. C
19. D	19. A	19. C	19. A		19. C	64. B
20. B	20. D	20. C	20. A		20. D	65. D
21. C	21. D	21. D	21. C		21. B	66. A
22. D	22. C	22. C	22. A		22. A	67. C
23. B	23. A	23. B	23. B		23. D	68. D
24. C	24. B	24. C	24. A		24. C	69. C
25. C	25. B	25. A	25. C		25. D	
26. D	26. D	26. C	26. C		26. D	
27. C	27. B/D	27. A	27. A		27. C	
28. B	28. C	28. B	28. B		28. A	
29. D	29. B	29. A	29. A		29. C	
30. C		30. B	30. C		30. D	
31. D		31. C	31. C		31. D	
32. D		32. B	32. B		32. A	
33. A		33. D	33. C		33. B	
34. C		34. C	34. C		34. D	
35. A		35. C	35. A		35. D	
36. A		36. B	36. A		36. D	
37. A			37. A		37. A	
38. C			38. B		38. B	
39. D			39. B		39. D	
40. C			40. C		40. B	
41. C			41. B		41. A	
42. B			42. B		42. B	
43. C			43. A		43. B	
44. C			44. D		44. B	
			45. C		45. D	

PAPER C

VR		DM		QR		AR				SJT			
1.	B	1.	A	1.	C	1.	N	46.	N	1.	D	46.	D
2.	D	2.	D	2.	B	2.	N	47.	A	2.	D	47.	B
3.	C	3.	C	3.	C	3.	A	48.	A	3.	B	48.	A
4.	B	4.	A	4.	D	4.	N	49.	A	4.	A	49.	B
5.	C	5.	B	5.	E	5.	B	50.	B	5.	D	50.	C
6.	C	6.	D	6.	E	6.	N	51.	B	6.	D	51.	B
7.	B	7.	C	7.	D	7.	B	52.	N	7.	A	52.	C
8.	C	8.	B	8.	D	8.	A	53.	N	8.	B	53.	B
9.	B	9.	C	9.	B	9.	B	54.	N	9.	C	54.	A
10.	C	10.	A	10.	D	10.	A	55.	A	10.	B	55.	A
11.	A	11.	C	11.	B	11.	B			11.	C	56.	A
12.	C	12.	C	12.	C	12.	B			12.	C	57.	D
13.	C	13.	D	13.	D	13.	A			13.	A	58.	A
14.	C	14.	B	14.	A	14.	B			14.	D	59.	B
15.	A	15.	C	15.	D	15.	A			15.	D	60.	A
16.	A	16.	B	16.	A	16.	B			16.	C	61.	A
17.	A	17.	D	17.	C	17.	N			17.	D	62.	A
18.	A	18.	B	18.	B	18.	N			18.	A	63.	A
19.	C	19.	C	19.	B	19.	N			19.	B	64.	C
20.	C	20.	D	20.	C	20.	A			20.	D	65.	A
21.	B	21.	C	21.	A	21.	A			21.	D	66.	B
22.	A	22.	D	22.	C	22.	N			22.	A	67.	D
23.	A	23.	C	23.	B	23.	B			23.	D	68.	D
24.	C	24.	B	24.	D	24.	N			24.	D	69.	D
25.	B	25.	D	25.	C	25.	B			25.	A		
26.	B	26.	C	26.	A	26.	B			26.	D		
27.	B	27.	D	27.	D	27.	D			27.	A		
28.	C	28.	C	28.	C	28.	A			28.	C		
29.	C	29.	A	29.	D	29.	C			29.	D		
30.	B			30.	B	30.	B			30.	A		
31.	B			31.	B	31.	A			31.	D		
32.	C			32.	C	32.	D			32.	D		
33.	A			33.	D	33.	A			33.	A		
34.	D			34.	E	34.	C			34.	C		
35.	A			35.	E	35.	D			35.	D		
36.	B			36.	C	36.	A			36.	C		
37.	C					37.	A			37.	C		
38.	A					38.	D			38.	C		
39.	A					39.	C			39.	B		
40.	B					40.	B			40.	A		
41.	A					41.	N			41.	D		
42.	C					42.	N			42.	A		
43.	C					43.	A			43.	A		
44.	A					44.	B			44.	C		
						45.	B			45.	D		

PAPER D

VR	DM	QR	AR		SJT	
1. B	1. E	1. B	1. N	46. B	1. B	46. D
2. C	2. B	2. C	2. A	47. N	2. A	47. C
3. A	3. C	3. E	3. B	48. B	3. B	48. D
4. C	4. B	4. B	4. A	49. N	4. A	49. A
5. D	5. A	5. D	5. A	50. A	5. C	50. C
6. A	6. D	6. C	6. B	51. B	6. C	51. D
7. A	7. C	7. C	7. N	52. B	7. C	52. C
8. A	8. A	8. A	8. A	53. D	8. B	53. C
9. B	9. A	9. E	9. A	54. B	9. C	54. C
10. B	10. A	10. C	10. N	55. A	10. B	55. B
11. B	11. E	11. D	11. B		11. D	56. A
12. A	12. D	12. A	12. B		12. C	57. D
13. A	13. B	13. E	13. A		13. B	58. A
14. C	14. C	14. E	14. A		14. C	59. C
15. C	15. C	15. B	15. A		15. A	60. D
16. D	16. B	16. B	16. A		16. C	61. B
17. D	17. C	17. A	17. N		17. D	62. C
18. B	18. C	18. D	18. N		18. C	63. C
19. D	19. C	19. B	19. B		19. A	64. B
20. C	20. E	20. E	20. A		20. D	65. A
21. A	21. C	21. C	21. B		21. D	66. A
22. A	22. C	22. D	22. A		22. C	67. C
23. A	23. C	23. C	23. A		23. B	68. D
24. C	24. E	24. B	24. N		24. A	69. A
25. C	25. A	25. C	25. B		25. D	
26. C	26. C	26. B	26. B		26. C	
27. C	27. B	27. A	27. D		27. A	
28. D	28. D	28. E	28. C		28. B	
29. C	29. E	29. D	29. B		29. C	
30. C		30. C	30. A		30. D	
31. D		31. C	31. D		31. A	
32. A		32. B	32. N		32. A	
33. B		33. A	33. A		33. B	
34. C		34. C	34. N		34. B	
35. A		35. D	35. A		35. C	
36. B		36. D	36. B		36. D	
37. C			37. B		37. A	
38. C			38. A		38. D	
39. D			39. N		39. B	
40. B			40. A		40. D	
41. A			41. B		41. D	
42. B			42. A		42. C	
43. D			43. N		43. C	
44. D			44. B		44. D	
			45. B		45. D	

PAPER E

VR		DM		QR		AR				SJT			
1.	D	1.	A	1.	E	1.	A	46.	C	1.	C	46.	B
2.	B	2.	A	2.	D	2.	N	47.	C	2.	D	47.	A
3.	C	3.	E	3.	C	3.	B	48.	C	3.	A	48.	A
4.	D	4.	C	4.	D	4.	N	49.	D	4.	B	49.	C
5.	A	5.	A	5.	B	5.	B	50.	C	5.	D	50.	D
6.	B	6.	B	6.	A	6.	A	51.	D	6.	D	51.	D
7.	C	7.	C	7.	C	7.	A	52.	A	7.	A	52.	A
8.	A	8.	B	8.	C	8.	B	53.	C	8.	C	53.	A
9.	B	9.	B	9.	B	9.	N	54.	A	9.	B	54.	D
10.	A	10.	A	10.	A	10.	A	55.	C	10.	D	55.	D
11.	A	11.	E	11.	E	11.	A			11.	D	56.	B
12.	C	12.	A	12.	B	12.	B			12.	D	57.	A
13.	B	13.	A	13.	A	13.	A			13.	A	58.	A
14.	A	14.	B	14.	C	14.	B			14.	C	59.	A
15.	D	15.	A	15.	A	15.	A			15.	D	60.	D
16.	D	16.	C	16.	D	16.	N			16.	B	61.	D
17.	A	17.	D	17.	C	17.	N			17.	A	62.	A
18.	C	18.	D	18.	A	18.	B			18.	D	63.	A
19.	B	19.	C	19.	C	19.	A			19.	D	64.	D
20.	C	20.	D	20.	A	20.	A			20.	C	65.	C
21.	C	21.	A	21.	B	21.	A			21.	A	66.	A
22.	A	22.	E	22.	A	22.	N			22.	D	67.	D
23.	A	23.	A	23.	D	23.	B			23.	C	68.	A
24.	C	24.	A	24.	E	24.	N			24.	B	69.	B
25.	B	25.	C	25.	B	25.	A			25.	D		
26.	C	26.	C	26.	A	26.	N			26.	B		
27.	A	27.	A	27.	C	27.	A			27.	D		
28.	D	28.	C	28.	B	28.	A			28.	A		
29.	A	29.	A	29.	A	29.	A			29.	D		
30.	A			30.	D	30.	B			30.	D		
31.	D			31.	C	31.	B			31.	A		
32.	D			32.	A	32.	B			32.	B		
33.	A			33.	C	33.	N			33.	C		
34.	C			34.	A	34.	A			34.	D		
35.	D			35.	A	35.	N			35.	B		
36.	C			36.	C	36.	N			36.	A		
37.	A					37.	A			37.	D		
38.	D					38.	N			38.	B		
39.	D					39.	B			39.	C		
40.	A					40.	A			40.	D		
41.	D					41.	B			41.	C		
42.	B					42.	A			42.	A		
43.	B					43.	A			43.	D		
44.	C					44.	N			44.	C		
						45.	B			45.	B		

PAPER F

VR	DM	QR	AR		SJT	
1. A	1. B	1. D	1. B	46. N	1. D	46. A
2. B	2. D	2. A	2. N	47. B	2. D	47. A
3. C	3. C	3. C	3. A	48. N	3. A	48. A
4. B	4. D	4. C	4. A	49. A	4. C	49. A
5. D	5. B	5. C	5. N	50. N	5. B	50. D
6. D	6. C	6. E	6. A	51. A	6. D	51. A
7. C	7. B	7. E	7. D	52. B	7. D	52. A
8. C	8. E	8. D	8. A	53. A	8. A	53. B
9. B	9. E	9. D	9. D	54. A	9. B	54. B
10. C	10. C	10. E	10. A	55. N	10. B	55. C
11. A	11. C	11. E	11. A		11. D	56. A
12. D	12. C	12. B	12. C		12. C	57. A
13. D	13. C	13. E	13. B		13. A	58. D
14. C	14. C	14. B	14. A		14. B	59. D
15. A	15. B	15. C	15. B		15. D	60. D
16. D	16. D	16. A	16. A		16. C	61. A
17. A	17. C	17. C	17. A		17. A	62. B
18. C	18. D	18. A	18. B		18. C	63. A
19. B	19. C	19. D	19. A		19. D	64. A
20. B	20. E	20. B	20. A		20. D	65. C
21. C	21. D	21. C	21. A		21. D	66. A
22. C	22. B	22. E	22. N		22. A	67. B
23. B	23. B	23. B	23. B		23. C	68. C
24. C	24. E	24. A	24. A		24. D	69. A
25. D	25. C	25. A	25. N		25. B	
26. C	26. C	26. D	26. N		26. C	
27. B	27. B	27. C	27. N		27. D	
28. C	28. D	28. E	28. A		28. A	
29. D	29. A	29. C	29. B		29. A	
30. A		30. A	30. N		30. D	
31. C		31. D	31. N		31. B	
32. C		32. E	32. N		32. A	
33. C		33. D	33. A		33. A	
34. C		34. C	34. N		34. C	
35. B		35. C	35. N		35. A	
36. C		36. D	36. A		36. A	
37. B			37. N		37. A	
38. B			38. B		38. B	
39. C			39. A		39. A	
40. C			40. B		40. C	
41. A			41. B		41. C	
42. B			42. A		42. D	
43. C			43. N		43. B	
44. D			44. B		44. A	
			45. N		45. C	

Mock Paper A Answers

Section A: Verbal Reasoning

Passage 1

1.   **B** The passage states that the existence of atoms 'has been suggested since ancient Greece'. Though there have been recent developments in our understanding of atoms, the idea of its existence is not recent or 19th century but 'ancient', and so, old.

2.   **D** It was 'concluded' that the nucleus was 'positively charged' (and so *not* negatively charged) in 1909, so well within the 20th century. Dalton only theorised about tiny units, he did not prove their existence, and Thomson's and Bohr's ideas described in **B** and **C** are stated in the passage to have been proved false.

3.   **D** Quarks make up neutrons, protons and electrons (not the other way round, so **B** is false), and so must be smaller than atoms, which are made up of these three things. It is said we will discover more on atomic structure, so **C** is incorrect, and that 'no less than six' types of quark have been discovered - not 'more than six', so **A** is also false.

4.   **B** It is said that electrons' behaviour is 'difficult to predict'. The other statements are verified by information in the passage. Quarks make up electrons, so electrons contain quarks. Electrons 'do not behave like other particles'. Electrons orbit the positively charged nucleus.

Passage 2

5.   **D** The passage simply states the powers were 'at odds' after WWII, but does not at any point explicitly state why.

6.   **C** The Soviet Union were the first to launch a man-made satellite into space, and so achieved the first success. There is no stated 'winner' of the Space Race, nor are there comments made on which side was superior/inferior - they each received different 'victories' in the race, so A and D are false. B is not stated anywhere in the passage: no link is drawn explicitly between the two in this manner.

7.   **D** The president of the US campaigned for the moon-landing, giving it a political element. Spacewalks happened before a man landed on the moon, so **C** is incorrect, and the reasons the Soviet space program was 'plagued by difficulties' is not stated, so **A** and **B** are also incorrect.

8.   **A** Only one man is explicitly mentioned in the passage to have landed on the moon, and there is no mention of a political truce or research *achieved*, only the potential for research. That 'humans' orbited the moon is stated, and the plural 'humans' equates to 'multiple humans', making **A** the only correct statement.

Passage 3

9. **C** The service is important for 'most' families, and parents 'generally' cherish their kids above all else, but these statements allow for there to be exceptions to this general trend, so **A** and **B** are incorrect. Both physical and emotional safety is stated as something parents need to feel assured of when entrusting their child to someone else, and as the emotional safety is not stated to be preeminent, **D** is wrong.

10. **C** Physical needs including food are mentioned, as are emotional needs and the desire for a child to feel respected. A tidy appearance is not stated as something to keep in mind when caring for a child.

11. **D** If the carer suspects abuse, they should contact the appropriate authority, and this is a 'procedure'. Unusual aggression and depression, as well as bruises, may suggest serious problems at home (including abuse) but they are not said to directly prove them.

12. **B** Holiday camps are not mentioned in the above passage. The other statements are all corroborated by the passage.

Passage 4

13. **A** Original settlers were criminals, and if they reproduced, modern Australians may be able to trace back to their criminal ancestors.

14. **C** The English settled, and so spread their language. The Dutch did not, and so could not spread theirs. The English did not steal the land, nor did the natives show a distaste for the Dutch or a preference for the English language.

15. **D** The passage states nothing about the effect this has on England, Great Britain or native Australians, only that Australia was then a major world power, and so the creation of an independent Australia created a 'new' major world power.

16. **D** The fictional character Sweeney Todd's fate is not to be seen as proof of many falsely-convicted men being forced to move to Australia. There is no mention about the reproduction habits of immigrants or natives, and neither **B** or **C** necessarily explain the current population demographic. 250 years has changed this country from predominately Aboriginal to predominately European decedents, so **A** is correct.

Passage 5

17. **A The theory of** Spontaneous Generation is said to be 'ancient', and so before the middle ages, and not conclusively disproved until 1859, so **C** and **D** are incorrect. The passage states that spontaneous generation was the idea that "certain types of life could appear on their own, without coming from a parental source." Therefore, A is correct.

18. **C** Pasteur is referred to as 'he', is said to be French, and is known as the 'father of microbiology' (and who conducts biological experiments, and who is implicitly called a biologist in the phrase 'Pasteur and other biologists'), so **A, B** and **D** are not the answers. The passage does not ever call him a genius, it simply states he made important discoveries.

19. **D** The methods are explicitly stated to be 'preservation' methods, and so stopped food from going off. The other statements may be true, but are not stated in the passage.

20. **C** The drinks are ruined by boiling, not pasteurisation. Pasteurisation gets rid of many microbes in a fluid, but the passage does not state it gets rids of them all- and the guarantee mentioned in **D** is not mentioned in the passage, so **A, B** and **D** are incorrect. The structure of the fluid is said to remain 'intact', so **C** is correct.

Passage 6

21. **A** The passage states few innovations took hold even though 'Farming has been common in Europe for thousands of years', so the practice remained roughly the same.

22. **A** The passage states changes in agriculture spurred an industrial revolution, linking agriculture and industry.

23. **C** Medieval farmers were practicing leaving lands fallow, but the passage states it was not until the 1700s farmers knew to plant crops to help return nutrients to the fields. It does not state when the discovery was made that crops could leech fertilisers and nutrients from the soil, and though we may assume Medieval farmers knew this as they knew to leave their lands fallow, it would not be relevant to the question unless the opposite was stated as true, which it is not.

24. **D** Both help restore the soil, but they do not do this in the same way: turnips are said to get nutrients from the Earth and improve the topsoil, whereas clover does not get nutrients from the atmosphere and we are not told what it does to the topsoil.

Passage 7

25. **B & A** The passage clearly states that advances in agricultural technology allowed farms to be operated by fewer labourers, creating a surplus of labour. This means there were fewer jobs for *farm* labourers, although many of these individuals found work in mills and mines. Arguably, C & D are also correct, as increased food production is linked, albeit indirectly, to the improvement of mills and migration to the colonies. B may be the *most* correct, but strictly speaking, they are all true according to the passage.

26. **C** Though it helped, the passage does not state if this was the main factor which led to colonisation - there could be other, unmentioned but more important causes.

27. **D** The passage does not make an explicit positive or negative judgment as to the Industrial revolution, it just states what happened because of it. It similarly does not say things got 'simpler', and the increased poverty described sounds like a complication. It says that overpopulation in the city led to 'disease, poverty and crime', a higher population thus led to problems and so **D** is correct.

28. **C** The passage states employers could 'overwork and underpay their employees', an exploitative act, making **C** correct. They are not described as possibly doing anything kind, nor is anxiety mentioned, nor is it stated they would cause pain in others for their own pleasure, so **A, B** and **D** are wrong.

Passage 8

29. **B** For 'most' of 100,000 years does not equate to for 'all' of them.

30. **A** What is now known as Scotland *used* to be covered by the sheet, but is not now.

31. **C** A piece of ice left to melt caused Loch Fergus, not ice-water that was flowing, or newly deposited sand, gravel or rocks.

32. **A** 'Sea levels rose to separate…the British Isles from Europe' means that the British Isles were once *not* separate from Mainland Europe.

Passage 9

33. **B** Germany is a 'republic' and so cannot be a monarchy.

34. **A** Over 100 million people are 'native speakers', whereas the other 80 million speakers learned it as a foreign language. The passage does not state it is the dominant language in Austria and Switzerland, only that it is widely spoken there. The possibility of **D** is not mentioned anywhere in the passage.

35. **C** It is ranked the 'fourth largest economy in the world', well within the top ten. Migrants are attracted to the strong economy, but that does not mean A or D are true - migrants may not strengthen the economy and there may not be jobs for all those who seek them. B is not mentioned. The strong economy does not necessarily guarantee each individual's income.

36. **C** Although the Croatians are excluded from the EU's Freedom of Movement treaty Article 39, there is not enough information on their migration rights comparative to Middle Easterners in the text to confirm or deny the statement.

Passage 10

37. **C** There is only a short extract from the book, and we can't tell whether the rest of the text might speak on Biblical matters.

38. **C** The passage does not state Hobbes' thoughts on war from birth, nor does he explicitly state a dislike of the Spanish. His father being a clergyman does not mean he was meant to be a clergyman. The conditions of his birth are described as stressful (premature birth upon hearing of a frightening invasion), so **C** is correct.

39. **C** Hobbes was born in the sixteenth century, so **A** and **B** cannot refer to him, and the passage does not speak of any other political philosopher. Though the extract is downbeat, one cannot necessarily deduce Hobbes is depressed at this point. *Leviathan* was published in 1651, so **C** is the correct answer.

40. **C** The enemy or war's necessity are not mentioned, just how war impacts negatively on society, so that there is 'no society' - it has been shut down. The illiterate poor are not mentioned: the stopping of literature and arts is not limited to the poor.

Passage 11

41. **C** The passage does not describe the entire century, so we cannot know the state of affairs after 1913.

42. **D** The passage states they fought for 'male and female equality', not female superiority. There is no mention of either the vote or alternative sexualities.

43. **C** The passage does not state whether hunger strikes did or did not have an effect on how the public saw feminism, nor does it state when they ended, so **A** and **D** are wrong. They may well have gained something, despite being countered with force-feeding (for example, publicity for the feminist movement), and as there is nothing to state the contrary **B** is an incorrect answer for this question. Force-feeding was awful treatment that answered hunger strikes, so **C** is correct.

44. **B** We do not know if she was pre-eminent, suicidal or a genius from the passage: only that she sacrificed much fighting for her cause.

END OF SECTION

Section B: Decision Making

1. **D** The question asks what the largest number you CANNOT make is, therefore work through the numbers from the largest. $52 = 20 + 20 + 6 + 6$. Correct answer is d as there is no combination that will add up to 43.

2. **B** Statement a is wrong because although it gives a reason as to why some people may not apply, it does not give a reason why 18 is the optimum age to start medical school. Statement c does not relate directly to the statement. It is possible to live at home and go to medical school in the same city. Statement d is close but it assumes that 18 is the age of an adult and we know that in certain situations different ages can assume a more mature role, e.g. contraception for 16 year olds. Statement b is the only statement that directly addresses the statement and provides a reasonable answer.

3. **A** Dr Smith can prescribe antidepressants.

4. **A** Lucas goes to school on Monday – Friday. Last week there were two sunny days, and three rainy days. Therefore, the probability he walks on a sunny day with an umbrella is 0.3 x 0.4 x 2 = 0.24, and the probability he walks to school in the rain without an umbrella is 0.2 x 0.6 x 3 = 0.36. He is more likely to walk to school in the rain without an umbrella.

5. **E** Statement a - Karen is a musician so she must play an instrument but we do not know how many instruments she plays. Therefore statement a is incorrect. Although all oboe players are musicians, it does not mean all musician play the oboe. Similarly, oboes and pianos are instruments but that does not mean that they are the only instruments. So statements b and c are incorrect. Karen is a musician but that merely means that she plays an instrument, we do not know if it is the oboe. So statement d is incorrect.

6. **E** The key to this question is *most likely* to be correct. There is a possibility that any one of these statements could be true or all could be false but one is most likely. There is no information to suggest that we can predict the baby's eye colour based on past babies. Each one presents a new event and probability so statement a is wrong. There is no information on the parents' eye colours and how it is linked to the baby's so we cannot be sure statement b is correct. We are not told how many boys and girls James has so we cannot work out probability and statement c and d are incorrect. Statement e – none of the answers are correct.

7. **C** If Millie starts the game with 4 pennies, she is guaranteed a loss. If she takes 1 penny, then Ben will take 3. If she takes 2, then Ben will take 2. If she takes 3, then Ben will take 1. In each scenario, Ben takes the last penny. This works for all multiples of 4. No matter how many Millie takes at the start, Ben can take enough to reach a multiple of 4. This rules out a, b and d. For statement c, Millie can take one penny and this will leave 12. This is a multiple of 4 so now Ben is in the situation where any number he chooses will lead to loss - as explained above.

8. **B** Statement B directly rejects the assumption of the statement and so is the strongest answer. The other options may weaken the argument but do not directly refute it.

9. **D** A sky view of the arrangement leads to:

C	A	B		*or*		A	C	B
D		E					D	E

    In both, D is to the left of E, thus is the only correct answer.

10. **A** If Arnold is correct then their father owns at least 4 cars, this would also make Eric right because he would own at least 1 car. If Eric is correct then their father owns at least 1 car, this would make either Carrie (less than 4 cars) or Arnold (at least 4 cars) also correct. Therefore, Carrie must be the correct one and their father owns less than 4 cars. If Eric is incorrect then he must own less than 1 car so therefore, zero cars.

11. **A** Statement a clearly states that there is an increase in a carbon monoxide to harmful levels with traffic which provides a clear link to poor health and traffic. Statements b and c rely on assumptions; that people will use their spare time to exercise, and that road tax reduces traffic, respectively. This makes them weaker supporting statements than statement a. Acid rain damages buildings and the environment so is not directly damaging to health so statement d is incorrect. Statement e weakens the conclusion.

12. **D** Work from the worst case scenario – Fred initially picks one sock of each colour. When Fred picks his 5th sock, it must make a pair. The worst case would be that this, and every other sock Fred picks, is an orange sock. This is because he cannot make all three pairs from any other colour, and therefore would require only one sock to complete the second and/or third pair. If you work this through: when Fred picks 1r 1g 1b 2o = 1pair, he next gets a pair when he takes out one green or blue, or two orange socks so must take 7 socks to guarantee 2 pairs, again for the third pair he needs one green or blue, or tow orange socks thus meaning he needs 9 socks to guarantee three pairs.

13. **D** We know how old Tom is in relation to the others so we cannot say who is oldest or youngest. So statements a and e are incorrect. We do not know is Olivia wins more or less than anyone else other than Tom and Gabby so statement b is incorrect. No opinions can be made from the statements so statement c is incorrect. Tom loses most often so everyone wins more than Tom so statement d is correct.

14. **B** There are two directions: clockwise and anticlockwise and rats will only collide if they pick opposing directions Rat A - clockwise, Rat B - clockwise, Rat C - clockwise = 0.5 x 0.5 x 0.5 = 0.125 Rat A - anticlockwise, Rat B - anticlockwise, Rat C - anticlockwise = = 0.5 x 0.5 x 0.5 = 0.125 0.125 + 0.125 = 0.25. So the probability they **do not** collide is 0.25.

15. **E** Statement a is clearly wrong. "Plums are not sweets."
We do not know the correlation with sweet and tasty. It is never explicitly said that sweet and tasty coexist or are interchangeable. Therefore, we cannot make a statement on the taste of plums. This makes statements b, c and d incorrect. If only *some* plums are sweet, then there must be a subset that are not sweet, so statement e is correct.

16. B The info as a Venn diagram is shown below:

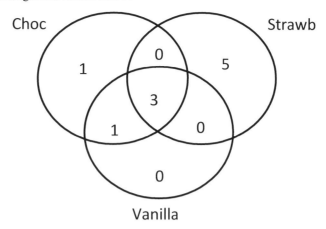

Remember the children can make numerous responses so ensure each is only counted once. Start with the total number of children who like each ice cream in the large circle, then minus the number who also like other ice cream when they are placed in the overlapping parts of the Venn diagram. You will end up with the above, and finally add in the 2 children who don't like ice cream, giving 12.

17. C We start be working backwards so filling in the centre first with 4 children liking all types. We also know that 4 like vanilla only so we can put that in. We get:

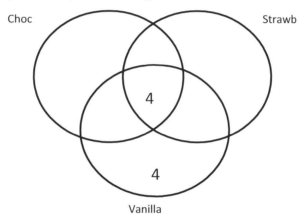

We know 1 child likes chocolate and vanilla and 4 children like strawberry and vanilla so we get:

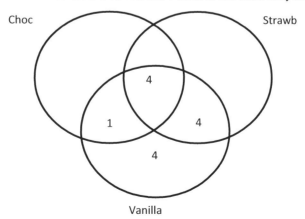

That leaves 12 that like chocolate. 12 - (4+1) = 10. We have no information on strawberry by itself.

So the answer is:

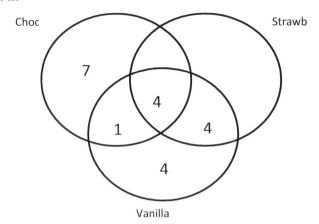

18. **B** "Every dress she bought was blue." Do not be distracted by the extra information and the different sentence structure. Statement a and e - blue dresses were not necessarily the only dress she saw. She could have seen but not bought other dresses. Statement c - there were not necessarily no other dresses there Statement d - Tasha bought all the blue dresses she saw, not every dress she saw.

19. **C** Scenario 1: John picks Kelly, Kelly picks Lisa, Lisa picks John Scenario 2: John picks Lisa, Kelly picks John, Lisa picks Kelly Any other scenario will result in starting the game over. As it is not possible to pick one's own name, there is a 50/50 chance of picking each other person. So statement c is correct.

20. **E** Helpful to draw a tree diagram and build up from the information.

    The woman's brother's father is also her father. As the father is the ONLY son of her grandfather, it follows that this must also be Mary's father. The woman is therefore Mary's sister and so her daughter's aunt.

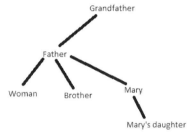

21. **A** If society disagree that vaccinations should be compulsory, then they will not fund them. So Statement a is correct. It attacks the conclusion. Statement b - society does not necessarily mean local so this does not address the argument. Statement c strengthens, not weaken, the argument for vaccinations. Statement d – the wants of healthcare workers do not affect whether vaccinations are necessary.

22. **A** Statement a - if the pain is from a blood test, Tim assumes needles will cause pain. Statement b - This can only be true via the mechanism of the needle, hence answer (a) is a more proximal cause and therefore a better response Statement c - experiencing pain with this doctor before doesn't mean Tim always will. Statement d - leaving a bruise does not necessarily inflict pain. Statement e - similar to (b), trying different needles is not the primary cause of pain - using the needle in the first place is.

23. **B** The runners aren't apart at a constant distance; they get further apart as they run. Xavier and Yolanda are less than 20m apart at the time William finishes. Each runner beats the next runner by the same distance so they must have the same difference between speeds. When William finishes at 100m and Xavier is at 80m. When Xavier crosses the finish line then Yolanda is at 80m. We need to know where Yolanda is when Xavier is at 80m. William's speed = distance/time = 100/T. Xavier's speed = 80/T. So Xavier has 80% of William's speed. This makes Yolanda's speed 80% of Xavier's and 64% (80% x 80% = 64%) of William's. So when William is at 64m when William finishes. 100m - 64m = 36m, thus William beats Yolanda by 36m.

24. E The statements can be written as: E > C, E > M > J, and N > E > J. Therefore, Naomi is the tallest.

25. A Erica - Ferrari, Mercedes, BMW
Chris - Audi, BMW
Harry - Ferrari, Ford
Lily - Volkswagen
Michael - Ford, BMW

26. D At 99% water, the watermelon is 99g water and 1g matter. After drying, the watermelon still has 1g matter but now 98% water. So 1g is now worth 2%. This means that the total weight is 50g (1 / 0.02 = 50)

27. C We only have information of Emmanuel's time. We know that Jon and Saigeet are faster but we do not know the definitive times so we cannot make statements on their ability to qualify. Emmanuel's best time is greater than the qualifying time so he will not qualify. Though statement e could be true, it is not necessarily so.

28. A Statement a - all doctors are handsome and some are popular so some must be handsome and popular. Statement b and d - We are not told that Francis and Oscar are doctors. We are not told about the general public, only doctors, so statement c is wrong. We do not know who Oscar is popular with so statement e is incorrect.

29. B Lucy must live between Vicky and Shannon, as we know Vicky and Shannon are not neighbours. Either Lucy or Shannon have a blue door, however as Vicky lives next to someone with a red door, Lucy must have the red door. This leaves Shannon with the blue door and Vicky with the white. There is a house with a green door is across the road, which is the forth house on the street and so does not belong to any of them.

END OF SECTION

Section C: Quantitative Reasoning

Data Set 1

1. B 10 m/s. Speed is distance/time, which will be the slope of the car line on the graph. Distance/time = (20 meters – 10 meters)/(1 second – 0 seconds) = 10 m/1 s = 10 m/s.

2. B 25 m/s. Speed is distance/time, which is the slope of the motorcycle curve on the graph. The maximum speed will be at the point of maximum slope. The slope is steady and greatest between 4 and 5 seconds or 50 and 75 meters. Speed = distance/time = (75 meters – 50 meters)/(5 seconds – 4 seconds) = 25 m/ 1 s = 25 m/s.

3. A 0 m/s^2. The car keeps a steady speed throughout, as seen by its unchanging slope on the graph. Therefore the car never accelerates and the acceleration is 0 m/s^2.

4. D True – but only initially. The car travels at a constant speed throughout, and thus has no acceleration. The motorbike decreases its speed in the first few seconds, as seen by the increasing slope of the curve, indicating it has a positive acceleration. It reaches a steady speed and zero acceleration like the car at about 3 seconds.

Data Set 2

5. C £ 3.19. A seven minute call is in the 1-10 mins category, so in peak time will be charged at £ 0.42 per minute. The UK-France connection charge is £ 0.25. Cost = £ 0.25 + 7 min x £ 0.42 = £ 3.19.

6. B £ 0.10. There is no per minute or connection charge on an unanswered call, but there is a £ 0.10 surcharge on Europe-Australia unanswered calls.

7. C 11%. 463 seconds = 7.72 minutes, which rounds up to 8 minutes. 15:43 is a peak time. The per minute charge of a 1-10 mins call is £ 0.42. Duration charge = 8 min x £ 0.42 = £ 3.36. No surcharges apply to this call. The USA-China connection charge is £ 0.43. Percentage of fixed charges/overall cost = 100% x (£ 0.43)/(£ 0.43 + £ 3.36) = 100% x 0.11 = 11%.

8. D £ 14.37. 1935 is off peak. The 75 minutes will include minutes charged at 4 different per minute rates.

(10 min x £ 0.25) + (10 min x £ 0.18) + (40 min x £ 0.16) + (15 min x £ 0.15) = £ 2.50 + £ 1.80 + £ 6.40 + £ 2.25 = £ 12.95.

There is an off peak over-an-hour surcharge of £ 0.88 and a China-France connection charge of £ 0.54. Total cost = £ 12.95 + £ 0.88 + £ 0.54 = £ 14.37.

9. C £ 6.09. This is a peak time call that is 763/60 = 12.72 minutes long, which rounds up to 13 minutes. No surcharges apply, but there is a France-Australia connection charge of £ 0.78. The first 10 minutes will be charged at a per minute rate of £ 0.42 and the 3 final minutes at a per minute rate of £ 0.37. Total cost = £ 0.78 + (10 minx £ 0.42) + (3 min x £ 0.37) = £ 0.78 + £ 4.20 + £ 1.11 = £ 6.09.

Data Set 3

10. C 155%. From the graph, in January 2004 the price of gold is approximately $ 400 per ounce and $ 620 in January 2007. Percentage change = 100% x new price/old price = 100 x ($ 620 per ounce/$ 400 per ounce) = 100% x 1.55 = 155%.

11. B £ 486. In January 2004 gold cost $ 400 per ounce. The conversion rate is $ 1=£ 0.68, so $ 1 x 400 per ounce = £ 0.68 x 400 per ounce = £ 272 per ounce. 1 ounce = 28g, so 50g gold = 50g/28g per ounce = 1.785 ounces. 50g gold costs 1.785 ounces x £ 272 per ounce = £ 485.52.

Data Set 4

12. C 31.5 g/kg. A lasagne has 190 g sugar per 450 g total mass. 1 kg lasagne therefore has 1000 g/450 g x 19 g sugar = 42 g sugar. The beef noodle dish contains 35 g sugar per 475 g total mass. 1 kg beef noodles has 1000 g/475 g x 35 g sugar = 73.5 g sugar. 73.5 g/kg sugar is 31.5 g/kg more than 42 g/kg sugar.

13. E More information required. To answer this question the energy content of sugar and the other ingredients per weight would need to be known. If the rest of the meal is made of high calorie ingredients, the relative proportion of sugar calories will be smaller than if the rest of the mass is comprised of low calorie ingredients.

Data Set 5

14. C £5.80. Average error = (|wine 1 retail value – estimate respondent 1| + |wine 1 retail value – estimate respondent 2| + |wine 1 retail value – estimate respondent 3| + |wine 1 retail value – estimate respondent 4| + |wine 1 retail value – estimate respondent 5|)/5 = (|8-13| + |8-17| + |8-11| + |8-13| + |8-15|)/5 = (5 + 9 + 3 + 5 + 7)/5 = 5.80.

15. C Respondent 3. Average error = (|wine 1 retail value – estimate 1| + |wine 2 retail value – estimate 2| + |wine 3 retail value – estimate 3|)/3. Respondent 1: (|8-13| + |25-16| + |23-25|)/3 = (5 + 9 + 2)/3 = 5.33. Respondent 2: (|8-17| +|25-16| + |23-23|)/3 = (9 + 9 + 0)/3 = 6. Respondent 3: (|8-11| + |25-17| + |23-21|)/3 = (3 + 8 + 2)/3 = 4.33. Respondent 4: (|8-13| + |25-15| + |23-14|)/3 = (5 + 10 + 9)/3 = 8. Respondent 5: (|8-15| + |25-19| + |23-29|)/3 = (7 + 6 + 6)/3 = 6.33. Respondent 3 has lowest average error and thus highest average accuracy.

Data Set 6

16. C £ 252.00. Weekly profit per product = (Sale price – Cost per unit) x Number sold per week.
Weekly profit Gobstopper = (£ 0.40 - £0. 22) x 150 = £ 27.00.
Weekly profit Bubblegum = (£ 0.50 - £ 0.35) x 180 = £ 27.00.
Weekly profit Everton mints = (£ 0.90 - £ 0.45) x 300 = £ 135.00.
Weekly profit = (£ 0.65 - £ 0.50) x 420 = £ 63.00.
Total weekly profit = £ 27.00 + £ 27.00 + £ 135.00 + £ 63.00 = £ 252.00.

17. C £ 267.00. Value of sales per product per week = Sale price x Number sold per week.
Value of sales Gobstoppers per week = £ 0.40 x 150 = £ 60.00.
Value of sales Everton mints per week = £ 0.90 x 300 = £ 270.00.
Purchase price per product per week = Cost per unit x Number sold per week. Purchase price Bubblegum per week = £ 0.35 x 180 = £ 63.00.
Gobstopper sales + Everton mint sales – Bubblegum purchase price = £ 60.00 + £ 270.00 - £ 63.00 = £ 267.00.

18. A Gobstopper. Gobstoppers and Bubblegum both generate the same profit of £27.00, which is the lowest of the sweets in his shop. However Bubblegum requires half the shelf space that either Gobstoppers or the new product require.

Data Set 7

19. D 38.6 million. Country A : 140% Country B. Country C : 70% Country D. Country D : 120% Country B.
Population Country A/1.40 = Population Country B = 45 million/1.40 = 32.1 million people.
Population Country B x 1.20 = Population Country D = 32.1 million people x 1.20 = 38.6 million people.

20. B $ 1.45 bn. Country A : 140% Country B. Population Country A/1.40 = Population Country B. 45 million people/1.40 = 32.1 million people. Health initiative cost = $ 45 per capita x 32.1 million people = $ 1.45 billion.

21. C 12.3 million. Country C : 70% Country D.
Population Country C/0.70 = Population Country D. 25 million people/0.70 = 35.7 million people.
48 % population are men = 0.48 men x 35.7 million = 17.1 million men. 72% men are adults = 0.72 adults x 17.1 million men = 12.3 million adult men.

Data Set 8

22. E 393 euros. Difference in advertising costs = Most expensive advertising costs – least expensive advertising costs = 576 euros – 183 euros = 693 euros.

23. B Civil service.
Trade and industry = 447 euros/5370 euros = 0.083.
Civil service = 431 euros/4380 euros = 0.098.
Liberal professions = 75 euros/3001 euros = 0.025.
Crafts and skilled trades = 139 euros/2895 euros = 0.048.
Agriculture = 168 euros/2311 euros = 0.073.
Civil service has the greatest proportion lost working hours/overall recruitment cost = 0.098.

Data Set 9

24. D 207. Y321 had 230 offences in 2011. 2013 is 10% lower than 2011 rate = 2011 rate x 0.90 = 230 offences x 0.90 = 207 offences.

25. A 373. There were **more than** 1837 Y offences in 2012 – 210 Y321 offences – 490 Y632 offences – 754 Y230 offences = **less than** 383 Y115 offences in 2012. The only option lower than 383 is 373.

26. D Y632. Percentage change per crime type = number of 2013 offences/number of 2012 offences.
X632 = 2670/2453 = 1.088 = 8.8% increase.
X652 = 3231/3663 = 0.882 = 21.8% reduction.
Y321 = 207/210 = 0.986 = 1.4 % reduction.
Y632 = 432/490 = 0.881 = 21.9% reduction.
Y115 = 431/383 = 1.125 = 12.5% increase.
Y230 = 714/754 = 0.947 = 5.3% reduction.
Y632 has the greatest reduction of 21.9% between 2012 and 2013.

Data Set 10

27. C 38 minutes. The arrival times for each bus stop listed in the left-hand column are shown in that grey or white row. The bus moves forward in time, down the column of stops, starting at the station at the top of the column at each of the times listed in that row, and going toward its end station listed at the bottom of the column. Different times apply on Saturdays. On Monday the 1225 bus to Winchester Bus Station from Petersfield Tesco stops at East Moon All Saints Church at 1247 and Itchen Abbas Trout Inn at 1325. There are 38 minutes between these stops.

28. E 165 minutes. At 1321 on Tuesday the bus to Winchester Bus station from Petersfield Tesco has just left at 1306. The next bus at 1506 only goes one stop further to New Alresford Perins School. The 1606 bus goes all the way to Winchester Bus Station past Winchester City Road at 1638. There are 165 waiting minutes between 1321 and 1606, 2 hours and 45 minutes.

Data Set 11

29. C 7%. July 1987 saw 66 mm of precipitation. The long-term July average is 71 mm precipitation. The percentage difference = 100% x (long-term average – 1987 value)/long-term average = 100% x (71 mm – 66 mm)/71 mm = 100% x 0.07 = 7%

30. A 1985. Average seed moisture content = (slight + moderate + severe)/3. By visual appraisal the answer is clearly 1985, as it has the lowest value for each of the three rows. By calculation: 1985 = (22.6 + 24.8 + 24.7)/3 = 24.03.
1986 = (22.7 + 26.0 + 25.9)/3 = 24.87.
1987 = (26.5 + 25.9 + 27.2)/3 = 26.53.
1988 = (26.0 + 25.1 + 31.3)/3 = 27.47.
1989 = (31.6 + 30.4 + 31.6)/3 = 31.2.
1985 has the lowest average of 24.03%.

31. D 1760. Corn grain yield in slightly eroded Marlette soil in 1986 was 8150 kg/ha. Corn grain yield in moderately eroded Marlette soil in 1989 was 9910 kg/ha. Difference = 1989 – 1986 = 9910 kg/ha – 8150 kg/ha = 1760 kg/ha.

32. E September. Average monthly precipitation = (monthly precipitation 1985 + monthly precipitation 1986 + monthly precipitation 1987)/3.
May = (69 mm + 89 mm + 38 mm)/3= 65.3 mm.
June = (56 mm + 119 mm + 64 mm)/3 =79.7 mm.
July = (51 mm + 71 mm + 66 mm)/3 = 62.7 mm.
August = (104 mm + 99 mm + 127 mm)/3 = 110 mm.
September = (89 mm + 203 mm + 102 mm)/3 = 131.3 mm.
September has the highest average precipitation of 131.3 mm.

33. C 1684 cm^2. In lightly eroded Marlette soil in 1986 there were 59 400 plants/ha. 1 ha = 10 000 m^2 = 10 000 x 100 x 100 cm^2. 100 000 000 cm^2 x 1 ha/59 400 plants/ha = 1684 cm^2/plant.

Data Set 12

34. D 25th. The point on the lower weight versus age plot on the vertical 2 month line (2nd point from bottom left) also lies on the 25th percentile line.

35. C 71.5 cm. This point lies in the upper length section of the graph on the vertical 24 month line, just below the intersection with the 72 cm line.

36. D 19.2 lbs. The three final weight measurements in pounds are 18.2 lbs, 18.5 lbs and 20.8 lbs as read off the table in the bottom right corner for 17.5, 24 and 36 months of age. Average weight = (18.2 lbs + 18.5 lbs + 20.8 lbs)/3 = 19.2 lbs.

END OF SECTION

Section D: Abstract Reasoning

Rules:

Set 1: Edges in Set A= 14; edges in Set B = 15.

Set 2: Sum of Arrow directions; Set A -Point left/right, Set B- point Up/down. N.B. – the double headed arrows 'cancel out', which is why the answer to 7 is B

Set 3: In Set A there are two points of intersection between the lines in total; in set B there is one point of intersection between the lines in total

Set 4: Set A has at least one White triangle. Set B has at least one Black Quadrilateral.

Set 5: Number of shapes is Odd in Set A and even in Set B.

Set 6: Number of corners in shaded shapes is 8 in Set A and 9 in Set B.

Set 7: Arrow points to corners in set A, arrows point to edges in set B.

Set 8: Number of edges Inside circle minus number of edges outside circle = 4 in set A and 3 set B.

Set 9:

41. A white shape with an increasing number of sides is added each time. The colour of each shape alternates between white and black each tile.
42. Central pentagon rotates 90° clockwise. Dots rotate one space clockwise.
43. Tiles alternate between reflecting in the horizontal and rotating 90° clockwise
44. Number of triangles increases by 1 each time. Colour and size of shape is irrelevant.
45. Number of black circles increases by 1. Number of white circles decreases by 2.

Set 10:

46. Shapes with an even number of sides turn black and gain 1 side. Shapes with an odd number of sides turn white and decrease number of sides by 1.
47. White shapes turn to grey with the number of black circles = the difference in number of sides between the grey and white shape.
48. The number of white shapes is double the number of black shapes. 2 white shapes are added.
49. The grid rotates 45° clockwise. The are n+1 black circles opposite n grey circles in the first tile; the number of grey circles decreases by 1 and the number of black increases by 2 (so n grey = n+3 black).
50. The image is reflected in the horizontal.

Set 11: Set A; the number of white circles = number of edges of black shapes. Number of black circles = number of edges of white shapes. Set B; the number of white circles = number of edges of white shapes. Number of black circles = number of right angles in black shapes.

END OF SECTION

Section E: Situational Judgement Test

Scenario 1

1. **Very inappropriate.** This is too much of a risk – she may end up doing worse than the original exam because of the time elapsed.
2. **Appropriate but not ideal.** Given the time and money spent on the trip, cancelling would mean she would lose a lot. However, she would have plenty of time to study.
3. **Inappropriate but not awful.** Her best friends might understand as they are all on the same course. However, this would also compromise their trips.
4. **Inappropriate but not awful.** This is a risky option – she may not enjoy the holiday because she is studying, or she may have very little time to study. However, bringing work is better than not doing any.
5. **Very appropriate.** This option ensures she gets to enjoy some of the trip whilst leaving enough time to revise for her exam properly.

Scenario 2

6. **Inappropriate but not awful.** His friends have to fund their own tours, but may be willing to contribute if they really want him there.
7. **Inappropriate but not awful** It is not really worth the debt just to go on a sports tour.
8. **Inappropriate but not awful.** It is probably better to ask his parents rather than his friends/the bank and they may be willing to help, however it is not a really important reason to ask for money.
9. **Appropriate but not ideal** He will have other numerous opportunities to make friends and have an active role within the club. He is only in his first year so there will be many other tours throughout his university career.
10. **Very inappropriate.** His medical studies should definitely not be sacrificed just for sports tour.

Scenario 3

11. **Very inappropriate.** If he were to be successful with his application, he would not have given his parents any warning about the costs. If they are unable to help him he will then be letting the other students down.
12. **Appropriate but not ideal**. This way he avoids the financial aspect of living away from home, but still has opportunities to make friends.
13. **Very inappropriate.** Clearly Ade is unhappy, so he should make an effort to rectify his situation.
14. **Very inappropriate.** It is unreasonable to put this pressure on his parents.
15. **Very appropriate.** This will allow Ade and his family to plan and make appropriate financial decisions.

Scenario 4

16. **Inappropriate but not awful.** This is risky and would have to be very carefully worded.
17. **Very appropriate.** This means that she can both inform a superior and take action.
18. **Very inappropriate.** Doing nothing would mean that the doctor would continue to upset patients in the future and no change would happen.
19. Inappropriate but not ideal. It could be deemed inappropriate to go behind the doctor's back and encourage patients to report him. It would be better to gain some advice first.
20. **Very appropriate.** This action means that a superior has been informed and they can decide whether or not to take the matter further.

Scenario 5

21. **Inappropriate but not awful** It would most likely make the patient uncomfortable to sit in silence, so some kind words should be offered. If the patient then says they don't want to talk, silence is appropriate.

22. **Very appropriate.** It is difficult to word this correctly, but the patient is likely to appreciate kind words.

23. **Inappropriate but not awful.** Although the doctor shouldn't really have left the room at such a poignant moment either, it would be inappropriate to leave the patient on his own at a time like this.

24. **Inappropriate but not awful.** Given the little knowledge she has about prostate cancer, this could do more harm than good. However, if the patient wishes to know the information and it is beneficial to them, it might be acceptable to share.

25. **Very inappropriate.** It is a good idea to offer support. However, the patient has been told he is terminal and thus saying he will be Okay seems like empty words.

26. **Appropriate but not ideal.** It is true that often talking to a medical student is less daunting for a patient and after receiving such shocking news the patient may wish to express themselves. However should she then proceed to answer any prognostic questions she would then be being inappropriate.

Scenario 6

27. **Very appropriate.** If she has not been taught how to complete the procedure, she should definitely not attempt it alone for the first time on a patient without any one to help.

28. **Appropriate but not ideal.** This way a trained professional can talk her through the procedure and be on hand if anything goes awry. However, this introduces a delay for a time pressured procedure and so it would be preferable for Larissa to go and directly seek a member of staff to assist, rather than wait.

29. **Very inappropriate.** There is no way of knowing if the skills she has learnt elsewhere are transferable. This is putting the patient at considerable risk.

30. **Inappropriate but not awful.** Whilst it is necessary to inform the patient that it would be the first time, it is still dangerous to carry out the procedure alone for the first time.

31. **Very inappropriate.** It would be very unprofessional to just leave without ensuring someone else can carry out the ABG, especially seeing as Larissa knows the patient needs it done ASAP.

Scenario 7

32. **Very important.** He has been told the chances of publication are very likely, and currently he has no other chances for publication. It is too much of a risk to reject the offer.

33. **Important.** A publication would make him more competitive when it comes to applying to jobs. Many of his friends already have publications.

34. **Important.** It is important for Romario to make the most of his last year as a student before the reality of becoming a junior doctor.

35. **Of minor importance.** It is important to continue to look professional, especially as his consultant could influence his grades.

36. **Of minor importance.** There is a group of friends going so no one would be left in the lurch. All of his friends already have publications so should understand his decision to stay.

Scenario 8

37. **Not important at all.** He has been told that if he is at all unwell he should not come in. Even a cold could put high dependency respiratory patients at risk.
38. **Important.** This would be a great learning opportunity for Mahood and probably his only chance to see such interesting cases.
39. **Very important.** This instruction has been given for a reason, as a minor illness could put such sick patients at serious risk. Mahood should follow these explicit instructions.
40. **Important.** As above, this is a one-off learning opportunity for Mahood.
41. **Very important.** It is part of Mahood's responsibility as a medical student to be professional and not put patients at risk.

Scenario 9

42. **Important.** The captain is responsible for organising the team. Theodore may be letting the rest of the team down if he is absent.
43. **Very important.** Extra-curricular activities should not compromise medical studies, especially if this will impact on Theodore's final grades.
44. **Very important.** Although important, Theodore's rugby matches should not count for more than his medical career.
45. **Very important.** This is Theodore's only chance to learn from his consultant and also impress him.
46. **Important.** Theodore has the chance to show leadership skills beyond medicine.

Scenario 10

47. **Important.** Both the essay and the clinic seem to be important to Malaika. Therefore, if she allocates an appropriate amount of time to her essay, she may be able to do both.
48. **Very important.** The essay should be her priority if it contributes towards her final mark, as the clinic is not compulsory and would be being attended out of interest.
49. **Of minor importance.** Malaika cannot tell how much she will gain from the clinic. The fact that she is interested in the field suggests she will learn something useful.
50. **Not important at all.** Although Malaika is interested in the field, this clinic is optional and so she will not look unprofessional for not attending. Not handing in her essay in time and getting a poor grade would be worse for her academic reputation!
51. **Very important.** If it is possible for her to attend another clinic, she need not worry.

Scenario 11

52. **Important.** It is important for students to look and act professional around both the staff and the patients.
53. **Important.** As they are clinical partners, it is important that they are on good terms so they continue to work well together. However, if they are good friends, Franklin should realise that Jean means well and is not trying to insult him.
54. **Of minor importance.** Franklin should realise this is not a personal attack and that Jean is ensuring they look professional as a pair.
55. **Very important.** It is important to look and act professional around patients, even as a student.
56. **Not important at all.** If Jean remains tidy and professional herself, there is no reason why the appearance of her colleague should influence opinions about her.

Scenario 12

57. **Of minor importance.** Although the project may feel important to Rory, taking patient notes home without permission breaks patient confidentiality and is unprofessional.
58. **Very important.** This is a serious breach of confidentiality and hospital policy, and Priya could get in a lot of trouble.
59. **Very important.** Lack of access to the notes could compromise patient care.
60. **Very important.** A member of staff is very likely to report the issue. Both Rory and Priya could then both be at risk of being in trouble, as the staff may not know who took the patient notes.
61. **Not important at all.** The project is not worth breaching patient confidentiality for.
62. **Not important at all.** Despite this, Rory has an obligation to put a stop to Priya's actions, which should not be ignored.

Scenario 13

63. **Not important at all.** Patient safety cannot be compromised for the sake of Nelson's embarrassment!
64. **Very important.** This is the most important factor of all – patient safety. Contaminated equipment could have devastating effects.
65. **Not important at all.** The duration/site of the contact is irrelevant – if the equipment is contaminated it cannot be used.
66. **Not important at all.** The staff would much rather carry out a safe and clean procedure than put their patient's life at risk.
67. **Not important at all.** Horatio and Nelson's learning experience is not worth risking the life of the patient.
68. **Not important at all.** We know that Nelson did touch the trolley and so the equipment is already contaminated. If the procedure shouldn't be carried out because of this then Horatio needs to inform them.
69. **Very important.** If Nelson were scrubbed in a sterile surgical gown then there would be no chance of contamination, indeed had he not been told to stand back he may have been expecting to participate in the surgery and handle the equipment regardless.

END OF PAPER

Mock Paper B Answers

Section A: Verbal Reasoning

Passage 1

1. **A** The Scottsboro boys case shows another example of racism in the USA, showing how race could lead to false indictments.

2. **B** The law did not state black people had to vacate seats, but conductors did. It is not mentioned if she was arrested at a previous protest, nor does it state how long she had been working - it may not have been for years.

3. **D** It treated white people and black people as different due to their skin colour, determining a 'coloured section' and a 'white' one, but it did not consider different members within the races. Rather, African-Americans were treated as one uniform group and white people as a (superior) other.

4. **A** The passage clearly states that Rosa was arrested "for her refusal [to move]". Therefore, the answer is A.

Passage 2

5. **A** He wanted prostitutes to leave their trade, not stay in it with better conditions. He is neither said to be in a romantic engagement, nor to have used prostitutes. The advert at the end, which he wrote, considered the condition of prostitutes, and so **A** is correct.

6. **B** He originally tried to 'dissuade' her, so initially opposed it, and **A** is incorrect and **B** correct. His general opinion of Burdett-Coutt, whether positive or not, is not mentioned, nor is his behaviour to her described as 'patronising'.

7. **D** It would be unlike other institutions with their harsh ways. It would actively be 'kind', so 'sympathetic', and not neutral, so **A** and **B** are wrong. No mention is made of religious behaviour within the refuge.

8. **D** The advert promises women they can gain back what they lost, so **A** is false, and it also states he does not see himself as 'very much above them' . He does not mention the 'devil', but promises that deserving women can gain back 'friends, a quiet home, means of being useful to yourself and others, peace of mind, self-respect, everything you have lost', making **D** correct.

Passage 3

9. **C** The passage states that values are relative, and does not claim objective right or wrong (so **D** is incorrect.) It does not tell as not to act charitably or that all attempts at doing good end up hurting the do-er, so **A** and **B** are incorrect.

10. **D** The passage does not state **A**, **B** or **C**, as there is nothing to link one view of happiness/morality to the entirety of the poor population or to all prostitutes. Dickens' expectations may have equalled some prostitutes, and not others, and the passage does not state what expectations are 'higher' than others. Instead, the passage states people are contented by different things.

11. **C** The passage states the unnamed writer was 'angered by those who would actively try to get rid of her means of earning', and so she accused 'do-gooders' of depriving those in her trade of work, not of making conditions worse. Conservatism and prudishness are not explicitly stated as accusations.

12. **C** It does not mention either social reformer by name, and was written in the 19th century (1858). She does not call on reformers to be ashamed, but questions their self-perception of being 'pious' with the phrase 'as you call yourselves', suggesting they merely *think* they are, not that they *truly* are, and thus questioning what piety is.

Passage 4

13. **C** The passage does not state the specific contents of either novel, only that they are dystopias. They may not be set in the future.

14. **C** The passage does not state how Orwell would feel about this, it simply states one may wonder how he would feel.

15. **A** The constant surveillance is not specified to be *camera* surveillance. The other three statements are contained within the passage ('children informing on parents', truth being rewritten corroborates and 'love being sacrificed to fear' shows how **B, C** and **D** are all described in the passage.)

16. **D** Concerns of the different dystopias are not stated, and though 'many' audiences are drawn to them, that does not mean all are. The passage does not state that dystopias can be used to make us feel better about the current state of things, but it does say that many are 'drawn' to dystopias, and so these horrible fictional realities 'do not necessarily repel people'.

Passage 5

17. **B** Though Wilde is a playwright who lived during the Victorian era and wrote comedy, none of this is mentioned in the passage. The fact he wrote a novel is mentioned, so he is a writer of fiction.

18. **D** The critic is not denounced as a 'beast', and Wilde does not state how a critic should react to things that aren't beautiful. He describes two types of critic: one who is 'corrupt', the other 'cultivated', so **D** is correct.

19. **D** Books cannot be 'moral' or 'immoral', and books are not specified as being the 'highest form of art'. As a book is either simply 'well written or badly written', a beautifully written book is simply beautifully written.

20. **B** There is no claim that the writer himself will be able to produce a 'well-written' book, or that he will be able to achieve the aim of 'concealing the artist'. There is no guarantee either that critics will like this work. However, as 'all art is useless', this book, as art, must be useless and so **B** is correct.

Passage 6

21. **C** It is said 'Eugenists seem to be rather vague' about what Eugenics is, not that they are too blunt/pithy or badly define their opponents.

22. **D** Though he states he would be justified in his actions, Chesterton acknowledges 'I might be calling him away from much more serious cases, from the bedsides of babies whose diet had been far more deadly'.

23. **B** The passage states 'we know (or may come to know) enough of certain inevitable tendencies in biology' that we can understand an unborn baby as we would an existent person. This mention of the use of science to predict the reality of 'the babe unborn' supports **B** as the correct answer.

24. **C** The passage states that 'The baby that does not exist can be considered even before the wife who does. Now it is essential to grasp that this is a comparatively new note in morality.' So, the baby who is not yet born can be considered more important than a woman who exists, making **C** correct. The other statements are not mentioned in the passage as ethical truths belonging to anyone.

Passage 7

25. **C** The source of the curiosity is not explicitly stated, and may or may not have anything to do with the gender of the individuals working there.

26. **D** She states 'one's first thought' of seeing the workers in the kitchen is of 'some possibly noxious ingredient that might be cunningly mixed in the viand'. She does not make a mention of their safety of comfort whilst working in a kitchen, but only of the potential for poisoning.

27. **C** They call out 'fie! A dirty skirt!', but we are not told about them at work, nor does the passage discuss the other inmates' apparel. No reference is made to them cleaning the writer's skirt.

28. **B** She wonders about their 'intellects; particularly as some of them employed in the grounds, as we went out, took off their hats, and smiled and bowed to us in the most approved manner.' Her specific reason is that they show politeness, not that they speak well or demonstrate particularly high intelligence.

Passage 8

29. **D** She has self-immolated by the writing of the passage, and so was dead.

30. **C** Hydrophobia (fear of water) and being next to a river is the cause stated for his agony.

31. **D** He was to light the pile, and so burn his father and enable his mother to jump into the flame. He was to lose both his parents, and so become an orphan. But the passage does not state he would go on to support his siblings.

32. **D** She was 'apparently', but not definitely, praying, and though she did not move that does not mean she was unable to. She was 27 years younger than her husband, which is less than three decades. She had, however, chosen to leave her children: a choice which is registered by the fact she prepared for her suicide by leaving her children in her mother's care.

Passage 9

33. **A** The passage states many qualities and actions a boy must have and do to be 'respected'.

34. **C** There is no mention of what girls can achieve in terms of respect.

35. **A** He should 'never get into difficulties and quarrels with his companions'.

36. **A** These boys, who will be respected, will 'grow up and become useful men'.

Passage 10

37. **A** The revenge tragedy genre in which Shakespeare wrote was very popular, and so populist.

38. **C** Though the other revenge tragedies are not named after characters, there may be unmentioned ones that are.

39. **D** Biting off one's own tongue, a raped woman and a poisoned skull are named, meaning **A**, **B** and **C** are not the correct answers for this question. Suicide is not explicitly mentioned, so **D** is correct.

40. **C** One play involves an act of cannibalism, but this does not mean all others of the genre are 'obsessed' with it. We can't tell for certain where all plays are set, as this is not mentioned as a general rule. The tragedies are described as 'dramas', but not as poems. The ethical code, where killers die, is mentioned and described as a general rule for the genre, so **C** is correct.

Passage 11

41. **C** The passage makes no reference to the modern existence of wild cats, but nor does it deny their existence.

42. **B** Cats may have been originally used to hunt rodents, have a gut and tongue suited for raw meat consumption and may find food outside their owner's home. The passage does not state that cats are still used to kill things.

43. **C** 'Washington University's Wesley Warren' states a relationship between the academic body and the individual, so **C** is correct. The passage does not state the individual's gender, whether they are a student or whether they are an acclaimed professor.

44. **C** The passage does not talk in terms of all humanity: though humans are said to have affection towards cats, and previously humans wanted rats dead, it does not state that all humans are unified in affection or loathing. Though originally the relationship with cats might have been one of gratitude for a job well done, this does not mean that human love is predicated on this exchange. For thousands of years, however, cats have been domesticated and kept as pets, so **C** is correct.

END OF SECTION

Section B: Decision Making

1. **C** We are not told whether any of the chemistry exams constitute the paper double sided exams, so I is not correct. Likewise, we are not told whether non-chemistry papers can make up paper exams, but it is likely they can. Therefore, only statement III is true.

2. **E** The Government wants to improve health by encouraging walking so we can conclude that walking is a good form of exercise. The other statements do not explain the Government's initiative.

3. **B** Our Venn diagram should look like the one shown to the right. The total is 30 and we have (6+5+3=) 14 people represented. We need to account for 16 students. We know 14 like Pharmacology and 19 Biochemistry. From the data already on the Venn diagram, there are 11 students who like biochemistry +/- Pharmacology, and 9 who like Pharmacology +/- Biochemistry, however as we know there are only 16 student in total, there must be an overlap of 4. (11+9=20, 20-16=4)

    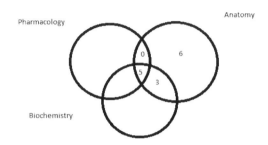

4. **B** The conclusion is: "developing countries need better medical care". Supporting statements must directly affect the conclusion. Statements a, c, and d are not the focus of the conclusion and are not mentioned in the question. Statement e may be true but it does not address the conclusion which is about developing countries.

5. **B** All mothers are women, but not all women are mothers, therefore:

    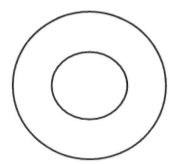

    Some women and mothers are doctors but not all doctors are women, which looks like:

    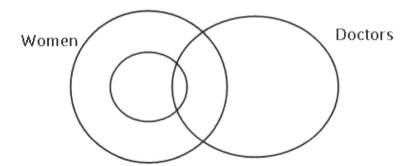

6. **A** Statement a - Percentage change = percentage difference/original percentage so 20-15/20 = 5/20 = 25%. So it is correct.

Statement b - we are not given any information on food costs. Although there are more dogs we do not know how expensive it is to feed one.

Statement c - dogs + rabbits + cats = 65% but some reptiles have four legs so they must be included.

Statement d - Although the same number of reptiles and rodents are purchased, it does not mean that they are purchased together.

Statement e - rabbits were the most owned animals at 25% and dogs are at 20%

7. **B** Statement b – The formula is: (difference in amounts/ original amount) x 100% So Jake has the greatest percentage change at 40%.

Statement a - 2012-2013: The overall loss is £15. 2010-2011: The overall loss is £35. 2010 - 2011 has the greatest loss.

Statement c - we can see that Lauren earns more than Nathan in 2010, 2013, 2014 and 2015. So she earns more than him for 4 years.

Statement d is clearly wrong.

8. **C** Statement c - Anna's average is (90+35) 62.5, Emily (65+75) 70, Isabelle (20+55) 37.5, Olivia (75+40) 57.5, and Uri (55+35) 45. Emily has the highest average.

Statement a - Anna beats Olivia by 50 marks and Olivia beats Anna by 35.

Statement b and e - There is no way of knowing this from the data given. Statement d - Isabelle and Uri's combined score - 35+55 = 90. Anna also scored 90 so they share the highest score.

9. **D** Statement d - BMWs contribute the greatest percentage of annual revenue but have the least number of cars sold. This means that each car must have high value. Mazda cars are the opposite contributing the least to total annual revenue but sold the most so they must be cheaper.

Statement a - Audi is included in the total revenue so Millie has made money.

Statement b - there is no data that suggests a change in price during the year.

Statement c - BMW and Ford sold (10 + 20+40 +35) 105. Audi sold (60+35) 95 cars

10. **E** There are 20 students in Y10 and Y11 and 160 in Sixth Form. We convert the table from proportion to numbers. Remember that being female and being a swimmer are not mutually exclusive.

	Female	**Swimmers**
Year 10	0.3 x 20 = 6	0.65 x 20 = 13
Year 11	0.7 x 20 = 14	0.5 x 20 = 10
Sixth form	0.45 x 160 = 72	0.3 x 160 = 48
Total	92	71

We can see from the table that this is correct

A - 10 swimmers in Y11 and 71 swimmers in total gives a percentage of 10/71 = 14%. Even if you cannot work this out, we can see that 65% is the percentage of people in Y11 that can swim and not the percentage we want.

B - there are 71 swimmers in the school (0.355 x 200)

C - there are (20 - 14) 6 males in Y11 and (160 - 72) 88 in Sixth Form.

D - being female and a swimmer are not mutually exclusive so cannot be worked out from the data present.

11. C If we work back from the final totals of all having 32p. Queenie loses the first, Rose the second, and Susanna the third game. Each time, they keep half their money, and the other two get one quarter each. Therefore at the beginning of the round they lose, they will have double the money.

Start game 3: Q 16 R16 S 64 – Susanna has double (64p) as she has lost half her money, and therefore Q and R 16p less. The 32p lost is split equally

Start game 2: Q 8 R 32 S 56 – Rose has double (32p), so the other two 8p less

Start game 1: Q 16 R 28 S 52 – Queenie has double (16p), so the other two 4p less

Queenie therefore has the lowest starting amount and Susanna the most.

12. A
  A. Correct as it addresses the fundamental challenge to the communist social idea.
  B. Correct, but not necessarily a problem of communism.
  C. Correct, but also not necessarily a problem of communism.
  D. Irrelevant for this question.

13. A

Although we are not told about Joe by name, we can exclude the other options from the information available to conclude the name of the last place finisher is Joe.

14. D
  A. False – we are not told her profession
  B. False – no information regarding this
  C. False – he only wears white shirts with black trousers, but we can't assume he doesn't wear other colours of shirt with black trousers
  D. True – he owns white shirts as per the text

15. D
  A. False – we are not told about the healthcare directly but are told that injury and disease posed a threat
  B. False – the terrain was difficult and mapping was poor
  C. False – outlaws were a significant threat
  D. True – as the text states, there was a marked lack of bridges.

16. B
  A. Potentially correct, but extreme sports also carry higher risks of injury.
  B. True.
  C. True, but irrelevant for the question.
  D. Potentially correct, but irrelevant to the question.

17. D
  A. Incorrect. The text clearly states that the exercise routine is resistance training based.
  B. False. Both groups contain equal numbers of men and women per the text.
  C. False. Both groups are age matched in the range of 20 to 25 years.
  D. Correct. As the only difference between the two shakes is the protein content.

18. B

    A. Incorrect. As the text states, often lead animals are the only ones to mate, not always.

    B. Correct answer.

    C. False as per text – they are born in late summer in order to be stronger by the winter.

    D. False.

19. A

From weakest to strongest:

Fisher's Ale	Knight's Brew	Monk's Brew	Ferrier Ale	Brewer's Choice

Even though Brewer's Choice is not mentioned in the question, as none of the beers mentioned are the strongest, this must be the strongest. We are not told enough information about which beers are German or are made from spring water to deduce any correlation.

20. D

    A. False, there are total 80 plants

    B. False, he plants the same amount of potatoes as heads of salad.

    C. False. There must be 40 tomato plants

    D. True. There are 40 tomato plants and 10 salad heads.

21. D

    A. Incorrect. According to the text, the majority is due to inhaled substances and due to food intolerances.

    B. Incorrect. The majority of diagnoses are being made from 5 and younger.

    C. False as irrelevant for the question.

    D. Correct. The study demonstrates that this increase has occurred.

22. C

    A. Incorrect - the text describes the technical skills required.

    B. Incorrect – the text explains the physical forces that need to be optimised.

    C. Correct – the text describes both aspects of rowing.

    D. Incorrect as per text.

23. A

    A. Correct. By piecing together the two pieces of information about restaurant density and its relation to obesity, it can be deduced that obesity is higher in low-income neighbourhoods.

    B. Incorrect as per text, there are more around schools.

    C. Irrelevant for the question.

    D. Beyond the scope of the question.

25. B

Three friends only support the dress, so this must stand alone. Adding in the other preferences you see that the 5 represents recommending earrings, overlapping also with 3 necklace and 4 shoes, and 1 handbag.

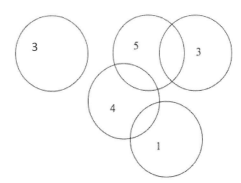

26. B
- A. Correct, but irrelevant for the question.
- B. Correct answer, it provides a sound rationale for the policy.
- C. Depends on the perception of moral responsibility, so weaker than (B).
- D. Irrelevant for this question, it argues in the wrong direction.

27. D
- A. Incorrect, the text states it is an important problem.
- B. Incorrect, this is described in the text.
- C. Incorrect, this is described in the text.
- D. Correct.

28. B & D
- A. False – he has Canadian stamps
- B. True – as Germany is in Europe Stan may have German stamps
- C. False – Half a dozen = 6 and therefore this is more than the 3 Canadian stamps
- D. True – as France is in Europe Stan may have French stamps

29. C

Communication skills	Mental flexibility	Intelligence	Dependa-bility	Empathy

Communication skills: 5%
Mental flexibility: 15%
Intelligence: 18%
Dependability: 28%
Empathy: 34%

We can see from that diagram that A and B are incorrect. Although empathy is not mentioned, we can conclude it is 34% so D is incorrect.

30. B
- A. True, but not far-reaching enough.
- B. Correct answer. Sugar does indeed have an addictive potential as it causes the release of endorphins and the health concerns are well known. This characteristic makes it like alcohol and smoking and potentially suitable for similar policies.
- C. True, but similar to option a) and thus too limited.
- D. Potentially true, but also too limited.

END OF SECTION

Section C: Quantitative Reasoning

SET 1

1. **B** The cost of the order is (£2.00x2)+ £5.00 + £4.30 + £6.00 + £3.40 = £22.70. The rice comes to £9.00, so the proportion is 9:22.7 = <u>1:2.5.</u>

2. **B** The most expensive option is to buy the Roast Pork and the Sichuan pork, for £11.60, and then two egg fried rice for £4.00, so the cost is £15.60 without the deal.
    She thus saves £5.60, giving the proportion: 5.6/15.6 = <u>36%.</u>

3. **A** The cost of Anne's order is £20, and the cost of the items separately is £24.60. So, she saves nothing on delivery, as for both orders the delivery charge is a flat rate of £3.50.

4. **B** Her previous bill would have come to £39.4x1.10 = £43.34. Her new bill comes to £41.40 with free delivery, so she saves £1.94: 1.94/43.34 = <u>4.5%.</u>

SET 2

5. **D** Bob would pay £300 with Red Flag; with Chamberlain he has a 12% discount, so would pay £308; with Meerkat Market he has a £40 discount so would pay £260, and with Munich he pays £250. Therefore Munich is the cheapest insurance provider for Bob.

6. **B** Normally it would cost £300 per year. With the 10% discount this comes to £270, and with the two free months this comes to (10/12) x£270 = £225.
    225/300 = 75%. If 75% of the original cost is paid, the total saving is <u>25%.</u>

7. **C** Laura has 3 years of no claims discount so she has a 3 x 3% = 9% discount. The cost with 3 years no claims discount from Chamberlain is £350 x 0.91 = £318.50
    The cost with 3 years no claims discount from Meerkat Market is £300-£30 = £270, so the saving = £318.50 - £270 = <u>£48.50.</u>

8. **A** Delia's current policy is £270 and Elliot's current policy is £350x0.85 = £297.50. The total old cost = £270 + £297.50 = £567.50 Together their new cost will be £500x0.90 = £450, thus, the proportion is 450:567.5 = <u>0.79:1.</u>

SET 3

9. **C** The number of students playing at least one game = 341 (the sum of every individual box)
    Therefore the number of students playing none of the games = 500 – 341 = 159
    The number of students playing exactly one game = 27+23+31+18 = 99
    The number of students playing at most one game
    = number playing none + number playing one
    = 159 + 99 = <u>258</u>

10. **A** This includes only students who play cricket or basketball, not both. Therefore the number of students who play either cricket or basketball but not football = 27 + 18 + 16 + 31 = <u>92.</u>

11. **B** Students playing at least three games = students playing exactly three + student playing exactly four
    = 8 + 13 + 36 + 31 + 37 = <u>125.</u>

12. **B** The number of students playing at most one game = 258
    Therefore the number of students playing at least two games = 500 – 258 = 242
    Difference = 258 – 242 = <u>16</u>

13. **C** The number of students who do not play cricket, football or hockey is the number of students who play only basketball plus the number of students who play no sports at all = 31+ 159 = <u>190</u>.

14. **C**

The number of students playing at least one game = 341

The number of students playing none of the games = 159

The number of students playing exactly one game = 27+23+31+18 = 99

The number of students playing none of the games **outnumber** the number of students playing exactly one game by = 159 – 99 = <u>60</u>

SET 4

15. **A**

Number of daily train users = 300,000

Since 41.67% use trains, the total number of travellers is 300000/0.4167 = 719,942 people

Since 16.67% use private cars, the number is 719,942 x 0.1667 = <u>120,000</u> people

16. **B** In 1999, the number of people using taxis as means of public transport = (150000 x 0.0833)/0.4167 = approximately <u>30,000</u> people.

17. **B**

Number of people using public transport in 1998 = 200,000/0.25 = 800,000

Number of people using public transportation in 2000 = 300,000/0.4167 = 720,000

Therefore there is a decrease of 80,000 passengers per day from 1998 to 2000

Percentage decrease = 80,000/800,000 = 0.1 = <u>10%</u>

18. **A** Number of people using taxis as a means of transport in 2000 = percentage using taxis x total travelling = 0.0833 x 720,000 = 60,000

Therefore the number of people using taxis as a means of transport in 1998 = 60,000/1.5 = 40,000

Percentage of total in 1998 = number using taxis/total travelling = 40,000/800,000 = 0.05 = <u>5%</u>

19. **C** Number of people using buses in 2000 = 0.33 x 720,000 (total travelling) = 240,000

Therefore the number of people travelling by buses in 1999 = (240,000 x 5)/6 = 200,000

And the number of people travelling by buses in 1998 = (240,000 x 8)/6 = 320,000

So the total fare collected is the sum of the total fares for each of the two years of interest (1998 & 1999), which is the daily fare multiplied by the number of passengers multiplied by the number of days.

1999: 200,000 x £1.5 x 358 = £107,400,000

1998: 320,000 x £1 x 358 = £114,560,000

Total amount collected in 1998 and 1999 together = <u>£221,960,000</u>

20. **C** In 1999, the percentage using private cars and buses combined is 16.67 + 33.33 = 50%

Therefore the number this corresponds to is the total number travelling in 1999 multiplied by 0.5 = (720,000 x 0.5) x 0.5 = <u>180,000</u> people

SET 5

21. D

The network from the given data is:

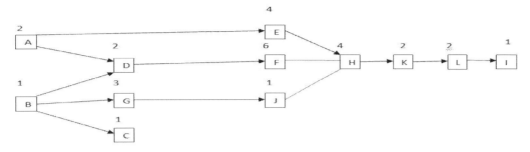

False ceiling (H) can start at earliest in the 11th week. Before this work begins, E, F, J, A, D, G and B must be completed, as shown by the flowchart. Work can take place simultaneously, so simply adding these durations is not the correct solution. The longest individual path is A > D > F which takes 2+2+6 weeks = 10 weeks. Therefore the false ceiling work can begin in the 11th week.

22. C

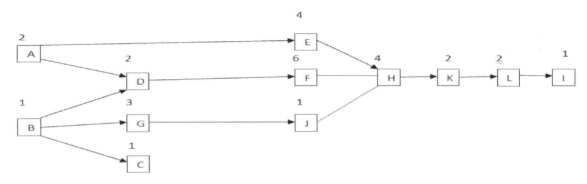

J must start before H. The latest it can start is by the 10th week. It could start as early as week 5 once G and B are completed, but in order not to hold up construction work it doesn't have to begin until week 10, so it is completed at the same time the air conditioning system is finished, allowing the false ceiling work to begin in week 11.

23. B

If no more than one job can be done at any given time, the minimum time all the work can be done, one after the other, is in a total of 29 weeks. Simply add together the duration totals to arrive at this figure.

24. C

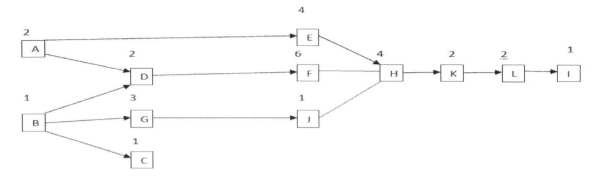

Maximum time gap between F and L is F+E+G+C+J+H+K = 21 weeks.

25. A

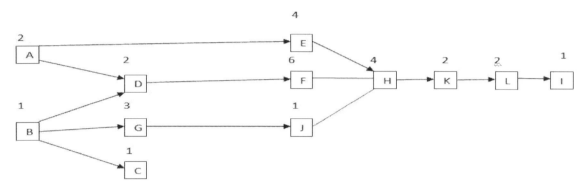

F must finish before H can start. Since F takes 6 weeks and must be completed by the end of week 10, the latest it can start is by week 5 to keep the building work on schedule.

SET 6

26. C Target food grain production in 1999-2000 = 1989 production x 1.2 = 175 x 1.2 = 210 million tons
Actual food grain production = 190 + 13 = 203 million tons
Therefore the absolute deficit = 210 – 203 = 7 million tons
Deficit as a percentage of total production target = 7/210 = <u>3.3%</u>

27. A Firstly calculate the percentage decrease in the production of pulses from 1989-90 to 1999-2000
= (15 – 13)/15 = 13.33%
Therefore the average price of pulses for the given years increase by the same 13.33%
Therefore the price in 1999-2000 = 1.8 x 1.133 = £2.04, approximately <u>£2.00</u>

28. B Relative percentage = 1999-2000 production/1989-90 production = 190/160 = 1.1875 = <u>118.75%</u>

SET 7

29. **A** To solve this you need to calculate the time taken by each team to complete the race. Whilst it might be tempting to attempt to find the average speed of each team, this cannot be done arithmetically as different runners run for different durations depending on their speed.
Time to complete = time per metre x number of metres = 1/speed x distance
USA = (1/8 + 1/12 + 1/9 + 1/11) x 400 = 164 seconds
Kenya = (1/9 + 1/14 + 1/6 + 1/11) x 400 = 176 seconds
Russia = (1/10 + 1/15 + 1/9 +1/6) x 400 = 178 seconds
Australia = (1/10 + 1/13 +1/10 +1/7) x 400 = 168 seconds
Therefore USA finishes in the shortest time and is therefore the winner.

30. **C** Once again calculate the time each team takes to complete the course.
Time to complete = time per metre x number of metres = 1/speed x distance
USA = (1/8 + 1/12 + 1/9 + 1/11) x 400 = 164 seconds
Kenya = (1/9 + 1/14 + 1/6 + 1/11) x 400 = 176 seconds
Russia = (1/10 + 1/15 + 1/9 +1/6) x 400 = 178 seconds
Australia = (1/10 + 1/13 +1/10 +1/7) x 400 = 168 seconds
The Russian team took the greatest length of time to complete the course and therefore finished last.

31. **B** The third Runner from Russia and Kenya did not meet. The first, second and third Russian runners were all faster than their corresponding Kenyan runners, so the Russian is ahead at the start of the third runner and the gap only increases during the third leg.

32. **B** This is a simple case of adding up the lap times:
First Runners: 8 + 9 +10 +10 = 37
Second Runners: 12 + 11 + 15 +13 = 51
Third Runners: 9 + 6 + 9 + 10 = 34
Fourth Runners: 11 + 14 + 6 + 7 = 38
Thus, the second runners were the slowest on average.

SET 8

33. **D** In 1999, the total weight of exports = overall cost/cost per tonne = 140000000/7000 = 20 000 tonnes
Total weight of imports = overall cost/cost per tonne = 180000000/6000 = 30 000 tonnes
Exports are less than imports by = (30 − 20)/30 = 33.33%

34. **C**
Trade surplus in 1998 = Exports − Imports = 130 − 110 = 20 Million USD
Trade surplus in 2001 = Exports − Imports = 160 − 150 = 10 Million USD
Therefore the percentage decrease = (20 − 10) / 20 = 50%
Trade surplus in 2002 = 50% of 2001 value = 0.5 x 10 = 5 Million USD
Imports in 2002 = 150 + (0.2 x 150) = 180 Million USD
Exports in 2002 = Imports + Trade surplus = 180 + 5 = 185 Million USD

35. **C** This question tests compound growth – the 2001 value must be increased by 10% three times.
Imports in 2004 = Imports in 2001 x $(1.1)^3$ = 150 x $(1.1)^3$ = 199.65, approximately 200 Million USD

36. **B** The compound annual growth rate is the rate that describes the growth when applied each year.
Therefore 2001 exports = 1996 exports x $(CAGR)^5$
Substitute the numbers in: 160 = 100 x $(CAGR)^5$
Rearrange: CAGR = $(160/100)^{(1/5)}$ = 1.098
Therefore the compound annual growth rate is 9.8%, approximately 10%

END OF SECTION

Section D: Abstract Reasoning

Set 1: In Set A there is always a quadrilateral that is cut by two circles. In Set B there is always a quadrilateral cut by one circle.

Set 2: In set A all the triangles have a black dot above them; in set B all the circles have a black dot below them

Set 3: In Set A, there is at least one circle that is cut by two lines. In Set B there is always at least one circle that has 2 tangents.

Set 4: In set A the sum of the dots is always odd, whilst in set B it is always even.

Set 5: In Set A there is always one more dot than there are rhomboids; in set B there is always one more rhomboid than there are dots.

Set 6: In Set A, the number of rhomboids and crosses are equal; In Set B, the number of rhomboids and black dots are equal

Set 7: In set A there is always one more black shape than white shape. In set B there is always one more white shape than black shape.

Set 8: In Set A there is always one square that overlaps with a circle. In Set B there is always a circle inside a larger square.

Set 9:
41. Black shapes increase number of sides by 1 and white shapes decrease number of sides by 2. There is at least one grey shape.
42. If a shape has an odd number of sides, it turns black in the next tile and stays black for subsequent tiles. Size and orientation is irrelevant.
43. Number of sides increases by 1 each time. Colour, shape and size is irrelevant.
44. Total number of right angles increases by 1 each time. Colour is irrelevant.
45. Number of intersections increases by 1 each time.

Set 10:
46. The number of intersections decreases by 1.
47. The total number of sides increases by 2. Size and colour of shape is irrelevant.
48. The number of acute angles increases by 2.
49. Each shape rotates 90° anticlockwise. Positions are irrelevant.
50. There is an added curved line.

Set 11: Set A; total number of white edges = even. Total number of black edges = odd. Always 1 grey shape.
Set B; total number of black edges = even. Total number of grey edges = odd. Always 1 white shape.

END OF SECTION

Section E: Situational Judgement Test

Scenario 1

1. **Very appropriate.** This way Arthur is not speaking for the patient but inviting him to share his feelings so that they can be resolved.
2. **Inappropriate but not awful.** Suicidal thoughts are a 'red flag' and so need to be mentioned to the doctor. However, going against the patient's request in front of them jeopardises their future engagement with the GP.
3. **Appropriate but not ideal.** Although this is good to inform the doctor, and true that confidentiality can be broken in this instance, Arthur should not have let the patient go home without exploring the matter further.
4. **Very inappropriate.** Although the patient has asked Arthur not to say anything, suicide risk is a very serious matter and one that needs to be reported, even if it risks breaking patient confidentiality.
5. **Inappropriate but not awful.** Although this is better than doing nothing, Arthur needs to explicitly tell the doctor the issue, there is no point just hinting about something so serious.

Scenario 2

6. **Very appropriate.** His personal tutor is there to help him out in difficult situations and together they may be able to come up with a solution.
7. **Very inappropriate.** Nico cannot break the rules of the medical school simply because he is feeling underprepared.
8. **Very appropriate.** It is important that Nico maximises the remaining time he has, all he can do is his best! He should not be embarrassed at the prospect of failing.
9. **Very inappropriate.** As a medical student, this would be very unprofessional and not solve Nico's problem.
10. **Very inappropriate.** Nico should not hide from the problem, he must make the most of the time he has left and put in as much effort as he can.

Scenario 3

11. **Very appropriate.** This is a very polite and appropriate way of acknowledging the patient.
12. **Inappropriate but not awful.** Encountering patients can be a difficult situation to navigate. If Mrs Hamilton has not spotted Alan there might not be an issue – Alan is not obligated to approach her.
13. **Very inappropriate.** This would be a breach of patient confidentiality. Mrs Hamilton's health issues are private and she most likely wouldn't want them broadcast on the bus.
14. **Inappropriate but now awful.** This may be seen as rude and unprofessional of Alan. However, he is not obligated to talk to her.
15. **Very appropriate.** Again Alan is not obligate to talk to her, nor may she even want to talk to him. Alan should follow her lead, not actively seek conversation, but respond politely and professionally if confronted.

Scenario 4

16. **Very inappropriate.** Sabrina should put a stop to the obvious breach in patient confidentiality. As an older student she should remind the younger students of their professionalism.
17. **Inappropriate but not awful.** Sabrina should talk to the girls before reporting them. Perhaps if they had behaved in this way before then reporting them straight away would be more appropriate.
18. **Very appropriate.** Discussions about patients should occur in a private and professional manner, and Sabrina should remind her fellow students of this.
19. **Inappropriate but not awful.** This may slightly help the situation, however it is still a public environment and they could easily be overhead.
20. **Very inappropriate.** This leaves the problem unresolved and the patient's confidentiality is still at risk. Observations of wrongdoing should be stopped if possible.

Scenario 5

21. **Appropriate but not ideal.** This may be unfair to the rest of the group who are prepared, however it is not Maxine's fault that she did not receive the email.
22. **Very appropriate.** The best option is for Maxine to be honest to the doctor.
23. **Very inappropriate.** It would appear as though Maxine had just not bothered to prepare if she does not turn up – she needs to be honest to her consultant.
24. **Inappropriate but not awful.** Although she may gain some information, this does not give Maxine adequate or equal time to understand the contents of the test and she may not achieve her full potential or achieve a safe level of understanding.
25. **Very inappropriate.** Copying another student's work does not achieve anything for Maxine, as she will not have learnt from preparing for the test. Cheating is not the solution!

Scenario 6

26. **Very inappropriate.** This is not Marcus' position as a student. The consultant may have had valid reasons for phrasing it how he did. It is also an entirely inappropriate way for the patient to hear such news.
27. **Inappropriate but not awful.** This could come across very rudely, so would need to be phrased differently. As above, the consultant may have had a valid reason.
28. **Very appropriate.** This will give Marcus' a greater understanding as to why the consultant did what they did, and he may gain greater insight into the skill of delivering bad news.
29. **Inappropriate but not awful.** Marcus should not ignore something that he strongly believes in, however as a student it is not necessarily his place to ensure the patient understands everything the doctor tells them.
30. **Very inappropriate.** Marcus should not go behind the doctor's back, especially as he is likely to have a limited understanding of the patient's condition. This also has implications for patient confidentiality.

Scenario 7

31. **Very inappropriate.** Tunde himself should not lie.
32. **Very appropriate.** This answer is honest, and does not put the other student's marks at risk.
33. **Appropriate but not ideal.** Whilst he is being honest to the other students, it does not provide them sufficient time to adequately prepare for the test.
34. **Very inappropriate.** This again is dishonest and it is likely that the consultant would give all of the students the test anyway.
35. **Very inappropriate.** This does not solve the problem in any way.

Scenario 8

36. **Very inappropriate.** This would be very inappropriate given Delilah's position as a medical student in a clinic that he is a patient of.

37. **Very appropriate.** This is a polite way of refusing the patient's offer whilst maintaining professionalism.

38. **Appropriate but not ideal.** If Delilah feels uncomfortable because of the patient's behaviour, she should inform her seniors as she does not have to put up with him. However, initially a simple polite refusal to the patient may end the situation there without the need for escalation via a complaint.

39. **Very inappropriate.** Despite what she may be feeling, Delilah must remain composed and maintain her professionalism in the face of adverse patient behaviour.

40. **Appropriate but not ideal.** Whilst this solves the solution for Delilah, this could put another student at risk of the patient's behaviour.

Scenario 9

41. **Very important.** Medical studies should be prioritised over extra-curricular activities.

42. **Important.** Chad has only been offered a trial and should recognise that, although this is a good opportunity, it by no mean guarantees him becoming a professional footballer. Therefore this may not be worth compromising his exams for.

43. **Important.** Conversely to the above factor, although a small step in a long process, if this is something Chad wants to pursue this may be a very good opportunity for him.

44. **Important.** Chad may have ample time to study for his exams as well as trial for the team.

45. **Not important at all. Whilst ideally Chad's parents would agree with his chosen profession, the final decision should remain with Chad.**

Scenario 10

46. **Important.** Although the situation is unpleasant, Troy must consider the underlying reasons for the patient's actions, as he does not know the patient's reason for being there. However whilst explaining why this has occurred it does not excuse aggressive and verbally abusive behaviour towards members of the medical team.

47. **Of minor importance.** Whilst the patient's diary is not Troy's responsibility, and there is nothing Troy can do about the delay, he must consider that this could be a very important reason as to why the patient has become aggressive.

48. **Of minor importance.** All of the other patients have to wait too yet have remained calm, suggesting this patient is being unreasonable.

49. **Important.** Although it has inconvenienced the patient, the doctor was delayed due to an emergency, so Troy can explain this to the patient.

50. **Very important.** Troy needs to maintain his professionalism in front of the patients who he will later be seeing.

51. **Of minor importance.** Troy is there to help out but should be supported by other professionals in the clinical environment.

52. **Important.** There is no excuse for violence towards healthcare professionals but those patients who have a diagnosis that pre-disposes them to violence should be treated differently as it is expected.

Scenario 11

53. **Important.** This factor will greatly influence Leroy's decision even though to an impartial person the answer may seem obvious. He must balance the implicit trust in a relationship with his duty to prevent plagiarism.

54. **Important.** Although he does not want to betray Bryony's trust, he now knows she has done something wrong and should do something about it.

55. **Very important.** Plagiarism is a serious offence and if Leroy has the chance to put a stop to it he should.

56. **Not important at all.** Plagiarism should be reported no matter how prestigious the environment in which it occurs.

57. **Important.** This could cast an embarrassing light on the university, which could make Byrony's punishment more severe.

Scenario 12

58. **Very important.** Benson should do all he can to do his best in his studies.

59. **Very important.** If Benson is able to borrow the books from the library he does not necessarily need his own copies.

60. **Important.** Students tend to get into lots of debt in these scenarios, which could create extra stress for Benson and compromise his studies.

61. **Very important.** It is not worth putting his studies at risk for a job, as there are other ways of obtaining funds for medical books.

62. **Very important.** A contribution from his parents could make all the difference when buying these books.

63. **Of minor importance.** His medical studies should be put before going on a sports tour. However, this may be a meaningful part of his University experience, and so should not be completely discounted.

Scenario 13

64. **Important.** As Albert-Clifford placed this as his first choice, whilst Anna-Theresa put it as her last choice, it seems there may have been some sort of irregularity in the allocation system which may be worth querying as part of an appeal.

65. **Not important at all.** Anna-Theresa was not allocated her first choice of project either, so it does not seem fair that Albert-Clifford should be re-allocated and no one else.

66. **Very important.** Albert-Clifford should realise that he is not the only one who wanted this project and it may be unfair of him to demand his place on the project over other people.

67. **Of minor importance.** He has put extensive time and resources into the project and has the opportunity to take this further. However, unfortunately putting in that amount of effort prior to a random allocation process was always going to be a risk.

68. **Not important at all.** Albert-Clifford should commit to whichever project he has been allocated. Many other students are likely to be in a similar position.

69. **Of minor importance.** This may help Albert-Clifford if he decided to appeal as he has demonstrated enthusiasm to the project supervisor. However, similarly to above, demonstrating enthusiasm does not have any influence in a random allocation process.

END OF PAPER

Mock Paper C Answers

Section A: Verbal Reasoning

Passage 1

1. **B** The text states that the pest-control business as a whole profits $6.7 billion of which bird-control only comprises one offshoot. Therefore, this statement is false.

2. **D** The text states that cryptococcosis is exceedingly rare and the $1.1 billion a year of excrement damage is the reason there is a need for pest-control in the first place irrespective of bird lover views. C is difficult to evaluate but it can be assumed to be false due to the "lucrative" bird control industry, leaving the correct answer D which is directly implied through the trade journal article the text mentions.

3. **C** The text states that pigeons can spread more than 60 diseases among humans however it does not offer any insight on how the significance of this issue compares with say the $1.1 billion a year due to excrement damage. Therefore, more information is needed to appraise this statement.

4. **B** The text explicitly states that culling remains a common fall-back method in response to public complaints. Our growing concern for pigeon welfare questions the ethics of culling but the text does not imply that it has caused its outlaw.

Passage 2

5. **C** At no point does the author categorically state or imply that he is or isn't the creator. He makes several vague references to time throughout the passage but without knowing the date the text was written/age of the author it is impossible to make any definitive conclusions.

6. **C** We know that the gun can hit a toy soldier at 9 yards 90% of the time but that doesn't necessarily mean that the accuracy deteriorates 1 yard further away. More information is needed.

7. **B.** The author states that the breechloader supersedes spiral spring makes thus implying it uses an alternative mechanism.

8. **C** This is the only conclusion which has direct evidence within the text. In the middle of the second paragraph where the author explains how the breechloader shoots wooden cylinders about an inch long. B seems like it could be an answer, however the text states 'the game of Little war, as we know it, became possible with the invention…' which may imply that the game is only possible with this gun however only states that the game at its current form needs the gun. This doesn't rule out the possibility that an inferior version of "little wars' was played prior to the invention of the spiral spring breechloader.

Passage 3

9. **B** Whilst much of the article describes the benefits to trading vessels, the second sentence states that the canal was opened to strengthen Germany's navy.

10. **C** The article does directly imply that bypassing the difficult navigation at Kattegat and Skager Rack would avoid the half million moneys loss of trading vessels each year. It doesn't, however, state a currency for that amount. Therefore the statement in the question may or may not be true.

11. **A** Recall that Germany's primary motivation for opening the canal was to strengthen their navy, trade benefits are just a coincidence, which is what the article is trying to demonstrate.

12. **C** Benefits to the navy are not described in the article therefore more information is needed. Furthermore, the text describes the potential future benefits of the new canal and doesn't describe any benefits that have come from the new canal (as it has only just opened).

Passage 4

13. **C** The statement in the question implies that each of the books in the trilogy was discussing the geology of South America. If you look back at the text you will see that the author only states that one specific volume of the trilogy focused on South America. It is of course possible that the other two volumes discussed South America as well but it is not mentioned.

14. **C** It is in a quote that Darwin himself calls his work dull. In the very first sentence the author describes Darwin's work as remarkable which we can assume does not mean boring. However, the author presents no opinion of whether they find Darwin's work boring or not so we are unable to answer this question with any certainty.

15. **A** The author states that there was a year delay in the completion of Darwin's work and the publishing of his manuscript. In a subsequent quote from Darwin himself the author implies ill health as a reason.

16. **A** In the last quote Darwin refers to his 240 pages much condensed. Implying that initially his draft was much longer.

Passage 5

17. **A** There is not a single argument in the text to argue against this, indeed one could argue that this is the main argument of the text through the examples that it uses.

18. **A** The article not only states that cattle are not indigenous to America, it also describes how the Conquistadores first introduced them to South America. Whilst you're not expected to know the Conquistadores nationality, references to the Spanish national sport and old Madrid certainly imply that they were Spanish.

19. **C** Whilst all of the conclusions are to a certain extent true on inference C is the most important. Indeed, answers A, B and D are actually all used to exemplify answer C in the text.

20. **C** The text states that was uptake of the Spanish national sport in all these locations. That does not however mean necessarily that the Conquistadores landed in each of those locations.

Passage 6

21. **B** The author explicitly states that in his opinion this belief is a fallacy.

22. **A** The author states that white pine has grown in New England "from time immemorial", which does not necessarily mean it grows best there. However, the use of the author's rhetorical question in conjunction with this statement suggests a degree of sarcasm and as such heavily implies that the statement in the question is true. The phrase 'Where do you find white pines growing better' also supports this statement.

23. **A** All the other answers are used as part of the authors argument towards A. Using examples such as the English oak and white pine to demonstrate that soil nutrient depletion is not sufficient to prevent the growth of established forests, unless the entire population is destroyed in which case new species move in.

24. **C** Although almost word for word this is the concluding sentence of the passage there are subtle differences in the wording which means it is used more as an observation or metaphor in the text. It is therefore impossible to know whether the author believes this.

Passage 7

25. **B** The only thing the text states about conflict is that in the event of war the cables may be cut by each of the aggressors with no implication that America has the right to seize overall control.

26. **B** The text states that "seven are largely owned, operated or controlled by American capital". As there are nine cables in the question it is impossible for America to own them all.

27. **B** Although if taken on face value this statement is not 100% true with the cables also residing above 40 degrees north, it is in itself not false – just half the truth. A and C directly conflict with passages of the text and D is not mentioned at all.

28. **C** The passage tells us that France owns only one of the cables, that however does not allow for any inference of usage. More information is required.

Passage 8

29. **C** The author presents this as one possible theory whereas the question presents it as fact. There is not enough evidence in the text to either accept or reject the statement.

30. **B** The text explicitly states that the excavations occurred during winter 1894-95, i.e. December of 1894.

31. **B** The text states that Libyans may have resided in Egypt between the old and middle kingdoms, i.e. before the middle kingdom ended not after it.

32. **C** The text refers to 3,000 "graves". 3,000 bodies is therefore a good estimate however one cannot assume that there was exactly one body to every grave and as such more information is needed.

Passage 9

33. **A** The text opens with "the best feed" and at no point does it mention any meat, so this statement can therefore be assumed to be true.

34. **D** At first glance B may appear the correct answer but note the text only describes a change in diet depending on wet/damp and cold weather and does not specify a specific time of year. The sentence "water and boiled milk, with a little lime water in each occasionally, is the best drink..." directly implies a turkey's need for distinct items of food and drink.

35. **A** The text directly states not to combine water and rice since the chicks are unable to swallow the paste, which therefore may cause them to choke.

36. **B** The author writes that the whites should only be added after the chicks are several days old.

Passage 10

37. **C** The author writes the passage from a relatively neutral standpoint, describing Lintner's opinion on the matter. We cannot be sure the author believes this himself.

38. **A** The text directly implies this in the penultimate line where it states the closed system labours under the disadvantage of making it difficult to maintain cleanliness.

39. **A** The text describes one of the flaws of the current malting system being a problem with the removal of carbonic acid, thus implying that carbonic acid has a negative impact on the malting process. As germination ceases at 20% carbonic acid, we can thus assume that 30% carbonic acid, which is greater than 20% and would occur if you can't remove the carbonic acid, would also make germination not possible.

40. **B** Although this sentence is not explicitly reiterated in the text it is implied through the mentioning of at least two types of pneumatic malting apparatus: closed vs. open.

Passage 11

41. **A** The question does not state that it ONLY flows north of Cork and is therefore at least in part true.

42. **C** Whilst the final paragraph does describe how the new bridge displays many features dictated by local conditions, it does not mention anything about practicality or difficulty.

43. **C** The new Angelsea bridge is certainly a swing bridge however the mechanism adopted by the previous bridge is not mentioned.

44. **A** In the second paragraph the text states that Cork train station is on the southern back of the southern branch of the river Lee.

END OF SECTION

Section B: Decision Making

1. **A** The statement currently presented demonstrates a jump of logic. The only way to establish logical flow is if a reason is given for not reading biased journals. It does not matter whether President Trump or the magazine are "good" or not.

2. **D** If a Venn diagram is drawn with the above information, such a picture should emerge. 25-13= 12, which is the number of students who like Beer/Spirits but not Wine. 12- 6 (Students who like spirits) = 6. Thus the number of students who liked beer alone is 6.

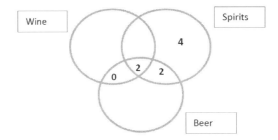

3. **C** Statements A and D are unrelated to the need for sleep, (which refutes the lack of sleep statement) and thus can be eliminated. Statement B is wrong because the functionality of the immune system is not a logical indicator of life/death.

4. **A** This argument contains the slippery slope fallacy. Statements B and C are irrelevant to the argument because they would not change the logic within it. Statements A and D are quite similar, however the focus of the statement is on gay marriage in particular, not gays and thus A is the answer.

5. **B** The word to focus on in this passage is that the factory "claimed" …. This indicates that none of the conclusions about the casualties can be confirmed, but also this does not indicate that there is some ulterior motive as in statement A.

6. **D** With the information given to us, it is impossible to work out the probability of full attendance of students, hence the answer is D.

7. **C** Pervill's number cannot be formed by adding multiples of 7, 28 or 16. 7 and 28 are both divisible by 7. As 7 and 28 are divisible by 7, and thus we can get to any total in the 7x table, the easiest method is to keep subtracting 16 from the potential answers to see if we reach a number in the 7x table which indicates that this total number of marbles can be achieved. 55-3x16=7, so this total can be reached. 51-16=35. 67-2x16=35. The odd one out therefore is C, as taking multiples of 16 away from 63 gives us 47, 31 and 15, thus this total is impossible to reach with combinations of 7, 16 and 28.

8. **B** Within the Venn Diagrams, reading the information should make it obvious that 17% fulfils all three categories and is placed at their intersection. Since this is included in the 60% of students who get into medical school, to find those that ax§re not included in the middle should be 60-17=43%. The 8% should be obsolete of any junctions as these are the students who do not "get in". Option B fulfils all these categories.

9. **C** If a diamond is drawn with all of the names placed as described, Hans is eliminated for standing in the front. Amelia can be eliminated because she stands at the very back, being 2 steps behind Hans in the diamond. That leaves Jake, Lola and Macy. Jake has been positioned to the far left and thus can be eliminated. As Lola is right next to Jake, she is the one in the middle. The formation is shown below.

```
             Amelia
    Jake    Lola    Macy
             Hans
```

10. **A** The total number of shoe pairs available is 14. The probability that Stan smiths are selected is $\frac{4}{14}$ and the probability that BAIT are selected is $\frac{3}{14}$. To select both we can either have Stan Smiths AND BAIT or BAIT AND Stan Smiths. Thus $\frac{4}{14} \times \frac{3}{14} \times 2 = \frac{6}{49}$

11. **C** The candidate has to identify which of the options would support the statements in the question. They have to choose the answer that explains which lung cancer patients should be charged. Options A and B do not explain why the smokers are charged, but rather introduces new patient categories. Thus option C is the right answer.

12. **C** There are n+1 people at the conference so hugs given = 1+2+3+ ... + n. Since this sum is n(n+1)/2, we need to solve the equation n(n+1)/2 = 28. This is the quadratic equation n2+ n -56 = 0. Solving for n, we obtain 7 as the answer and deduce that there were 8 people at the party.

13. **D** It is important to realise that the past does not predicate the present, thus decisions of the past will not affect establishment of deals now. Thus answers A and B are unlikely. Option C can be ruled out because there is not enough information to draw such a conclusion about Trump's negotiating abilities from the question.

14. **B** Statement A & C refers to individuals whom we cannot assume from the question are in any other category than those assigned to them (good grades/ party hard). The problem with statement D is that it generalises to include everybody, but not all medical students get good grades. Thus B which states the possibility of being in the good grades and party hard category is the right answer.

15. **C** It is important to realise that the Swimmers aren't a constant distance apart. They get further apart as they swim. Holly and Alex are less than 15m apart when Jon wins. Since they beat each other by the same distance, the difference between speeds must be equal.
When Jon reaches 200m, Alex is at 185m. When Alex finishes, Holly is at 185m. We need to know where Holly is when Alex is at 185m.
Alex is at has 92.5% ($\frac{185}{200}$) of Jon's speed. Holly's speed is 92.5% of Alex's and 85.5% (92.5% x 92.5% = 85.5%) of Jon's.
Thus 85.5% of 200 = 171m

16. **B** While option A might be true, it does indicate not why action is better than inaction.
Option B refers to improvement, the result of an action which is better than inaction, and is thus the correct answer.
Option C refers to success but the question only refers to betterment or improvement and is thus wrong

17. **D** The concept of greater good is irrelevant to whether medicolegal services should be hired and thus option A can be eliminated. The expense of the services does not directly explain why doctors do not need to hire them thus eliminating B. It is true that hospital lawyers can protect the doctors but that does not automatically exclude the need for medicolegal services. D is the only option that gives a reason and fits between the two statements in the question.

18. **B** All four of the options could be considered true facts. The question is looking for the assumption associated with the cybersecurity which gives reason for pornography being banned. Only statement 2, about pornography being a risk factor explains why it should be banned.

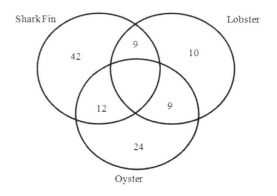

19. **C** A Venn diagram can help to calculate how many restaurants keep only one of each seafood type.

As n=130, adding up all the numbers in the Venn diagram (=106) and subtracting from 130 should give us 24.

20. **D** From the information given, the hexagon represents the group that can use most rooms but within a larger group. The only one at the intersection of all the groups is Women and thus D is the answer.

21. **C** The question might seem like a probability calculation, but actually just requires careful reading. The probability of completing the assignment if the brain stem is dissected is not given, while the probability of finishing the assignment or even completing 85% is quite high if the cerebellar cortex is dissected. Thus the *best* reason for choosing the cerebellar cortex as the region of choice is option C.

22. **D** Several snippets of information are provided but many of the options have a gap in logic. The question does not talk about all Dalmatians, so concluding anything about the whole group would be wrong. The same applies for the statement on all dogs, because the question only refers to dogs that bark loudly. Thus options A and B can be eliminated. Option C could be a solution in a real life situation, however according to the question we only know about one Dalmatian and cannot generalise it to even a few others in the breed. Thus option D is the most accurate answer.

23. **C** From the question, on Saturday, airline F is the cheapest. On Thursday, airline F is not the cheapest but airline B is the cheapest. That should already allow us to conclude that option C is wrong. With option D, we know airline N has good service. The first statement in the question says that Airlines F, N and B provide cheap fares so option D is not wrong.

24. **B** The number of students who get to watch the play is $0.6 \times 83 = 50$. Since the probability of them not winning the voucher is $1-0.25 =0.75$, the chance of both of these instances occurring is $0.75 \times 50 = 37.5$ (rounded up to 38)

25. **D** Methane is a subset of greenhouse gases. It shares all characteristics of greenhouse gases but not the other way around. Thus options A and B can be eliminated for generalising flammability and pollution to all greenhouse gases. Option C may be correct but the question does not state a link between flammability and pollution, thus the answer would be a reach. Since greenhouse gases are released in cow faeces, and methane is one of them, it has to be released in cow faeces too.

26. **C** Any answer that is a multiple of 4 will lead to her losing.
If she starts with 4 marbles, she is guaranteed a loss. If she takes 1, Tim will take 2, leaving the last for her. If she takes 2, then Tim will take 1, again leaving the last one.

For statement c, Jess takes one marble, leaving 12. This is a multiple of 4 so the opposite of what has been happening before will take place, allowing Jess to win.

27. **D** Armstrong's probability of doing all three actions was $0.14 = 0.7 \times 0.4 \times P(\text{NASA communication})$
Thus $P(\text{NASA communication}) = 0.5$
Thus Buzz Aldrin too has a 50% chance of communicating with NASA.

28. **C** Option A is wrong because of the false generalisation. While it is true as a fact, the focus of the question is the logic, where Aspirin is a subset of NSAIDs. The discussion of herbal medicines does not arise in the questions and is thus irrelevant to the answer.

29. **A**
If three columns are drawn like below:

Monday	Chicken	Mike
Tuesday	Fish	Leia
Wednesday	Lentiles	James
Thursday	Noodles	Carrie

Leia and Mike can be ruled out as they do not make noodles. Option D can be ruled out because it exactly contradicts the sentence 4 "*On Monday, chicken was cooked but not by Carrie*". Lentils were not cooked on Thursday, so the only day left for them is Wednesday. This leaves only the option A.

END OF SECTION

Section C: Quantitative Reasoning

SET 1

1. **C** There are multiple ways to approach this question but the most important step is recognizing that Chris is cooking 10 portions when the recipe only provides 6. Therefore calculate the amount of squid per portion (450/6 = 75g per person) and multiply by 10 giving the correct answer of 750g. Alternatively the problem can be solved in one step using the ratio of desired portions to the actual number provided by the recipe (10/6 x 450).

2. **B** In this question it is easiest to total up the cooking time before subtracting from 8.30pm. Which is a total of (10 x 7) 70 minutes plus the required 10 minutes early. Subtracting 80 minutes from 8.30pm leaves Chris needing to start his preparations at 7.10pm or 19:10.

3. **C** There are a few ways of approaching this question; the quickest is probably noting that a single portion is a quarter of Chris' recommended daily intake and therefore he must first eat 4 whole portions to exceed it. Leaving the calorific content of the remaining 6 portions in excess (6 x 575 = 3450). Alternatively it is possible to calculate the total calorific content of the meal (10 x 575) and subtract Chris' recommend daily intake like so: 5750 – (4 x 575).

4. **D** In this situation it is first necessary to calculate the total cost of the meal, which for 10 portions is £18.00. Then divide this amongst the remaining 8 friends giving the correct answer of £2.25. Or, total up the cost of the uneaten portions (2x £1.80) and divide this by the 8 remaining friends (3.60/8=£0.45). Now add that to the normal cost (£1.80 + £0.45 = £2.25

5. **E** At this stage it is important to look back to the recipe and recall that it only serves 6. Therefore the recipe requires 12/6 = 2 prawns per person. Meaning for the now 11 people eating the email 22 prawns are required.

SET 2

6. **E** In this question there is lots of information to consider. To achieve the correct answer the full £3 surcharge (£2 for premiere and £1 for 3D) must be added to peak prices for each ticket. Hence the new prices are £14 for an adult and £10 for a child or concession. Adding up these cumulative charges (2 x 14) + (2 x 10) + 10 gives the correct total of £58.

7. **D** Although on a Tuesday all tickets are £5, the 3D surcharge must still be added making each individual ticket £6. Therefore 3 admissions out of the £50 leaves 50 – (3 x 6) = £32 change.

8. **D** As a standard non 3D film Joe pays the basic rate of £5 per ticket, therefore 4 x 5 = £20 in total. Next the interest must be considered at 30%. For Joe's transaction this is calculated as 0.3 x 20 = £6. As the question asks for the total cost to Joe this must be added onto his initial expenditure of £20 producing the correct answer of £26. A quicker method would be to simply calculate 20 x 1.3.

9. **B** First ascertain the price per child: 5.50 + 2 = £7.50. Next divide this into the total amount of money available giving 100/7.50 = 13.3333. As it is not possible to take a third of a child to the cinema this number must be rounded down to the correct answer of 13.

10. **D** Once again recall that all tickets are £5 on a Tuesday but the correct surcharges (of £3) must be applied. Therefore the total cost on a Tuesday would be (5 + 3) x 5 = £40. Alternatively the total cost at peak time can be calculated as 11 + 3 + [2 x (7 + 3)] + [2 x (5 + 3)] = £50. Hence this leaves the correct difference of £10.

SET 3

11. B First recognize that the 1 bedroom flat has 0kWh gas usage and the same electricity usage as the 2 bedroom flat. Therefore it is not necessary to calculate the household electricity rate just yet. Instead isolate the gas component of the 2 bedroom flat bill like so (700 – 300 = £400). The question asks for the answer as per kWh therefore 400/8,000 = £0.05 is the correct answer.

12. C First calculate the household electricity rate from the 1 bedroom flat (300/2000 = £0.15). Next use the given business gas rate of £0.03 per kWh in any of the business rows to isolate a business electricity component – probably easiest with the microbusiness like so: 635 – (4,500 x 0.03) = £500. Therefore per kWH the business electricity rate is 500/5,000 = £0.10. Finally remember to calculate the difference as asked in the question 0.15 – 0.10 = £0.05

13. D Recall the gas rates for both businesses (£0.03) and households (£0.05) are now known. Therefore Kate's expenditure on business gas is £270 (0.03 x 9,000) and on household gas is £900 (0.05 x 18,000). Giving a total gas expenditure of £1170 per year. Notice the question asks "every two years" so don't forget to double your answer to give £2340.

14. A First calculate Luke's share of medium business energy costs as 0.5 x 3360 = £1680. Add on the energy costs of his 1 bedroom flat giving a total annual cost of 1680 + 300 = £1980. Here however the question asks for a monthly expenditure so divide by 12 to give the final answer of 1980/12 = £165

15. D Recall that household and business electricity rates were £0.15 and £0.10 respectively. Business rates are 0.1/0.15 or two thirds the cost of household rates. With a difference in price of a third, business rates are approximately 33% cheaper. As 1m² of photovoltaic cells equates to a 1% saving, 33m² are required.

SET 4

16. A Note that both quantities are less than 99 and therefore taking pricing from the first category it is £9 for a rugby ball and £5 for a pack of 3 tennis balls. Therefore although 60 tennis balls are being purchased in total this equates to only 60/3 = 20 packs. As such the correct answer is given as (50 x 9) + (20 x 5) = £550. As this is collection in store there is no delivery charge to add on.

17. C As 8 footballs costs (8 x7) £56, that leaves a remaining £19.01 to cross the minimum spend figure for free delivery. As footballs can only be sold as a whole this must be rounded up to the nearest integer number of footballs – in this case to £21 and the correct answer of 3 footballs.

18. B Recall that the data set says "home delivery is free on orders *exceeding* £75" and that therefore with a budget of £75 exactly the delivery charge must be paid. For a home 36 miles away this is £20 leaving a remaining £55 for tennis balls. Assuming this cannot purchase more than 99 items use the price of £5 per pack allowing for a total of 55/5 = 11 packs to be purchased. Recall that these are packs of 3 tennis balls and therefore the correct answer is 11 x 3 = 33 tennis balls.

19. B Notice here that when the quantities of items are halved the individual pricing of items changes and therefore one unfortunately cannot simply calculate the total price and halve it. Instead each order must be calculated separately and the difference taken as so: [(1000 x 4) + (500 x 7)] – [(500 x 5) + (250 x 8)] = £3,000

20. C First calculate the actual number of tennis ball packs being purchased (6,000/3 = 2,000). From the graph we know this equates to a price of £2 per item before the inflation. The new unit price can be calculated as 2 x 1.2 = £2.40. Therefore 2,000 tennis ball packs cost £4,800. At this point do not forget to add the cost of delivery which is also subject to the 20% price increase (20 x 1.2 = £24) Giving a total cost of 4,800 + 24 = £4,824.

SET 5

21. **A** The quickest way to solve this problem is to recognise that the plan illustrated effectively consists of 3 squares (with one broken into two right angle triangles). Secondly note the use of different units in the question and answer so straight away convert 200m to 0.2k to avoid a more difficult conversion later. Calculate the length of a square's side as 0.2/2 = 0.1km. Then calculate the total field area as 0.1 x 0.1 x 3 = 0.03km².

22. **C** For this question it is quickest to calculate the area of field A in metres as follows 100 x 100 = 10,000 m². As each cow requires 2m² the correct answer is given by 10,000/2 = 5,000 cows.

23. **B** First ascertain the total number of cows within the fields. As described previously this can be achieved using the idea that the total area is three times that of field A or D. Therefore there are 3 x 5,000 cattle present. First consider water expenditure per day as 3.5 x 15,000 x 0.02 = £1050. Next it is simplest to convert the pellet food price given into a per kg form like so 10/20 = £0.50. Daily food expenditure is therefore given by 4 x 15,000 x 0.5 = £30,000. Leaving a total of 30,000 + 1050 = £31,050 per day.

24. **D** First calculate the total volume of milk produced upon milking as 200 x 1.5 = 300L. As a cow can be milked only once every two days, if milking were to begin on Monday, milking could then only occur on Wednesday, Friday and Sunday. Thus there is the opportunity for 4 milking periods giving a total of 4 x 300 = 1,200L.

25. **C** Note the new daily cost of feed at £2.93 per cow. However don't forget that each cow also required 3.5L of water (3.5 x 0.02 = £0.07 per day). Therefore on the new feed (with water) the new daily cost of keeping a cow is £3. On a £1,000 daily budget this leaves the farmer able to house 1,000/3 = 333.33 cattle. As it is impossible to keep a third of a cow this must be rounded down to the nearest integer of 333.

SET 6

26. **A** In this question the first important thing to notice is that although the track time has been given in minutes; all other units use hours. Therefore begin by converting the lap time into an hourly figure like so: 3/60 = 0.05 hours. Next recall that distance = (average) speed x time. Therefore the correct length of the race track is given by 150 x 0.05 = 7.5 miles.

27. **D** As we are given the cost of fuel in terms of pence per litre, first convert car C's fuel consumption into a value in terms of miles per litre. Given that one gallon = 4.5 litres, a fuel consumption of 36mpg is therefore the equivalent of 36/4.5 = 8mpL. As the distance required is 10 miles, car C therefore requires 10/8 or 1.25 L of fuel which costs a total of 1.25 x 102 = 127.5 pence.

28. **C** The use of the term "average track speed" is a red herring in this question; it is actually not necessary to use car D's average car speed in any calculation. Instead the question is telling us that we can assume car D's fuel consumption will remain constant at 13.5mpg as it is driving at its average track speed. As the fuel tank capacity is given in litres first convert fuel consumption to miles per litre like so: 13.5/4.5 = 3 miles per litre. With a total fuel tank capacity of 40 litres, on a full tank car D can travel a total of 3 x 40 = 120 miles

29. **D** Assuming both cars at travelling in the same direction as it is a race track, car B has a speed of 180 – 150 = 30mph relative to car A. That is car B is gaining on car A at a speed of 30mph. As car B must first travel the 1.8 miles, and time = distance/speed, it will take car B 1.8/30 = 0.06 hours to overtake car A. As speed is given with units miles per hour, the correct answer has units of hours.

30. **B** As fuel tank capacity is given in litres, first convert the fuel consumption into units of miles per litre like so: 54/4.5 = 12 miles per litre. Therefore on a full tank car A can drive a total distance of 60 x 12 = 720 miles. On a track 8 miles long this equates to 720/8 = 90 laps

SET 7

31. B From the graph 4 bedroom house prices have risen by £40,000 pounds between 2010 and 2015 (240,000 – 200,000 = £40,000). As the question asks for the percentage RISE, we need to calculate what percentage this difference is out of the original house price in 2010. Which is (40,000/200,000) x 100 = 20%.

32. C First note the difference in price between 2 and 4 bedroom houses in 2014: 225,000 – 115,000 = £110,000. As they are able to save a total of £50,000 in the first 10 years this leaves them with a remaining £60,000 to save. By saving £10,000 a year this should take them 6 years. Leaving a total of 10 + 6 = 16 years to reach their minimum savings to upgrade. Don't forget to subtract these 16 years from 2014 giving the correct answer of 1998.

33. D After the price increase a 4 bedroom house in 2010 now costs 125% of the 2009 price. Therefore 1% is the equivalent of 200,000/125 = £1,600 and the price of a 2009 4 bedroom house can be calculated as 100 x 1,600 = £160,000. A quicker method is to spot that the 2010 house is 125% (1.25x) the price of the 2009 house so 200000/1.25=160000.

34. E The quickest method here is to first calculate the difference in 2 bedroom house prices in 2012, 3 bedroom house prices in 2013 and 4 bedroom house prices in 2014 compared to 2010 prices. Then calculate the sum like so: (105,000 – 95,000) + (135,000 – 125,000) + (225,000 – 200,000) = £45,000

35. E First calculate the maximum price that a 2010, 3 bedroom house could be let for each year: 125,000 x 20% = £25,000. Now consider the profit that would be received from selling the house in 2014 as 145,000 – 125,000 = £20,000. However when letting the house, the initial cost of the house must be first paid off before an investor would start making any profit. Therefore the total amount of money that must be earned to equate to a profit of £20,000 when letting is in fact 125,000 + 20,000 = £145,000. When receiving £25,000 a year this would therefore take 145,000/25,000 = 5.8 years.

36. C
(2 x 120) + (3 x 150) + (3 x 240) = 1,410
240/1,410 = 0.17 = 17%

END OF SECTION

Section D: Abstract Reasoning

Rules

Set 1: In set A number of white edges = 2(black edges); in set B number of white edges = 3(black edges). → quite a hard rule compared the standard ones you see in the test?

Set 2: In set A black edges > white edges; in set B white edges > black edges. There is always a grey arrow – its orientation/size is not relevant.

Set 3: In set A the angle is acute when total number of sides is even, whereas the angle is obtuse when the total number of sides is odd. In set B the converse is true – in both sets the colours are irrelevant.

Set 4: In set A the number of grey edges = the sum of black and white edges; in set B grey edges = the difference between black and white edges.

Set 5: In set A the number of edges within the largest shape is less than the number of edges of the largest shape; in set B the converse is true. Colours are irrelevant.

Set 6:

26. With each progressive frame the central star rotates 90° clockwise whereas the outer arrows individually rotate 90° anticlockwise.
27. Black shapes decrease by one side whereas white shapes increase by one side. Each progressive frame rotates 90° clockwise as well.
28. The number of black shapes increases by 2 each time. The number of white shapes is half the number of black shapes.
29. Shapes transform until they contain a right angle. Once a right angle shape is reached the shape alternates between white and black.
30. Each frame is rotated 45° clockwise and then flipped 180° along an alternating horizontal/vertical axis.

Set 7:

31. The number of grey sides is equal to the product of the number of black and white shapes.
32. If an arrow points to a shape it turns black. If an arrow points away from a shape, the shape transforms and the arrow itself becomes black.
33. The number of white shapes added is equal to half the number of black sides.
34. Shapes with an odd number of sides move inside those with an even number of sides and change colour. Size, orientation, and position within the second frame is not relevant.
35. Black shapes change colour to white. White shapes disappear. The size and position of the remaining shapes is not relevant, but their orientation is.

Set 8: Set A has more black shapes than white shapes. In set B the converse is true.

Set 9: All the frames are divided in half. In set A in any half there must be an even number of white shapes and an odd number of black shapes. In set B the converse is true with an odd number of white shapes.

Set 10: In set A there are less black and white shapes than there are grey sides; in set B there are more black and white shapes than there are grey sides. The ratio of black to white shapes does not matter.

Set 11: In set A when there is one grey shape there are more black than white shapes; when there are two grey shapes there are more white than black shapes. In set B the rule is reversed with more white than black shapes in the presence of one grey shape.

END OF SECTION

Section E: Situational Judgement Test

Scenario 1

1. **Very inappropriate** – although the patient has asked her not to say anything, it is something that is a cause for concern and would be better discussed with the GP, who may be able to persuade the patient to go to the police and therefore resolve the issue.

2. **Very inappropriate** – this breaches confidentiality and in this case, although domestic abuse is apparent, it is not in the student's remit to escalate this to the police without the patient's consent. Only the patient can escalate the issue (unless children are involved, then confidentiality can be broken).

3. **Appropriate but not ideal** – this reassures the patient that anything she says won't leave the practice walls, but also makes her aware of the fact that the GP may come to know of her situation and thus she is less likely to be surprised if the GP calls her in and talks to her about abuse. It is not ideal, however, as Maya has a duty to report all reports of abuse to her senior.

4. **Very appropriate** – the GP is likely to know of the protocols available in this kind of situation and how to go about discussing this issue with the patient. They will be able to take matters further if required.

5. **Very inappropriate** – this is a breach of confidentiality and it does not help the patient in any way. Maya's parents are unlikely to be able to help or provide any advice on this matter as they are not involved in the patient's care.

Scenario 2

6. **Very inappropriate** – there are still things he can do and people he can see and as it is a timetabled day, he should attend his activities, regardless of who is and isn't there.

7. **Very appropriate** – other members of staff on the ward are equally as helpful to observe or to tell him what to do and see. There are many things you can learn on a ward if you search hard enough.

8. **Appropriate but not ideal** – although Abdul is still doing something in his timetabled slot, there is a reason the university has given him that particular placement and so it would be better if he found something to do in the paediatric ward.

9. **Inappropriate but not awful** – The supervisor, while good to email, is not likely to be able to do anything as they may not be able to arrange an alternative option for placement with such short notice.

10. **Appropriate but not ideal** – This is a good idea, as Abdul can practice his communication skills, but it is often best to ask the doctors or nurses on the ward which patients would be the best to go and see as they are more likely to know which patients will be happy to have histories taken from them.

Scenario 3

11. **Inappropriate but not awful** – This is not the best way to deal with the situation. It is embarrassing for the students and doesn't address the problem of the upset relative. However, their supervisor has been alerted of their behaviour and this will allow them to deal with the students.

12. **Inappropriate but not awful**– This doesn't immediately stop the students talking about the patient, and they may say more in Mike's absence which may upset the relative even more, but their supervisor will be informed and thus further action can be taken.

13. **Very appropriate** – This reprimands the students and also makes sure that they apologise to the family member.

14. **Very inappropriate** – it is not Mike's responsibility to apologise for the students and does not stop the students from doing it again.

15. **Very inappropriate** – it doesn't address the situation and the students may continue with this behaviour.

Scenario 4

16. **Inappropriate but not awful** – although Alisha is has alerted someone else of the man, there is no guarantee that they will help and it is unlikely that they will be able to help in this situation better than Alisha can with the clinical skills she may have learnt. She will be able to administer CPR if needed and give the correct information to the paramedics when they arrive.

17. **Very inappropriate** – Alisha is likely to be trained in CPR and thus, is likely to be able to help until the ambulance arrives. It is also unmoral to leave an injured person in the middle of the road without stopping to help.

18. **Very appropriate** – Although Alisha may know CPR, it always helps to have a trained paramedic on the line who can guide her through the first aid, until the ambulance arrives, and by calling her date, she can explain the situation, and hopefully they will understand.

19. **Appropriate but not ideal** – although the ambulance is on the way, Alisha still has to wait for the paramedics so they find the right area and so that she can tell them what she knows.

20. **Very inappropriate** – the friend does not know where the patient is, and what has happened. The paramedics also need someone to guide them to the right place and may be able to tell Alisha what to do until they arrive.

Scenario 5

21. **Very inappropriate** – George is not competent enough to do the procedure and he may put the patient at harm.

22. **Very appropriate** – He has explained why he cannot do the procedure and has also not put the patient at risk.

23. **Very inappropriate** – it is likely the FY1 doctor is busy and it is unprofessional to both push a job onto someone else and leave without letting the nurse know.

24. **Very inappropriate** – this is not very professional as the nurse will return with the equipment and have to spend time finding someone else to do the procedure. She may be very busy, which is why she asked someone else to take the bloods in the first place.

25. **Very appropriate** – this means the nurse can get back to other jobs that she may have, and it means that the patient will have his bloods taken.

Scenario 6

26. **Very inappropriate** – although the patient does not know English, it is not very professional to make such a joke in a busy ward, where others can hear them. It reflects badly on the hospital and the rest of the staff working there.

27. **Very appropriate** – a senior has been alerted of this behaviour and it will be dealt with appropriately.

28. **Inappropriate but not awful** – It is best to lodge a complaint immediately, as action needs to be taken.

29. **Very inappropriate** – this joke should not be spread around, and a canteen is a very public place which means more people can hear it. This would damage the reputation of not only herself, but the entire organisation.

30. **Very appropriate** – This deals with the situation immediately, the FY2 will most likely apologise, and a senior can take this further if required.

Scenario 7

31. **Very inappropriate** – this could be harmful or even fatal to the patient if they have an allergic reaction to the drugs.
32. **Very inappropriate** – as a student, he should not be writing in or changing drug records unless told to do so and the patient has been prescribed the medication for a reason, so it is important a doctor knows so they can prescribe an alternative.
33. **Very appropriate** – an alternative can be proscribed for the patient.
34. **Inappropriate but not awful** – although the drug chart will be changed eventually, it may be signed off and the medication given to the patient in the space between they make that change, which could result in harm to the patient.
35. **Very inappropriate** – the drug chart will contain other medications that the patient needs so removing it may mean that they don't get the rest of the medication.
36. **Inappropriate but not awful** – patients should be aware of their medical treatment, however there is no guarantee the patient is fit to remember to inform the doctors. It is much better to alert the medical team directly as this poses a significant danger to the patient.

Scenario 8

37. **Of minor importance** – although the deadline is important, patient confidentiality and professionalism is more important.
38. **Of minor importance** – whilst this is obviously likely to be on Meena's mind, she should challenge Tom's inappropriate behaviour regardless of any potential repercussions for herself or Tom.
39. **Fairly important** – if their supervisor finds out, it could bring their marks down which could impact them on the final grade.
40. **Extremely important** – if Tom loses his phone, or shares the pictures, then patient confidentiality is broken. Patient information should be stored in a secure place and only be sent through encrypted pathways. It is unprofessional to take patient information out of the hospital without permission.
41. **Of no importance** – Tom does not have permission to take pictures of the notes.

Scenario 9

42. **Extremely important** – the hospital has a reputation to uphold and this post reflects badly on its values. It is a public post that anyone can access.
43. **Extremely important** – as stated above this is a public post and it can be linked back to Maria which can impact on her future.
44. **Of minor importance** – she has been linked to the post and therefore people can easily find out who she is.
45. **Of no importance whatsoever** – Maria has a responsibility in terms of her reputation and the hospital's reputation.
46. **Of no importance whatsoever** – the post will be available indefinitely so regardless of where she is placed, it will always affect her.

Scenario 10

47. **Fairly important** – this means that Muhammed will be able to go next year.
48. **Extremely important** – This is an avoidable situation, as there are many other opportunities to go to this conference.
49. **Fairly important** – supervisors are often busy and it is best not to rearrange too often, unless there is no other option.
50. **Of minor importance** – there are other ways to get work experience in those laboratories e.g. emailing the researcher.
51. **Fairly important** – Muhammed may not be able to get to the conference easily, if it is out of London

Scenario 11

52. **Of minor importance** – the patient can continue and resume her story.
53. **Fairly important** – if Melissa asks politely enough, the patient should be able to understand that it would take two minutes to return the sheet to the front desk.
54. **Extremely important** – patient confidentiality is very important
55. **Extremely important** – the person may not be a staff member and that would compromise patient confidentiality.
56. **Extremely important** – the nurse may need it to give medicines to the patient and thus without it, it will cause a delay.
57. **Of no importance** – The reason for the patient's admission to hospital doesn't affect the importance of listening to their story over the dropped documents.

Scenario 12

58. **Extremely important** – As they are clinical partners, it is important on good terms so they work well together, but if they are good friends then Tariq should not take offence.
59. **Fairly important** – Tariq may take offence at Jeremy's comments but he should realise that this is not a personal attack.
60. **Extremely important** – if they are to have so much contact with patients, they must look professional.
61. **Extremely important** – If Tariq is to make a good impression then he must remain professional.
62. **Extremely important** – professionalism is very important in medicine and must be maintained at all times.

Scenario 13

63. **Extremely important** – the patient must be treated immediately, or they could die.
64. **Of minor importance** – it is an inconvenience, but patient safety is extremely important.
65. **Extremely important** – if the patient contracts an infection, this could be more fatal than the initial reason for the surgery. It will also mean that the patient has to stay in hospital for longer which could have been avoided.
66. **Fairly important** – it is not fair to make others wait, but safety of the patient currently on the operating table is paramount.
67. **Of no importance** – as a student he is still part of the team and has a duty of safety to the patient.
68. **Of no importance** – everyone and everything in the room must be sterile, as even the smallest chance of infection could have big consequences for the patient.
69. **Of no importance** – she may touch the surgical equipment, or other utensils that come into contact with the patient.

END OF PAPER

Mock Paper D Answers

Section A: Verbal Reasoning

Passage 1

1. **B** The passage states that children have temporary teeth which appear in the 6th or 7th months. That means their teeth begin to appear before the age of one hence B is the correct answer.

2. **C** The second paragraph states that no teeth that come after the sixth year are ever shed. This means that all teeth that come after that age are permanent hence C is correct.

3. **A** The final sentence of the first paragraph states that temporary teeth require the same care that is exercised towards permanent teeth. Thus, statement A is correct.

4. **C** Statement A is correct as the third paragraph states that wisdom teeth appear in between 18 and 24 years of ages. Statement B is correct due to the answer for question 2. Statement D is correct as it states in the third paragraph that parents mistakenly suppose molars are temporary teeth. Hence C is the answer.

Passage 2

5. **D** The second paragraph states that magnetism may be cultivated and is inherent in every human being hence statement D is correct.

6. **A** The second sentence in the third paragraph states that self-preservation is the first law of nature hence statement A is correct.

7. **A** In the second paragraph when describing a person being hypnotized the first step mentioned is that the patient experiences a soothing influence which relaxes their muscles. Hence, statement A is correct.

8. **A** In the second paragraph it states that a pleasant, drowsy feeling is followed by a refreshing sleep. Hence A is the correct statement.

Passage 3

9. **B** The author states that coffee if rightly used is the most valuable addition to the morning meal. Hence statement B is correct.

10. **B** Tea and its active constituent theine is stated to be a pain destroyer, nerve stimulant and can produce hysterical symptoms. Thus by elimination B is the correct answer.

11. **B** It is stated in the text that when coffee is taken strongly in the morning is often produces dizziness and muscae volilantes hence statement B is correct

12. **A** The passage states that excess caffeine can cause a peculiar vision symptom whose name when translated is read as dancing flies, this is distinctly different from actual flies dancing making statement A incorrect and thus the answer.

Passage 4

13. **A** The first sentence states that pianos should avoid being exposed to atmospheric changes. Thus pianos must be exposed to similar atmospheres in winter and summer thus making the statement true.

14. **C** The passage states that the absence of frost in zero degrees weather is positive proof of a dry atmosphere. From this the converse can be humidity causes frost in cold temperatures hence C is correct.

15. **C** The passage states that an entirely dry atmosphere can cause physical effects which consequently puts the piano seriously out of tune. Hence C must be correct.

16. **D** The passage states that loosening of the glue joints produces clicks and rattles. Hence the converse is that's stiffening of these joints will make the clicks and rattles subside hence D is correct.

Passage 5

17. **D** The passage states that theosophy excludes all reasoning processes as they are considered imperfect. Therefore statement D is correct.

18. **B** In the passage it states that of the sects which practiced theosophy, one is the Hesychasts off the Greek Church. The Greek church is distinct to the Roman church hence statement B is correct.

19. **D** The passage states that the leader of the Theosophists had become fascinated with the doctrine of Buddhism. Hence statement D is correct.

20. **B** The passage states that the object of theosophical study is to understand the nature of divine things. This is synonymous with statement B.

Passage 6

21. **A** The first sentence of the last paragraph states that the dream must occur during healthy and tranquil sleep for it to have any significance. Hence the answer is true.

22. **A** The 3rd paragraph recites a passage about dreams written in the Holy Scripture. From this we can deduce that the answer is true.

23. **A** In the third paragraph it is written that Hippocrates said the body is asleep hence statement A is the answer.

24. **C** The passage says that modes of worship have been founded upon the interpretations of dreams which means statement C is correct.

Passage 7

25. **C** The passage states that deep sunken eyes are selfish. Selfishness is the opposite of selflessness, therefore a lack of selflessness is synonymous with selfish hence C is correct.

26. **C** The passage states that the normal distance between the eyes is the width of one eye and a distance greater than this intensifies the character of the person. Since the person could have any number of characteristics. Due to the ambiguity we cannot tell if erraticism is the characteristic that would be intensified.

27. **C** The passage states that eyes that are far apart and open indicate frankness. Frankness is synonymous to honesty therefore C is the correct answer.

28. **D** The passage states that long, almond-shaped eyes are indicative of genius, thus the answer is D

Passage 8

29. C The second paragraph says that probably the most important appliance for gym work is the wooden dumbbell. Due to the use of the word probably, we cannot be sure this is a fact. Hence the answer is C.

30. C The passage states that increasing circulation is done by exercising the extremities, the arms, the head and the feet. The abdomen is not mentioned hence C is the answer.

31. D The passages states that oxygen-hunger is only satisfied by breathing exercises hence statement D is correct.

32. A The passage states that strengthening the muscles of the back holds the body erect. This is synonymous with maintaining posture hence the answer is A.

Passage 9

33. B The passage states that a sponge in its natural state is different from what we see in commerce. Therefore the statement is false.

34. C The passage states that types of sponges are found in the Mediterranean Sea and are numerous in variety. Thus, the answer is C.

35. A The passage states that definite channels are constantly maintained and are essential to life of the sponge. A is therefore the correct answer.

36. B The passage states that the composition of the skeleton varies in different kinds of sponges. Hence B is the answer.

Passage 10

37. C Nowhere in the text is osteopathy defined as the study of bones, but also nowhere in the text is it said that osteopathy could be something else as a result of which it cannot be the study of bones. Since there is no evidence, the answer is C.

38. C In the text it says that what Dr Still asserted about osteopathy is not now maintained to any great extent by his followers. Hence C is the answer.

39. D The passage states that the constriction of an artery may be caused by a very slightly displaced bone. Thus the answer is D.

40. B The passage states that osteopathists do generally claim that all diseases arise from some maladjustment of the bones and ligaments that form the skeleton. This is synonymous with the answer B.

Passage 11

41. A The passage states that going over the body with a dry brush after bathing for two to three weeks will soften the skin. From this we can deduce that softening of the skin requires rubbing hence the answer is A.

42. B The passage states that the duty of pores is to carry waste matter off. Hence B is the right answer.

43. D The passage states that if you should quietly sit down in a tub of water and as quietly get up and dry off without rubbing, your skin wouldn't be much benefited. This supports answer D.

44. D The passage states that a little ammonia or any alternative should be used during bathing which makes the water soft.

END OF SECTION

Section B: Decision Making

1. **E** A is wrong because (18 x 10) + 8 =188. B is wrong because (18 x 2) + 8 = 44. C is wrong as 5 + 8 = 13. D is wrong because (8 x 2) + 5 = 21. This leaves answer E, which you can't reach using any combination of 5, 8 and 18.

2. **B** Statement A assumes all teenagers are immature. It also makes the argument that immaturity means that a person is unsafe at driving. Statement C does not provide an argument for why it is best for teenagers to start driving at 17. It simply provides a reason why most teenagers do not drive until they are 17. Statement D provides an argument as to why teenagers should drive at 17, however it is unknown what other factors are necessary to drive a car. Statement B provides a reason why it is illegal for teenagers to drive before the age of 17 and thus supports that 17 is the best age to start driving as this is when they are legally allowed to.

3. **C** Statement A is false as we know professor Moriarty can only lecture on physics. Statement B is false because identity is a philosophy topic and we know that professor Moriarty cannot lecture on philosophy. Statement D is a statement we can't say is true or false as no information has been given regarding this. Statement E is false as we know that professor Moriarty can only lecture on physics. Statement C is correct as we know all philosophy lectures are given by physics lecturers.

4. **B**

Table listing cars and their properties:

Colour of Car	Yellow	Red	Blue	Unknown
Seats	Fluffy pink cotton	Shiny black leather	Green cotton	
Tires	Orange	Silver	White	Gold

Table which indicates position of cars with regards to each other:

Red	Yellow	Blue
	Unknown	

By creating 2 tables and filling in information only when possible we can list all of the information and then can see clearly that the yellow car has orange tires.

We know that the yellow red and blue cars are adjacent and that the red car is not next to the blue car. From this we know that the yellow car is in the middle. We are told that a car opposite the yellow car has gold tires thus we know that this 4th car is not next to the red or blue cars. We know that a car next to the red car has orange tires, and since we know that the only car that can be next to the red car is the yellow car, we deduce that the yellow car has orange tires.

5. **A** It is never stated that Derek is intelligent so B is wrong. It is never stated that all funny people are intelligent. We can only infer that some funny people are intelligent as some engineers are funny and all engineers are intelligent. This makes C a wrong answer as well. It is never stated that Jamal is an engineer and it is never stated that all intelligent people are engineers hence D is wrong. A is proven correct as we know some engineers are funny. This makes E wrong as well, hence A is the answer.

6. **D** A is not correct as Prakash and Harry could still be faster than Ryan but set a time of above 12 seconds (e.g. both set a time of 12.1 seconds). B is not correct as Ryan's fastest time is 12.2 seconds which is above the qualifying time. C is incorrect as Prakash and Harvey may set a time of below 12 seconds, we do not know. D is correct as we know Ryan's fastest time is above the threshold time for qualification. Hence E is also incorrect and D is the right answer.

7. **C** If the coffee weighs 100g and it is 95% water and 5% coffee granules, then the water weighs 95g and the coffee weighs 5g. After drying the water content of the coffee is 75%. Assuming the weight of coffee granules has not changed, the 5g of coffee granules constitute 25% of the coffee. Therefore the total weight of coffee including the water is 20g. This added to the 10g weight of the cup make the total weight of the cup of coffee 30g. Hence C is the answer.

8. **A**
Table listing people and shops that they own.

	Julia	Polly	Fred	James	Harrison	Holly
Sweet shop	X	X	X			
Bakery			X	X		
Grocery store					X	
Pharmacy	X	X				
Sports shop						X
Shoe Shop		X		X	X	
Total	2	3	2	2	2	1

We can clearly see that Polly owns the most shops hence A is the answer.

9. **A** This question can be easily answered if written in inequalities and then the inequalities are combined.
Br<A.
Cl<M<A
A<Bi
Ch<Bi
From looking at these we know that Archith weighs more than Brian, Clarissa and Mike so these 3 can be discounted. We also know that Billy weighs more than Archith and Christy so those two can be discounted. Therefore, the answer is A – Billy.

10. **A** Sanjay thinks that a certain series of events (which we will call A) is more likely than a certain series of events which we will call B. Therefore, the probability for A is higher than B is what we are testing. The probability of A is 0.3 x 0.5 x 3 = 0.45. This is the probability of losing the match on a cold day (0.3) x probability of not eating ice cream (0.5) x the number of cold days (3). The probability of B is 0.1 x 0.5 x 4 = 0.2. This is the probability of losing the match on a hot day (0.1) x probability of eating ice cream (0.5) x number of hot days (4). Since 0.45>0.2, Sanjay is correct hence the answer is A.

A (much) quicker way to answer this is to look at the answer options before doing any maths. Without doing any calculations; answer a is possible, answer b is wrong (as the chance of winning the match when it's cold is lower than winning when higher, not lower, c is incorrect as he doesn't eat ice cream most days (only on half the number of days) and d is incorrect as he doesn't have an almost certain chance of losing his match on a hot day. Therefore, without needed to do any probability calculations, we know the answer MUST be A.

11. **E** A is false as we know all athletes play sports but we do not know which sports. B is false as we only know that Drogba is an athlete. C is false as we don't know which sports athletes play. D is false as we only know that Drogba is an athlete. Therefore, the answer is E.

12. **D** A is false as all events are independent and we can never be certain of the outcome as either outcome is equally probable. The same reasoning means B is also wrong. There is no information given to support C. D is correct as we know homer is flipping regular 10 pence coins thus both outcomes are equally probable. Hence, E is wrong and D is the correct answer.

13. **B** The cyclists aren't apart at a constant distance; they get further apart as they cycle. David and Alex are less than 50m apart at the time Alaric finishes. Each cyclist beats the next cyclist by the same distance so they must have the same difference between speeds. When Alaric finishes at 500m, David is at 450m. When David crosses the finish line then Alex is at 450m. We need to know where Alex is when David is at 450m.

    Alaric's speed = distance/time = 500/T. David's speed = 450/T. So David has 90% of Alaric's speed. This makes Alex's speed 90% of David's and 81% (90% x 90% = 81%) of Alaric's. So, when Alaric finishes, Alex is at 0.81 x 500 = 405m. 500 – 405 = 95. Therefore, Alaric beats Alex by 95m.

14. **C** A is false as we are not given any information about maths questions. B is false as we are not given any information about cheating. D is false as we are not given any information regarding how well Corrine is going to do in this exam. C must be correct as Corrine is assuming this exam will make her cry.

15. **C** The argument never states that only the local public should pay for the ceremony hence A is wrong. The fact that the MPs support Jed is completely irrelevant to the argument hence B is wrong. The fact that many politicians think the ceremony is outdated is irrelevant to the argument hence D is wrong. C directly challenges the fact that not all the public voted for Jed and thus not all of the public should have to pay for the ceremony. Hence C is the right answer.

16. **B** The only daughter of Isabella's grandmother is Isabella's mother. The person's sister's mother is the same as the person's mother. Therefore, the person's mother is the same as Isabella's mother. Therefore, the person depicted in the family tree must be Isabella's sister.

17. **C** We know that 5 children like all types of cake therefore B is wrong as the central overlapping segment is empty. We know that 2 like banana cake only so this also means that B is wrong. We know that only 1 child like all flavours except chocolate, therefore the region overlapping between carrot and banana must have 1 in it. This eliminates A and D as potential answers. C is the only diagram which satisfies all conditions hence C is the correct answer.

18. **C** If Harold says 1 number, then Fletcher can say "2 and 3". From this point onwards, Harold cannot win no matter what. If Harold says 4, then Fletcher will say 5,6,7 and Harold will only be able to say either 8, 8 and 9, or 8, 9 and 10. All of these result in Fletcher saying 11. If Harold says 4, 5 then Fletcher will say 6, 7 and the same outcome is achieved. If Harold says 4, 5, 6 then Fletcher will say 7 and once again the same outcome is achieved. Using this reasoning, the same can be said if Harold says 2 numbers as Fletcher will simply say 3 and thus win again. The only way for Harold to guarantee a win is to say 1, 2 and 3 when he starts. Whoever is able to say 3 is the person that can guarantee a win regardless of what the other person says. Hence C is the correct answer.

19. **C** Scenario 1: Imogen picks Lucy, Lucy picks Norden, Norden picks Imogen.
    Scenario 2: Imogen picks Norden, Lucy picks Imogen and Norden picks Lucy.
    All other scenarios result in the game restarting as Norden will have to pick himself. Since each person cannot pick themselves, they each have a 50:50 chance of dancing with either of the two-other people. Hence C is correct.

20. **E** Table depicting the position of each person:

Eric	David	
Andrew	Charlie	Barry

From this we can see that David is sat to the right of Eric. This is synonymous with Eric being sat to the left of David. Hence E is the correct answer.

21. **C** A is wrong as it simply states having more than 1 child is expensive, it does not state that this does not discourage people from having children. B is wrong as it talks about a certain group of the population which is not relevant to the argument. Statement D is wrong because it does not state why benefitting a certain group of people is bad. Statement C is correct as it disproves the assumption made I the argument that increasing cost discourages people from having children.

22. **C** If Peter is telling the truth, then Edwin is also telling the truth. If Edwin is telling the truth then either Max is also telling the truth, or Peter is also telling the truth. The only way that just 1 of them is telling the truth is if Max is telling the truth. If Max is being honest and both Peter and Edwin are lying, then they must have been to the swimming pool 0 times. Hence C is the answer.

23. **C** Statement A simply suggests that people living in cities are less fit than people living in rural areas. This does not necessarily mean that they are unhealthy or that living in a city is injurious to health. Statement B is incorrect since it doesn't necessarily mean that all people will buy unhealthy amounts of junk food if it is cheaper in cities. Statement D is incorrect as some vehicles may not produce as much pollution as others and we are not told if pollution is harmful. Statement E contradicts the conclusion; hence it is incorrect. High levels of carbon monoxide are dangerous to health and living in cities could be injurious if there are harmful amounts present. So, statement C is correct.

24. **E** There is a total of 17 shoes. In the worst-case scenario, Will picks 1 blue, 1 black, 3 orange and 3 red. That is 8 shoes in total. The next shoe he picks will be either black, orange or red and will complete his 3rd set. Therefore, Will has to pick at least 9 shoes to guarantee 3 matching pairs.

25. **A**

A. Bob is better at basketball than Chris.	**True**
B. Gerard is better at basketball than Bob.	**False**
C. Bob is the shortest.	**False**
D. Chris is the tallest.	**False**
E. Felix is the worst at basketball.	**False**

The height of the 4 boys can be written as: $G > F > B$. Statements B and C are incorrect as no information is given on Chris' height in relation to the others.

According to the given information, the level of proficiency in basketball can be written as: $F > G > C$. Statement B is incorrect as we cannot infer if Bob fits in between Felix and Gerard or Gerard and Chris in the above statement. Statement E is incorrect as Chris is the worst at basketball and Felix is better than Gerard. The question states that Chris is the worst at basketball, hence, statement A must be correct.

26. **C** The probability of the mice colliding is equal to 1 minus the probability of the mice not colliding. The probability of the mice not colliding is if all the mice travel in the same direction. If each mouse can travel either clockwise or anticlockwise, the probability of the mouse travelling in 1 direction is 1/2. Since there are 4 mice, the probability of all the mice travelling in the same direction and not colliding is: $1/2 * 1/2 * 1/2 * 1/2 = 1/16$. Since there are 2 directions that the mice can travel, the probability of them not colliding is: $2 * (1/2)^4 = 1/8$. The probability of the mice colliding is: $1 - 1/8 = 7/8$. $7/8 > 1/8$, therefore, the mice are more likely to collide than not.

27. **B** We only have information about whether lizards are poisonous, but we do not know any definitive facts about lizards and their offspring. Therefore, statements A, D and E are incorrect. The facts in the question clearly state that lizards are not mammals, therefore statement C is incorrect. We are told that some lizards are poisonous, therefore some are not poisonous. Hence, statement B is correct.

28. **D** Statement A is incorrect as we do are not told about all the goals scored in the match. We are only told about the goals Jacob scored. Statement B is also wrong based on this reasoning. Statement C is also wrong based on the reasoning. We are not told what goals Jacob saw during the match apart from the ones he scored so statement E is wrong. Statement D is correct as we are told every goal Jacob scored was from a free kick

29. **E** If 30 people liked all 3 sports and 80 people said they liked football, then 50 of these 80 people must like only football. If 10 people liked only rugby and cricket and 50 people said they liked rugby, then 10 of these 50 people must like only rugby (50 - (30 + 10) = 10). Out of the 40 people that said they liked cricket, 30 of these people said they liked all 3 sports and 10 liked rugby and cricket only. Therefore altogether, 50 people liked football, 10 people liked rugby and 40 people liked cricket, giving a total of 100.

END OF SECTION

Section C: Quantitative Reasoning

Data Set 1

1. **B** 98.4 miles per hour. Convert Jake's travel time into minutes: (2 x 60) + 8 = 128 minutes. Speed = distance/ time = 210/ 128 = 1.64 miles per minute. Convert this back into miles per hour: 1.64 x 60 = 98.4 miles per hour.

2. **C** 336 km. The conversion of kilometres to miles is given: 1.6 km = 1 mile. 210 x 1.6 = 336 km.

3. **E** 66%. To calculate this, convert both Jake and Martha's travel time into minutes. Jake: (2 x 60) + 8 = 128 minutes. Martha: (3 x 60) + 32 = 212 minutes. Martha's journey was 84 minutes longer than Jake's (212 - 128 = 84). Percentage change = difference/original so the percentage difference in travel time is calculated as follows: (84/ 212) x 100 = 66%, to the nearest whole number.

Data Set 2

4. **B** 16000m^2. Area = Length x Width. Area = 200m x 80m = 16000m^2

5. **D** 5min 36 sec. The distance Helen jogs is the perimeter of the field, which is calculated as follows: Perimeter = 2L x 2W. Perimeter = (2x200) + (2x80) = 560m. Time is distance/ speed, which will be how fast it takes Helen to jog the perimeter of the field. Time = (560/ 6000) x 60 = 5.6 mins. Remember to multiply .6 by 60 to give the answer in minutes. .6 x 60 = 36 so 5.6 mins = 5 min 36 sec.

6. **C** 161 rows. As the row intervals are given in cm, convert 80m into cm and divide by the given interval value: 8000/ 50 = 160. The question states that the first and last rows grow on the WX and YZ edges of the field. The trap to avoid is not to miss the final row; hence, there are 161 rows of cabbages on Sam's field.

Data Set 3

7. **C** 29300. Number of people older than 65 years = 7500 + 6450 + 6000+ 5700 + 3650 = 29300.

8. **A** 18.76%. If 65 years is the retiring age, then a person must be 51 years or above to be within 14 years of this age. The population of town X that is between 51-65 years = 8400 + 7850 + 8350 = 24600. Total population = 6900 + 6750 + 8200 + 7150 + 6350 + 6400 + 8750 + 7750 + 9950 + 9050 + 24600 + 29300 = 131150. 24600/131150 = 18.76% so A

9. **E** 80000. The working population consists of those aged 16-65, inclusive. 7150 + 6350 + 6400 + 8750 + 7750 + 9950 + 9050 + 8400 + 7850 + 8350 = 80000. A quicker way to do this is to subtract those not working age from the total population worked out previously i.e. 131150 – (6900 + 6750 + 8200 + 29300) = 80,000

10. **C** Approximate population under 30 years: 6900 + 6750 + 8200 + 7150 + 6350 + 6400 = 41750. Total population = 131150 so 41750/131150 = 0.318 so ~1/3 so C

Data Set 4

11. **D** 11:4. From the pie chart, it can be derived that the ratio of popularity between badminton and basketball is 22:8. This is 11:4 in its simplest form.

12. **A** 6/19. The difference between the popularity of tennis and football is 6%. Therefore, as a fraction, football is less popular than tennis by 6/19.

13. **E** 295. The racquet sports at this leisure centre consist of tennis, badminton and squash. If 570 people play a sport at the leisure centre a week, 108 play tennis (570 x 0.19), 125 play badminton (570 x 0.22) and 62 play squash (570 x 0.11), all calculated rounding down to a whole number. Therefore, the total number of people that play a racquet sport each week is 295.

Data Set 5

14. **E** 23000. From the graph, it can be read that the hot chocolate sales were £5000 in August and £28000 in December. The difference in the hot chocolate sales between these months is calculates as: 28000 − 5000 = 23000. Hence, the sales increase by £23000 from August to December.

15. **B** £17045. The sales of hot chocolate in October is £15000. £15000 represents 88% of company D's expected sales in October. The sum: 15000/ 88 gives the value of 1% of the expected sales. This can be multiplied by 100 to give the total expected sales in October: (15000/ 88) x 100 = £17045, to the nearest whole number. A quicker way to do this is to divide 15000 by 0.88.

16. **B** 10800. Company D made £27000 from the sales of hot chocolate during the month of January. The number of hot chocolates they made during this month is calculates as: 27000/2.50 = 10800.

Data Set 6

17. **A** 3/20. In focus group B, a total of 85% (55% + 30%) made some pronouncement, so 15% did not have a preference. This expressed as a fraction is: 15/100 = 3/20.

18. **D** 60. The number of people who preferred mango juice 2 in focus group A was 27 (108 x 0.25) and 33 in focus group B (110 x 0.30). Therefore, the total number of people who preferred mango juice 2 was: 27 + 33 = 60.

Data Set 7

19. **B** 67.7%. Average percentage obtained by the students = The sum of the students' percentages scored / the number of students. Average = (64 + 75 + 48 + 79 + 33 + 71 + 87 + 50 + 76 + 57 + 82 + 93 + 67 + 45 + 88)/ 15 = 67.7%.

20. **E** 88%. There are 15 students in the class. 15% of this is 2.25. This means that only the 2 students with the top scores will be able to obtain the maximum grade. From the bar chart, the 2 highest scores are 93% and 88%. Therefore, the minimum threshold mark value for this exam would be 88%.

21. **C** 4 students. The 2 lowest scores are 33 and 45, which combined = 78. If all the marks go up by 2% then 78 x 1.02 = 79.56.. If everyone else's marks went up by 2%, only 5 students would have marks greater than 79.56. Their marks would be (rounding to the nearest integer) 81, 89, 84, 95 and 90.

Data Set 8

22. **D** 15/17. Company R's April and September share prices are 85p and 75p, respectively. The calculation is as follows: 75/ 85 = 15/ 17.

23. **C** 96.05 p. Company R's share price in December 2007 was 85p. The new price in January 2008 would be: 85 x 1.13 = 96.05p.

Data Set 9

24. **B** B. BMI = Weight in kg / (Height in m)^2. The calculated BMI for each person is as follows:

A: 24.9   B: 20.4   C: 27.6   D: 21.0   E: 22.3

(however, you could have a reasonable educated guess by just eyeballing the data if you're low on time)

25. **C** 3kg. BMI = kg / m^2 so m^2 = kg/ BMI. m^2 = 45/ 17.6= 2.56. To have a BMI just within the normal range, a value of 18.5 is needed. BMI = kg / m^2 so kg = BMI x m^2. kg = 18.5 x 2.56 = 47.36 kg. Person F needs to gain just under 2.5kg.

26. **B** 1/10. BMI = kg / m^2 so kg = BMI x m^2. Person G's original weight is: 28 x 1.7^2 = 80.92 kg. The weight at a BMI of 24.9 is 71.96 kg. 80.92 − 71.96 = 8.96 kg needs to be lost. 8.96/ 80.92 is approximately 1/10.

Data Set 10

27. **A** $3x + 4y = 20.95$; $4x + 5y = 27.10$. In these simultaneous equations, x represents portions of chicken and y represents portions of chips, which equates to the price of the combined portions in £.

28. **E** £6.65. Multiply the first equation by 4 and the second equation by 3 to give 12x (12 portions of chicken) in both equations. Then subtract the two equations to eliminate x (chicken) and leave y (chips).
$4(3x + 4y = 20.95)$; $3(4x + 5y = 27.10) = 12x + 16y = 83.80$; $12x + 15y = 81.30$
$y = 2.50$. Use the calculated value of y and plug it into one of the equations to give the value of x (chicken): $3x + 4y = 20.95$, $3x + 4(2.50) = 20.95$. $x = 3.65$.
A 10% increase in the price of chicken would be $3.65 \times 1.1 = £4.02$, and a 5% increase in the price of chips would be $2.50 \times 1.05 = £2.63$. Therefore the cost of a chicken and chips would be $4.02 + 2.63 = 6.65$

29. **D** $27.52. At the original prices, a portion of chicken is £3.65, and a portion of chips is £2.50. The total cost of the order in £ is: $2(3.65) + 5(2.50) = £19.80$. This converted into $ is: $19.80 \times 1.39 = 27.52.

Data Set 11

30. **C** 180.4. The conversion factor for kg to lb is: 1kg = 2.2 lb. Thus, Kyran'a weight in lb is: $82 \times 2.2 = 180.4$ lb.

31. **C** 12 stone 4 lb. Alex weighs 78kg and the conversion factors are: 1kg = 2.2 lb, 1stone = 14 lb. Alex's weight in stones and lb is calculated as: $78 \times 2.2 = 171.6$ lb, $171.6 / 14 = 12.26$ stone. 0.26 stone is approximately 4 lb (rounded up).

32. **B** 417.9kg. To calculate this, the weight of Jay and Thomas needs to be converted firstly into lb and then to kg. Jay: $(13 \times 14) + 7 = 189$ lb, $189 / 2.2 = 85.9$ kg. Thomas: $(11 \times 14) + 11 = 165$ lb. $165 / 2.2 = 75$ kg. The sum of the boys' weights can now be calculated: $78 + 85.9 + 75 + 82 + 97 = 417.9$ kg.

33. **A** 1/5. Alex weighs 78kg and James weighs 97kg. 78/97 is approximately 4/5. Therefore, Alex is approximately 1/5 lighter than James.

Data Set 12

34. **C** 65%. The production of weed killer in 1998 and 1999 was 200,000 and 330,000, respectively. The increase in the production of weed killer from 1998 to 1999 was: $330,000 - 200,000 = 130,000$. Hence, the percentage increase is calculated as: $(130,000 / 200,000) \times 100 = 65\%$. (Hint: using the data from the table in 10,000s simplifies the calculations by making the numbers simpler to use).

35. **D** 4 years. The average production of weed killer in 10,000 litres between the years 1998- 2005 is: $(20 + 33 + 40 + 36 + 35 + 44 + 48 + 51)/ 8 = 38.4$. The years in which the production of weed killer was higher than the average calculated were: 2000, 2003, 2004, and 2005. Therefore, the answer is 4 years. (Hint: if short of time, you can take a fairly good guess by looking at the graph that the data is fairly evenly distributed around the mean so there's likely to be an even split between years above and years below the average so 4 would be a well-informed guess).

36. **D** The production of weed killer in 2005 and 2006 was 510,000 and 350,000, respectively. The difference in production between 2005 and 2006 is: $510,000 - 350,000 = 160,000$. The decrease in production as a decimal is therefore: $160,000 / 510,000 = 0.314$.

END OF SECTION

Section D: Abstract Reasoning

Rules:

Set 1: Set A: Circle intersects 1 shape. Set B: Circle intersects 2 shapes.

Set 2: Set A: At least 1 black square. Set B: Circle intersects 2 shapes.

Set 3: Set A: Even number of shapes. Set B: Odd number of shapes.

Set 4: Set A: Total number of sides is 16. Set B: Total number of sides is 11.

Set 5: Set A: Only curved shapes. Set B: Only shapes with straight edges.

Set 6: Darkly shaded segment rotates anticlockwise by one position and the lightly shaded segment rotates clockwise by one position.

Set 7: Number of shapes increases by 1 and the new shape is always shaded black.

Set 8: Shapes inside small square and big square swap. Position of shapes in big square rotates clockwise by 1 position after swap. Position of shapes in small square changes to the opposite position after swap.

Set 9: Outermost shape switches with second outermost shape. Innermost shape switches with second innermost shape.

Set 10: Set A: There is always a 4-sided shape in the top right corner. Set B: There is always a 3 sided shape in the bottom right corner.

Set 11: Set A: The total number of sides on white shapes is double the total number of sides on black shapes. Set B: The total number of sides on white shapes is equal to the total number of sides on black shapes.

Set 12: Set A: All the shapes have at least one line of symmetry. Set B: There are no lines of symmetry in any of the shapes.

Set 13: Set A: There is always one shape with one or more 90° angles. Set B: There is always a downwards pointing arrow in one of the 4 corners.

Set 14: Set A: Total number of sides of all shapes is 12. Set B: The number of straight edged shapes always equals the number of curved edge shapes.

Set 15: Number of sides increases by 2.

Set 16: Upward pointing arrow is always in a corner and moves anticlockwise through the corners.

Set 17: Set A: There are always 2 rectangles present. Set B: There are always 3 triangles present.

END OF SECTION

Section E: Situational Judgement Test

Scenario 1

1. **Appropriate but not ideal.** This is with good intention and will likely improve his performance in the exams, but it does not guarantee that he will pass.
2. **Very Appropriate.** This will ensure Afolarin has plenty of time to prepare for exams, and he will only miss out playing lacrosse for a month which is a short period of time.
3. **Appropriate but not ideal.** This is good as it will free more time for him to study however just 1 extra evening of work a week may not be sufficient.
4. **Very appropriate.** This is arguably the best option as it may enable him to still have time to play lacrosse as much as he wants whilst still finding enough time to study so his end of year exams are not compromised.
5. **Inappropriate but not awful.** This is not appropriate as lacrosse is important to him and maintaining a good work life balance is key to success in all parts of his life in the long term. This would however still ensure success in the exams so it is not extremely inappropriate.

Scenario 2

6. **Inappropriate but not awful.** This is inappropriate as it means he misses out on the opportunity of the project and the development it will grant him in his experimental technique. It will not jeopardise his exams however.
7. **Inappropriate but not awful.** This is not good as it risks Jamal not doing well in his exams as he may not have enough time to prepare. Although he will be able to continue with the project, it is not worth jeopardising his exams.
8. **Appropriate but not ideal.** This is a sensible approach to ensure equal work is done for both the exam and his project and that both will be done to an equal standard. The issue here lies with the fact that although a certain amount of time on the project will mean it is done well, the same amount of time spent preparing for exams may not guarantee passing.
9. **Inappropriate but not awful.** This will not necessarily cause any harm to Jamal's preparation for his exam or project however, it is unlikely to help either as there is no guarantee his peers have come up with a good solution and what may work for his peers may not necessarily work for Jamal.
10. **Appropriate but not ideal.** This would provide a solution as it would mean Jamal has enough time to prepare for his exams as well as work on his project. This is not ideal however because if Jamal receives an extension then it is unfair on his peers. It also may make Jamal look unprofessional in the eyes of his peers.

Scenario 3

11. **Very inappropriate.** The fact he failed by only 4% is largely irrelevant as the time between Jacob taking his exams and taking his resits means he may not be as prepared as he was when he took the exams as he may have forgotten some information. Thus, allocating only 3 days to prepare the exam is very inappropriate as it makes failing a likely outcome and will prevent him continuing with his studies.
12. **Inappropriate but not awful.** This is not an awful choice as he will have enough time to revise for exams, but it means he will not get a holiday. A holiday could be beneficial so he can relax and have some time off before diving into preparation for his resits. He has also taken a loan out so no going on the trip will cause financial issues too.
13. **Appropriate but not ideal.** This is good for Jacob as it means he can prepare for his exams and not miss out on the trip as he will be able to go the next year. This is not ideal however as it means his friends will have to miss out on going this year.
14. **Inappropriate but not awful.** Although this means Jacob is able to go on the trip and do some revision whilst there, this is not a great plan. Getting a lot of revision done on the trip is unlikely and after he returns home he only has 3 days to prepare further.
15. **Very appropriate.** This ensures that Jacob gets a break and is able to attend the trip with his friends which he has been looking forward to for a long time. It also means he will return home with enough time to prepare for his exam.

Scenario 4

16. **Inappropriate but not awful.** This is not helpful as it gets Ronit into trouble with the professor and this is a matter that can easily be managed by Shloke himself. Escalation of the issue is unnecessary.

17. **Very inappropriate.** This is very bad as it is compromising Ronit's safety and by doing nothing Shloke is responsible for any harm that may befall Ronit.

18. **Inappropriate but not awful.** This provides no help in resolving Ronit's safety and although this may pressure Ronit to wear his glasses it also embarrasses him in front of his peers.

19. **Very appropriate.** This will inform Ronit why it is important to take the proper safety precautions and will encourage him to wear the glasses.

20. **Very inappropriate.** This will compromise Shloke's own safety in the process of resolving Ronit's therefore this is completely inappropriate.

Scenario 5

21. **Very inappropriate.** This is terrible as it is compromising care of the patient as well as being very unhelpful to the nurse.

22. **Inappropriate but not awful.** This will likely help the patient and the nurse however this goes directly against strict protocol which Juan has been told to follow. This is inappropriate as it may result in Juan having disciplinary proceeding against him if the medical faculty find out.

23. **Appropriate but not ideal.** This communicates to the nurse why Juan shouldn't actively participate in care of the patient however it does not in any way help the nurse's dilemma

24. **Very appropriate.** This is ideal as it explains to the nurse why Juan can't do as she says, but also provides an alternative solution to the nurse's problem.

25. **Very inappropriate.** This involves directly lying to the nurse and compromising patient care which is a serious offence.

Scenario 6

26. **Inappropriate but not awful.** This is of no aid to Nicolas if he is unwell and thus is irresponsible of Jessica.

27. **Very appropriate.** This directly allows Nicolas to communicate what is wrong as well as providing water for him which may be beneficial if he needs it.

28. **Appropriate but not ideal.** This is good for Nicolas as it ensures that if he isn't feeling well that he will get the attention and care he requires, however it is unknown whether Nicolas requires help from the demonstrator and may be attracting unwanted attention. Jessica should see if she can help Nicolas before asking others to assist.

29. **Inappropriate but not awful.** This may cause no harm to Nicolas however this doesn't help Jessica find out more about what is wrong with Nicolas and thus opening the window may be of no benefit.

30. **Very inappropriate.** This would attract attention to Nicolas as well as stop the teaching session for other participants. This is an overreaction and Jessica needs to first assess how Nicolas is feeling herself by speaking to him.

Scenario 7

31. **Extremely important.** This is directly tied to Alvin's motivation for conducting the interview thus is of utmost importance.

32. **Extremely important.** This is an essential factor in Alvin deciding whether he can conduct the interview and write the essay as both require time for preparation.

33. **Fairly important.** If Alvin is likely to have another opportunity like this then that may sway him to focus on his essay and conduct the interview at another point, however there is no guarantee whether he will get another opportunity. Hence, this is a fairly important factor in his decision

34. **Fairly important.** This influences his motivation to write the essay however Alvin's assessment of the importance of writing the essay is probably not as accurate as his professor's assessment. Also, even if the essay does not help him much in his end of year exams, not handing it in on time will displease his professor.

35. **Of minor importance.** This factor partially motivates Alvin to write his essay however the benefit it provides him in terms of his academic development is the main driving factor. Thus, this factor is not very important.

Scenario 8

36. **Of no importance whatsoever.** With regards to the breach in patient confidentiality, completion of the project before the deadline isn't important, especially since stopping this breach of patient confidentiality shouldn't impact on their ability to complete the project. The most important factor is the breach in confidentiality and ensuring this doesn't happen again.

37. **Extremely important.** This is the most important factor for Damion to consider as this is a serious issue and needs to be resolved urgently.

38. **Of no importance whatsoever.** The breach in confidentiality is far more important than the grade that Damion and Philippa receive. This should not factor into Damion's decision at all.

39. **Fairly important.** This is important for Damion to consider as this may cause Phillipa to undergo disciplinary proceedings. It also means that how the breach occurred may be misconstrued and Damion may also then also be blamed. This factor is important in Damion understanding how urgently this must be resolved.

40. **Of no importance whatsoever.** The assessment of how important the data revealed is completely subjective and is not relevant to Damion at all. The breach in patient confidentiality is a very serious issue and thus needs to be addressed as such.

41. **Of no importance whatsoever.** Patient safety must be placed above any of his own personal gains and so the risk of losing all the time invested in the project is not important.

Scenario 9

42. **Of minor importance.** Although Carlos may care about his rapport with the consultant, being professional and looking professional in front of patients is more important.

43. **Of minor importance.** This is not very important is it is unlikely that the consultant will give Carlos a bad grade based on the decision Carlos makes.

44. **Of no importance whatsoever.** The consultant's appearance does not reflect on how presentable and professional Carlos looks himself. This should not be the reason why Carlos wants the consultant to look professional. The image displayed to patients should be his primary concern.

45. **Of no importance whatsoever.** Simply because the workshop was optional does not mean that the advice offered is any less important and that it should not be followed in a clinical setting.

46. **Of no importance whatsoever.** There is no relationship between the consultant's ability to do his job and his appearance.

Scenario 10

47. **Of minor importance.** Patient safety should be Arran's primary concern. The fact that this opportunity may not arise again is minor to compromising patient safety.

48. **Of no importance whatsoever.** His experience with tonsillitis is irrelevant when considering a breach in patient safety. Simply because he could focus with tonsillitis last time does not mean he is not compromising patient safety.

49. **Extremely important.** This is the main factor that Arran needs to consider when deciding whether to attend the surgery.

50. **Of minor importance.** Patient safety is far more important than whether the doctors conducting the surgery will ask Arran to sit in on other surgeries in the future.

51. **Of no importance whatsoever.** This fact is completely irrelevant. Simply because his peers have compromised patient safety does not mean it is okay and that Arran should do it himself.

Scenario 11

52. **Of minor importance.** Although being awarded the captaincy is something Abraham would like, this is not as crucial his scientific development which will assist him in his career.

53. **Of minor importance.** Simply attending the seminars will not mean the examiner will mark Abraham's papers more favourable and this should not be a driving factor in why Abraham chooses to attend the workshops. It is of minor importance however as it means that it is likely that the information taught in the seminars will translate into information that can be used in the exam.

54. **Of minor importance.** This is not a major factor in deciding whether Abraham should attend the seminars or play football. This is of minor importance as it may enable Abraham to build a better rapport with the lecturer which may the assist him his studies.

55. **Fairly important.** This is an important factor to consider as attending the seminars is largely motivated by the benefit it will grant Abraham in his exams. Although this is not the most important factor as the academic benefit derived from the seminars may not outweigh the benefits (e.g. mental wellbeing) derived from playing football.

56. **Extremely important.** This is extremely important to consider as it means that Abraham has the option of attending the seminars and still playing football on alternative dates thus reducing the proportion of training and football he is missing out on.

Scenario 12

57. **Of no importance whatsoever.** Although Brandon is more skilled at suturing and thus may perform alright in a tired state, this is not a valid factor when it may cause a breach in patient care and safety.

58. **Extremely important.** This is the main factor that needs to be considered and patient care is the upmost priority.

59. **Of minor importance.** Although embarrassing Brandon is unfortunate and not something Cameron wants to do, it is overshadowed by the importance of ensuring patient safety is not compromised.

60. **Of no importance whatsoever.** Assuming that Brandon is fine even though the information presented to Cameron suggests that Brandon isn't, is an irresponsible thing to do and thus this factor is irrelevant.

61. **Fairly important.** This is important in considering how to resolve the situation. Cameron needs to consider this factor because if he gets Brandon into further trouble he may risk Brandon undergoing severe disciplinary procedures. This matter can be resolved without alerting medical faculty to Brandon's tired state.

62. **Of minor importance.** Although this is unfortunate, it is overshadowed by the fact that patient safety could be compromised if Cameron does nothing.

Scenario 13

63. **Of minor importance.** This factor is directly involved with Jason's motivation for going on the trip, however it does not consider the negative impact it may have on his academics when considering the opportunity cost of a publication in his name.

64. **Fairly important.** This factor is key to Jason's decision as missing out on the tour may bring no benefit to Jason and thus would have been a wasted opportunity to go abroad with his peers.

65. **Extremely important.** A publication in his name is likely to be of tremendous use to Jason in his career and this factor may be far more beneficial to Jason than attending the tour.

66. **Extremely important.** This is a crucial factor for Jason to consider as it provides him with a solution of attending tour on an alternative year and still being able to undertake the extra work needed for the project this year. Although this may delay the tour, it still enables him to reap the benefits of both opportunities in the long term.

67. **Of minor importance.** This is important to Jason as he does not want to cause issues with his friends, however this is not so important when compared to the benefits granted by having a publication in his name. It also does not consider the fact that some of his friends may have publications already in their names which is why they are so keen to go on tour.

68. **Of no importance whatsoever.** The image Jason has in the eyes of the research team lead is irrelevant. The key factors at play are the benefits granted by either opportunity to Jason, not impressing the team lead is unlikely to have any negative consequences for Jason,

69. **Very important.** There is no obligation for Jason to work through his summer holidays and so the decision is purely down to him and what he would rather do.

END OF PAPER

Mock Paper E Answers

Section A: Verbal Reasoning

Passage 1

1. **D** The passage states that 'There were no chairs to be seen—the places of these useful articles being supplied by empty nail-kegs and blocks of wood'
2. **B** The passage states that 'of all the dinners that ever a white man sot down to' which would suggest that Godfrey is of Caucasian descent
3. **C** Based on the information given in the passage, the most likely option is that Godfrey is unemployed. Being a doctor, lawyer or university professor would mean that Godfrey would not live in poverty, as he does. He might be a farmer, but there is nothing in the passage to suggest so, hence the most likely option is unemployed.
4. **D** Based on the passage, one can infer that it is not set either in modern times, based on the state of the housing, nor is it set in the future. This leaves only option D as the plausible answer.

Passage 2

5. **A** The first line of the passage says that William Allen was a chemist, which is a scientist.
6. **B** It says in the passage that William Allen worked in Guy's Hospital, so option B is correct.
7. **C** William Allen discovered the proportion of carbon in carbonic acid; not the acid itself
8. **A** The only thing which can be properly inferred from the passage is that William Allen made a significant contribution to science, since he is quoted as an 'eminent scientist' and that the passage lists some of the achievements he had in the field of chemistry.

Passage 3

9. **B** Chaos is the name given to the mass of earth and sea and heaven before they were separated, not the name of a god.
10. **A** The passage states that 'The fiery part, being the lightest, sprang up, and formed the skies' so the skies are the lightest of the options
11. **A** The passage states that Prometheus 'made man in the image of the gods'
12. **C** The passage states that 'God and Nature at last interposed, and put an end to this discord, separating earth from sea, and heaven from both' which suggests that God and Nature together helped in the creation of the Earth from Chaos.

Passage 4

13. **B** The passage states that Merlin's mother was a woman
14. **A** The passage states that 'Merlin was the son of no mortal father' which would suggest that Merlin's father is immortal.
15. **D** The passage states that Merlin 'retained many marks of his unearthly origin', suggesting that he does have special abilities.
16. **D** Based on the information in the passage, that Vortigern is fearful for the return of the rightful heirs, one would suspect that Vortigern is a fearful character.

Passage 5

17. **A** The first line states that the author is born in Indiana, which is in the USA.
18. **C** The passage states that the author's father was James. P Mills, the grandfather was James Mills II and the great-grandfather was James Mills I
19. **B** The passage states that 'With his inheritance of $250, he and his brother Frank started West in a Dearborn wagon, crossing the Alleghenies.'
20. **C** There is no indication that the author is male or female based on the passage.

Passage 6

21. **C** The passage makes no mention of any siblings
22. **A** The passage states that Frank's face 'flushes with anger' when his friend suggests that Mr. Craven wishes to marry his mother, and refuses to think of it any more, suggesting that he is not comfortable with the idea of his mother marrying.
23. **A** Considering Frank is sitting outside on the lawn when Mr. Craven gets his attention, the most likely season would be summer.
24. **C** Considering Frank is worried about his mother getting married to Mr. Craven, the only plausible option is that his father is absent. Frank's mother may well be unmarried, but there is nothing in the text to distinguish whether she was ever married or not, so this option is not plausible.

Passage 7

25. **B** The passage states that 'no one in London who had a larger and more festive post than she', suggesting that she too lives in London
26. **C** The passage states that 'Even in her fiftieth year she retained with her youthful zest for life' but this does not give us any indication as to what age she currently is.
27. **A** This option correctly summarises the information in the passage. Although answer D seems to be appropriate, the passage focuses on Cynthia and her desires more than the benefits of sour milk injection.
28. **D** The passage states that 'but time was gradually lightening the heaviness of feature that had once formed so remarkable an ugliness' suggesting that Cynthia was ugly in youth, not good-looking

Passage 8

29. **A** The passage states that 'that other "Long Island"—the group of the Outer Hebrides—which, for an equal distance, extends along the Scottish coast from Butt of Lewis to Barra Head' so there is a Long Island in Scotland
30. **A** The passage does provide an account of the history of Long Island
31. **D** The passage states that 'the group of the Outer Hebrides—which, for an equal distance, extends along the Scottish coast from Butt of Lewis to Barra Head' so the Outer Hebrides are in Scotland
32. **D** The passage makes mention of all those animals, save for tigers

Passage 9

33. **A** The passage is set in the year 2126, which is in the future
34. **C** The passage talks about how the government and religion was a product of human behaviour, which would suggest that the passage is about human nature
35. **D** The passage states that England is 'under the absolute dominion of a female sovereign' so option D is correct
36. **C** The passage states nothing to suggest admiration for human compassion, however it also doesn't oppose this statement so the answer is C.

Passage 10

37. **A** The passage states that 'Moreover, he possessed only one eye, which was large and telescopic looking' which is proof of his one eye

38. **D** The protagonist describes the ghost as a 'horrid brute' which clearly demonstrates his lack of fondness for the ghost

39. **D** The description in the passage clearly brings out feelings of pity for the reader.

40. **A** The passage states that Ashton is interested in the story of the 'strange, dwarfish old man' and he 'resolved to look for him and see what his game really was' so clearly shows an interest in ghosts

Passage 11

41. **D** The passage states that they are based in Lytton Springs, India

42. **B** The passage states that the faith 'was a mixture of Catholicism and Hinduism' so cannot be labelled as either Hinduism or Catholicism

43. **B** The passage states that the Princes 'were all great Psychics', which would suggest that they had special abilities

44. **C** The passage makes no reference to Sita being the veiled princess, so the correct answer is can't tell.

END OF SECTION

Section B: Decision Making

1. A To calculate this one needs to find the lowest common multiple of both 73 and 104, and then add that value to 2007. The lowest common multiple of 73 and 104 is 7592, which when added to 2007 gives 9559 AD

2. A
    A. This is the most feasible option
    B. The ability to buy alcohol should have no semblance as to whether one can vote or not
    C. In a democracy, the more people that vote, the fairer the voting system becomes
    D. This is false. There will be plenty of people below the age of 18 who can still understand policies and the pros and cons of each

3. E When drawing out the whole arrangement as shown below we can clearly see that Caitlyn is directly opposite David.

Caitlyn

James   Simon

Adam   Ben

Joe   David

4. C 8 is the only common code amongst all 3 codes, as is the word 'you' so 8 must mean 'you' in this code.

5. A
    A. This provides the most sensible option as to why smoking should be condemned
    B. Though this is true, there is an element of free choice which must also be considered and therefore does not form a strong enough argument as to why smoking should be condemned
    C. The healthcare system being private does not form a coherent argument to this statement
    D. Though most people that smoke do also drink, there is no strict causation so this statement does not form a reasonable argument

6. B Jason is older than Peter and John, so let's assume Jason is the oldest. If Alan is younger than John, and Peter is older than John, this means that Peter is the next oldest, followed by John, followed by Alan.

7. C Based on the given information, S=Y, OILE can be derived from 'TOILET' as 'DJEB' and D=L, so option C is correct

8. B Assuming the public are proactive about their healthcare and have easy access to the information then by publishing mortality rates the public can have a more active role in their healthcare. This is especially relevant in the current era of medicine seeing a transition away from more paternalistic views to patient autonomy.

9. B If the number of girls is 40 more than the number of boys, and the boys make up 40% of the total number of students, then the discrepancy of 40 between boys and girls must represent 20%. Therefore, 1%=2 students and therefore the total number of students is 200.

10. A If Anna and May are in the same school year, and Isaac is May's younger brother, then Anna must be older than Isaac.

11. E If French is the third lesson of the day then maths has to be fourth so that there is a lesson between French and history and so that Maths is neither second nor fifth. The complete subject order can be seen below.

1 – Science
2 – English
3 – French
4 – Maths
5 – Geography
6 – History

12. A This statement is true-one of the main concerns about legalising euthanasia is that it provides a way by which families push their elderly family members into euthanasia to relieve themselves of the burden.

13. A Days pass at 2/5 of the speed on Mars so the gap between Olympic games would be 2.5x longer so 4 years x 2.5 = 10 years.

14. B

A. This statement is **false**. There is no reason as to why the illiterate members of society should be segregated from schoolchildren
B. This statement is **true**- such a scheme should not come at the expense of a household income for the illiterate members of society
C. This statement is **false**- providing unnecessary hardship benefits neither the schoolteachers nor the illiterate members of society
D. This statement is **false**- there does not appear to be any correlation between literacy and how crowded the streets are at night time.

15. A The probability that Jay will be late is the probability that Jay will walk (30%) multiplied by the probability that Jay will be late when he walks (60%), added to the probability that Jay will take the bus (70%) multiplied by the probability that Jay will be late taking the bus (20%), which gives 32%.

16. C 273546 rearranged would give 234567, which means that 2 and 5 retain their position within the original number, so the correct number is 2

17. D Based on the given information, a Venn diagram can be drawn as follows: (top circle = English; right = PhDs)

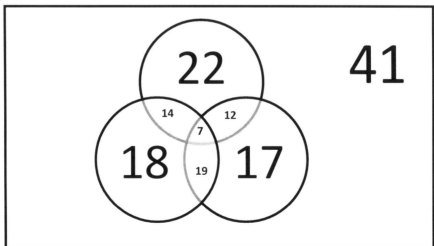

This means that the number of employees without any of the specifications is 41.

18. D The information provided states that motor vehicles include trucks, therefore it is fair for one to assume that all trucks are motor vehicles

19. C The information provided states that sharks are fish, and fish are aquatic animals. Therefore in turn sharks must be aquatic animals.

20. D If Alex starts off walking towards the sunrise, he is walking towards the east. If he then turns 90 degrees clockwise, he will be walking south. If he turns 180 degrees after this he will be walking north. Finally, by turning 90 degrees anticlockwise after this, he will be walking West for the last 4km.

21. A

A. This statement provides the strongest argument-if less people can possess a gun, then naturally there will be reduced murders because of gun crime

B. Although this argument is valid, it is not providing a strong case. Knife possession carries its own risks so carrying a knife instead of a gun does not really solve the issue

C. This statement is false-the point of black market sales is that it is already illegal-this will not really be affected whether possession is made illegal or not

D. This statement is false-using a gun in a public place is a crime anyway so people who are willing to use a gun in public are already not afraid of the consequences of possessing a gun.

22. E The Arabian Sea is a sea, which means that is must be a body of water partially enclosed by land.

23. A If 56% of the 300 students are boys, then 44% of the students are girls which =132 students. If 25% of these are Chinese that equates to 132 divided by 4 which = 33 students.

24. A

A. This option provides the strongest argument-the worst diets are often the cheapest and therefore tend to form a large proportion of the daily nutrition for the poorest of society. As a result, the poorest of society tend to be the most predisposed to obesity.

B. This option is false-obesity is associated with several co-morbidities and therefore is a very expensive morbidity

C. This option is false-contribution to the national health system is based on income, not weight

D. This option is false-although there is a genetic component to obesity and certain people are predisposed, unless one has a very rare mutation this is only a predisposition and can be controlled for.

25. C Karl is David's uncle.

26. C Based on the information, the school bus will get her to school at 09:01. The public bus arrives at 08:21, which she will miss, and the next bus will arrive at 08:38, which will take 18 minutes to arrive, meaning she will at school at 08:56, so the public bus at 08:38 will get her to school first

27. A Based on the information provided the seating arrangement can be drawn as shown below. Therefore, Neil must be sitting opposite Karen in the circle.

Neil

Richard   Opie

Jason   Mason

Pari   Liam

Karen

1992 = Neil; 1996 = Liam; 1997 = Karen; 2000 = Obie; 2001 = Jason; 204 = Pari; 2005 = Richard; 2010 = Mason

28. C

A. This statement is not a good enough excuse for the continued legalisation of zoos
B. This statement is false-there are plenty of methods by which children can learn about animals without using zoos
C. This statement provides the best argument for zoos-animals such as the giant panda are no longer endangered due to the conservation efforts made by zoos and other conservation organisations
D. Although this is true, this does not provide a good enough excuse for zoos. Animals should not be held in captivity so that humans can make a living, the same way animals should not be held in captivity for the sake of human entertainment

29. A We are told that all humans are apes and secondly that all apes are mammals. Therefore it follows that all humans are mammals.

END OF SECTION

Section C: Quantitative Reasoning

Data Set 1

1. **E** The mode is the most frequent group within the data set. Both 5 and 4 people are in the shop twice, making the made the average of these two so 4.5.
2. **D** The best time to visit the shop would be when there are the least number of people in the shop, which would be at 10am
3. **C** The range is the maximum value subtract the minimum value, which would give a value of 20
4. **D** The number of people who visit the shop throughout the day is 70. If each person spends £5, then the total income for the day is £350

Data Set 2

5. **B** If £1=1.8595 CAD, then 500 multiplied by 1.8595=$929.75
6. **A** The conversion table shows that 1 CAD=£0.5364, so 150 multiplied by 0.5364=£80.46
7. **C** One can see that 1 CAD=£0.5364, so $150 will only get Sally £80.46. However, if Sally received the same rate as when she was exchanging from pounds to Canadian dollars, she would've received £80.67 for $150 (150 divided by 1.8595), so she loses £0.21 in this exchange
8. **C** using the rate given in the table, for USD 700 Shoko can get JPY 82425. Using the rate given by the travel agency, Shoko can get JPY 83041. She therefore can get JPY 616 more using the travel agency as compared to the table.

Data Set 3

9. **B** The temperature today is 22.6°C which corresponds to a temperature anomaly of 1.3°C. Therefore the baseline is 21.3°C. In 1940, the temperature anomaly was 0.3°C, so the surface temperature must have been 21.6°C
10. **A** as mentioned above, the baseline is 21.3°C, which is taken as the normal surface temperature
11. **E** One can see that the gradient of the line of the graph is greatest in the period from 2010 onwards
12. **B** the temperature anomaly increases from 0.4 to 0.8°C in this period of 20 years, which gives a rate of 0.2°C per year.

Data Set 4

13. **A** Out of all of the students, Muskan had the highest overall average of 77.5%
14. **C** Rohit scored 60% in his Computer Science exam, which has a maximum of 40 marks, so Rohit got 24 out of 40.
15. **A** Out of all of the subjects, Maths had the highest average score at 82%
16. **D** Sajal scored 90% on his Maths exam, which is scored out of 150, giving him a mark of 135

Data Set 5

17. **C** The graph shows that the rainfall in 2010 was 45 inches
18. **A** The graph shows that the rainfall in 2009 was the most of any year
19. **C** the graph shows that the rainfall in 2007 was 32 inches, whilst the rainfall in 2009 was 70 inches. The difference between the two is therefore 38 inches
20. **A** The total rainfall over these 5 years was 259 inches, giving an average of 52 inches a year.

Data Set 6

21. **B** If the train travels 5km in 10 minutes, then in one hour the train will travel 30km, so the speed is 30km/h
22. **A** if the train travels at 30km/h, then in 2 minutes the train will travel 1km
23. **D** If Jamie misses the 08:33 train by 7 minutes, then he will be at the station at 08:40, so he must wait 8 minutes until the 08:48 train
24. **E** If he has to be at work by 9am, then he needs to reach Forest Road by 08:48, which means he has to take the 07:48 train from Central park, which means he has to set off from his house at 07:40

Data Set 7

25. **B** The surface area of a cylinder is defined as $2\pi r^2+2\pi rh$ where r=radius of the circle and h=the perpendicular height. This calculation gives, to the nearest centimetre, 1232cm²

26. **A** The volume of a cylinder is defined as πr^2h where r=radius of the circle and h=the perpendicular height. This calculation gives, to the nearest centimetre, 3233cm³

27. **C** the circumference of a cross-section is defined as $2\pi r$ where r=radius of the circle. This calculation gives, to the nearest centimetre, 44cm

28. **B** As the volume is defined as πr^2h, an increase in the radius by a factor of 2 will cause an increase in the volume by a factor of $2^2=4$

Data Set 8

29. **A** If the value depreciates by 15% each year, then the value after 1 year will be 0.85 of 9000. After three years hence, the value of the car will be 9000 multiplied by 0.85^3, which gives £5527

30. **D** If the value depreciates by 15% each year, then after 6 years the value is £3394

31. **C** The runner in the lead will take 12 minutes to complete a lap; the runner in the back will take 18 minutes to complete a lap. If one takes the lowest common multiple of these two values, then the answer is 36 minutes. This is the time where the lead runner will complete 3 laps whilst the runner in the back will complete 2 laps, so at 36 minutes the lead runner will lap the runner at the back

32. **A** In order to find this, one needs to find the highest common factor of 32 and 78, which is 2.

Data Set 9

33. **C** If the red represents 44% of the survey, this equates to 132 people

34. **A** If the green represents 6% of the survey, the angle this makes on the pie chart is 6% of 360° which to the nearest degree is 22°

Data Set10

35. **A** 10% of 9.325 + 15% of 28624 + 25% of 53949 of 28% of 84099 = £41461

36. **C** If he gets married, he goes from being in the 33% tax bracket to being in the 28% tax bracket, so this is a 5% tax reduction

END OF SECTION

Section D: Abstract Reasoning

Rules:

Set 1: Acute angles in Set A= Even; in Set B = Odd.

Set 2: The shape on the inside has more sides than the shape on the outside in Set A; the shape on the outside has more sides than the shape on the inside in Set B

Set 3: Number of lines in set A=1, number of lines in Set B>1.

Set 4: Set A has an odd number of right-pointing arrows and an even number of left-pointing arrows; Set B has an even number of right-pointing arrows and an odd number of left-pointing arrows

Set 5: Shapes in Set A only have straight edges; Shapes in Set B have at least 1 curved edge

Set 6: The arrow is pointing towards the shaded corner in Set A; the arrow is pointing away from the shaded corner in Set B.

Set 7: The number of corners in Set A is an even number; odd in Set B

Set 8: Set A has a vertical line of symmetry; Set B has a horizontal line of symmetry.

Set 9: Set A has less than half the circle shaded; Set B has more than half the circle shaded

Set 10: In increasing order from 1, the correct shape is highlighted

Set 11: The line crosses the shape by an increasing number of times in the sequence

Set 12: The shapes at the top have a horizontal line of symmetry, the shapes at the bottom have a vertical line of symmetry

Set 13: The shapes at the top have an even number of circles and an odd number of sides on the large shape; the shapes at the bottom have an odd number of circles and an even number of sides on the large shape

Set 14: To get from the left to the right, one must reverse the shading of the shapes and the shapes on the top must go to the bottom

Set 15: The number of vertical lines in Set A are greater than the number of horizontal lines; in Set B the number of vertical lines is less than the number of horizontal lines

END OF SECTION

Section E: Situational Judgement Test

Scenario 1

1. **Inappropriate but not awful.** This would not be an awful decision, because what James is doing is technically cheating. However, being James's friend, this is not the best way to approach the situation given the ramifications of cheating, and a solution in which James is not litigated by the medical school should be explored if possible.

2. **Very inappropriate.** The question bank is against the rules and undermines the validity of the end of year exams. Utilising the question bank is still cheating regardless of whether you were involved in the making of the question bank or not.

3. **Very appropriate**. This is the best way to explore the situation, such that James is not punished by the medical school but is not cheating.

4. **Appropriate, but not ideal.** This is a possibility and would be appropriate, but if James is at university this may be inconvenient for the parents, and is delegating responsibility which, if you were a good friend, you should take on yourself.

5. **Very inappropriate.** This option not only does not give James the opportunity to rethink his options about making the question bank, but also publicly exposes him and ruins your friendship with him. This is a highly inappropriate action.

Scenario 2

6. **Very inappropriate.** This would call into account Rashid's personal professionalism and if Rashid's plan was found out by the clinical school, it would certainly call into consideration Rashid's fitness to practice and whether he should continue with medical school or not.

7. **Very appropriate.** This would be the best response to the situation. Informing the clinical school means that Rashid is maintaining transparency as much as possible, and means that he can travel without worry of repercussions.

8. **Inappropriate, but not awful.** If Rashid does enough work on the placement before the holiday that he can afford to take the 2 weeks off and his supervisor is happy with it, it is not the worst decision. However, it would still be inappropriate as the placement is specified as 6 weeks and the supervisor would have to be dishonest with the clinical school by signing Rashid off early.

9. **Appropriate, but not ideal.** This would be an appropriate response, because taking a holiday during the clinical year is not technically allowed. However, if his parents have already pad significant amounts for his ticket, which they will not get back if the tickets are cancelled, then this is not the best way to handle the situation.

10. **Very inappropriate.** This would be very inappropriate. By asking the clinical school and then openly disregarding their decision would call into account Rashid's professionalism and make the clinical school more likely to take drastic measures with regards to repercussions.

Scenario 3

11. **Very inappropriate.** Taking the questions still qualifies as cheating-this would be a very inappropriate action.

12. **Very inappropriate.** It qualifies as cheating, regardless of how many questions he takes

13. **Very appropriate.** This would be the best course of action, as it is not cheating and prevents Jamie's friend from getting in trouble as well.

14. **Inappropriate, but not awful.** This would be justified-Jamie's friend is cheating and reporting a cheater to the exam board is the right thing to do. However, taking into consideration that it is Jamie's friend and she is trying to help him, reporting her without at least consulting her and exploring other options is inappropriate.

15. **Very inappropriate.** Although this may equalise the proportions within the classroom, the classroom still gains an unfair advantage compared to everyone else in the country writing the exam, so this still is cheating and is inappropriate.

Scenario 4

16. **Appropriate, but not ideal.** This is true, and HIV is a disease which you do need to inform your patients about if they have a risk of developing it. However, this is not really the time or the place to reveal this to the wife after she has received such news; a more sensitive approach is required.

17. **Very appropriate.** This is probably the best response to the situation. This way, you can maintain trust with the wife, as well as making sure that the husband is aware of the risk of HIV infection.

18. **Very inappropriate.** This abuses the trust that the wife has placed in you and such large news is not really your place to reveal as a professional unless you absolutely must.

19. **Very inappropriate.** This pressurises the wife into revealing the news in a position when she is not comfortable and is an abuse of the trust she has placed in you.

20. **Inappropriate, but not awful.** This is inappropriate, as it is involving yourself as a professional in the personal lives of your patients. However, if it means that the husband is aware of the risk of developing HIV, then it is not awful.

Scenario 5

21. **Very appropriate.** This is the most appropriate response to the situation. The registrar is the senior-most member of the team and will know the best way to proceed.

22. **Very inappropriate.** This puts patients at risk if the doctor goes into theatre and reflects very badly on the profession if a drunk doctor is doing the ward round

23. **Inappropriate, but not awful.** This would be a sensible response. If the doctor is at risk of endangering patients, then he should be reported to the board. However, reporting the doctor without consulting anyone would be an inappropriate thing to do.

24. **Appropriate, but not ideal.** This is an appropriate action, and is a perfectly sensible option. However, as the junior doctor, it is not ideal to send the consultant in charge home without consulting anyone senior.

25. **Very inappropriate.** This would result in massive distrust between that consultant's patients and the consultant, and bring the profession into disrepute.

Scenario 6

26. **Appropriate, but not ideal.** This is an appropriate response. Maybe the pastoral care can help Neil with regards to his drug use before finals. However, consulting the pastoral care before speaking to Neil is not ideal.

27. **Very inappropriate.** The drug use may underlie a more serious problem, which you are not solving by reporting Neil to the medical school. In addition, Neil has trusted you with this information as his friend. Telling the clinical school is an abuse of this trust.

28. **Very appropriate.** This is the best way forward. Perhaps the reason Neil's drug taking has increased is because of a more serious underlying problem, and by speaking to you he can address it and move forward into finals without drugs, which is the best outcome in this situation.

29. **Very inappropriate.** Stealing from Neil is a crime, regardless of whether it is in good intentions, and is hence always inappropriate.

30. **Very inappropriate.** This is an abuse of Neil's trust, may result in criminal charges against Neil and the end of his medical career, and does not solve the underlying problem of why Neil is taking drugs if there is one.

Scenario 7

31. **Very important.** If the teaching is likely to come up in her end of year exams, it is important that she gets good notes on it which she can consolidate closer to the exam.

32. **Fairly important.** If transplant surgery is something Janice is interested in, and this is a rare opportunity that she might get, then she should try her best to utilise it. Clinical experience is invaluable in addition.

33. **Of minor importance.** Although she is interested in transplant surgery, she still must pass the exams sufficiently such that she is able to become a transplant surgeon. It is important to have interests, but it is more important to know the basics.

34. **Of no importance whatsoever.** Whether the teaching is small-group or large-group should have no effect on Janice's decision, as she has a legitimate decision as to why she could not attend the teaching.

35. **Fairly important.** If Janice knows that she will pursue transplant surgery, then it is important to get as much experience as possible. However, if it is at the expense of priceless teaching which is guaranteed to come up in the end of year exams, then one should prioritise the teaching. Once Janice has passed her finals, then she can try and explore opportunities in transplant surgery.

Scenario 8

36. **Very important.** This is a very important factor to consider. If Johnathan does not achieve a suitable work-life balance, he may not be able to carry on with his medical career.

37. **Of no importance whatsoever.** Although a prize would be nice, it should not be an important factor in deciding what he should do in this situation. There are more important factors to consider than how much prize money he can get.

38. **Fairly important.** His father clearly is enthusiastic for him to join the university cricket team and has invested a lot into making sure he gets there; completely disregarding this would be unfair to the father. However, he must also consider the importance of these exams and the importance of balancing work with his hobbies to attain maximum reward.

39. **Of minor importance.** This is important to consider, Johnathan wants to make a good impression with his new teammates, but compared to whether he is competent enough to progress with his studies, this is a minor factor.

40. **Of no importance whatsoever.** Like the prize consideration; who wins the bet should have no influence on Johnathan's decision.

Scenario 9

41. **Of minor importance.** This is slightly important, as it may be difficult for Hayley to rearrange the meeting if her director of studies is on a tight schedule. However, simply the fact that her director of studies should not have that large an influence on whether Hayley goes on the conference or not.

42. **Very important.** The elective is a wonderful opportunity for the medical student to gain experience in something that they are interested in, and offers a different perspective before they become junior doctors. As such, consideration of the elective is very important.

43. **Of no importance whatsoever.** Which city the conference is in and whether it is an attractive tourist prospect or not should have little influence in Hayley's decision-making process.

44. **Of minor importance.** This is at least a little bit important, as the consultant has chosen Hayley because she believes that Hayley has the responsibility to carry out the task. This should be given some consideration, but is not of paramount importance.

45. **Fairly important.** Getting a publication adds a point to Hayley's foundation programme application, and in the long-run may give her an advantage over other applicants for training jobs, registrar jobs and even consultant posts. Getting a publication at this stage is hence important to the extent that it will affect your future, but not as important as gaining valuable experience from the medical elective. There will always be opportunities to get publications, but the medical elective is really the last chance you get to explore another healthcare system without repercussion.

Scenario 10

46. **Fairly important.** Latisha's reputation with the doctors is an important point to consider. All professionals, including the medical students, on the ward are a team and work together to deliver safe and effective care. If one member of the team is not working well with the other members then the efficiency of the team decreases. As such, Latisha's reputation within the team is important to consider.

47. **Very important.** If the consultant on the ward round is determining whether they pass or fail the placement, it is important for Tim to ensure that Latisha is doing all that she can to pass, as her friend and her project partner.

48. **Very important.** As professionals in the healthcare industry, the public place their trust in doctors that they will provide them with optimum treatment. As such, it is the duty of doctors to remain professional in the public eye. Latisha is no exception, and Tim should make her aware of that.

49. **Of minor importance.** Yes, Tim's friendship with Latisha is important because as clinical partners it is likely that they will spend a lot of time together on the wards. However, it is more important for Tim to make Latisha realise that she is compromising her position as a medical student in this current situation.

50. **Of no importance whatsoever.** That is a personal issue and this is a professional one. It is important for Tim to distinguish between the two and leave the personal side of things out of the workplace.

Scenario 11

51. **Of no importance whatsoever.** This shouldn't matter that much. This is a professional issue here and is very serious; Javad's friendship should not influence the action in this scenario.

52. **This is very important.** If Javad does choose to tell the clinical school, it would ruin his friend's career as a doctor, from which he cannot recover.

53. **This is very important.** The hospital will have to face the consequences of his friend's blunder.

54. **Of no importance whatsoever.** The history does not excuse the actions now, nor should it influence Javad's decision, especially when it comes to a situation this serious.

55. **Of no importance whatsoever.** The friends struggle does not excuse such as breach in confidentiality so this is of no importance.

Scenario 12

56. **Fairly important.** If Ross is working on the night, and the hospital depends on Alex for working the following day, then it is important that Alex is in a fit enough state to do so.

57. **Very important.** If there is already a shortage of beds in the emergency department, then Alex is placing more strain on it by admitting himself.

58. **Very important.** If the consultant is doing the ward round the following morning, then the consultant will see Alex in the Emergency Department, and Ross will have to explain himself.

59. **Very important.** If IV fluids require inpatient admission, then Ross as the admitting doctor takes full responsibility for Alex as a patient and will have to answer for his decisions.

60. **Of no importance whatsoever.** Hopefully Ross should never be in this situation, given the position of the Emergency Department in this situation, so this should not play a role.

61. **Of no importance whatsoever.** This could affect Ross's professional career if he decides to admit Alex-he cannot be allowing personal entertainment to be driving his decision making in this situation

Scenario 13

62. **Very important.** If MRSA requires strict isolation, then unless the medical students know how to approach the situation, they should not go in to see the patient.

63. **Very important.** If the nurses have told them not to go inside, and they disregard the nurses' warnings, then if something happens to the patient the medical students will be responsible.

64. **Not important at all.** The medical students have been notified that they should not enter the room unless necessary by the nurses, so the knowledge of the ward should not play a part in their decision.

65. **Of minor importance.** If they must present the findings back to the junior doctor, they may feel less comfortable in going against his wishes, should they choose not to see the patient.

66. **Very important.** If the patient is clearly unwell, then it makes even less sense for the medical students to see them, as it could clearly compromise the patient's health further if they go and see him/her.

67. **Of no importance whatsoever.** Opportunities to learn in hospital are governed by the state of the patient, not the other way around. Patient safety should never be compromised with the goal of a better learning experience for medical students.

68. **Very important.** There is a significant risk to the students in entering the side room, and if it is not essential for the students to examine the patient then they should question whether it is worth the risk.

69. **Important.** MRSA not only poses a risk to themselves but also those they interact with since it is a highly infectious disease. A party obviously presents significant opportunity for MRSA transmission but it is no more important than the risk of transmission to others they meet on a bus or train.

END OF PAPER

Mock Paper F Answers

Section A: Verbal Reasoning

Passage 1

1. **A** The passages states the English Sparrow 'wages war upon song birds, destroying their young' making **A** correct. It was 'first introduced into the United States' in the mid 19th century, not discovered then, so **B** is incorrect, and song birds sing soul-inspiring songs, making **C** incorrect. **D** is incorrect as 'with the change of climate, came a change of taste for insects'.

2. **B** The good English Sparrows accomplish is 'nullified' according to the passage. The other statements are verified in the text.

3. **C** English Sparrows are described as devouring the common white butterfly, and 'seed and vegetable eaters'. They are described as 'great insect eaters' but never as preferring insects to grains so **A** is incorrect. Song birds are described as singing 'soul-inspiring songs' and English Sparrows as 'inharmonious' but this does not necessarily mean that song-birds are more harmonious than English Sparrows, so **B** is incorrect. It says that English Sparrows devour white butterflies, whose caterpillars wreak havoc, but does not directly say they eat them, so **D** is incorrect.

4. **B** The Sparrows were observed to convey 'no less than 40 grubs per hour' not more than 40 grubs per hour, so **A** is incorrect and **B** is correct. They were only observed for a day, so **C** and **D** are both incorrect.

Passage 2

5. **D** 'The mental qualifications of their gods were of a much higher order than those of men'. **A** is incorrect as the passage states that they have sometimes 'earnestly prayed to be deprived of their privilege of immortality'. **B** is incorrect as it says that gods and goddesses become attached to mortals 'not infrequently', and **C** is incorrect as it states that they do need daily nourishment and refreshing sleep.

6. **D** All the other qualities are stated to be different in gods and humans except emotions or 'passions' as stated in the text.

7. **C** This is stated when the offspring of gods and humans are described as 'heroes or demi-gods'. Although demi-gods are renowned for their courage, it doesn't explicitly say they are more courageous than gods so **A** is incorrect. Hercules isn't mentioned in the passage which makes **B** incorrect, and **D** is wrong because the passage does not talk about the life expectancy of demi-gods versus humans.

8. **C** Gods are 'more commanding in stature, height being considered by the Greeks an attribute of beauty in a man or woman'. The other statements are stated as true in the passage.

Passage 3

9. **B** The passage states that digestion is unappreciated because 'few of us are aware of its happening in the same way we are aware of making efforts to use our voluntary muscles'. The other statements are not stated as a reason for digestion being under appreciated.

10. **C** The aspects of digestion described as hard work/effort are: chewing, making saliva, churning in the stomach, and manufacturing bile and pancreatic enzyme.

11. **A** It states that 'a large portion of the blood supply is redirected from the muscles in the extremities to the stomach and intestines' to aid digestion. Bile, which aids digestion, is produced in the gall bladder but digestion doesn't take place there (**D**) and although this is true, amylase is not mentioned in the passage (**C**). **B** is incorrect as the cabbage and carrots were just used as an analogy to explain how much effort digestion requires.

12. **D** Although the passage states that pancreatin solubilises proteins, this may not be the only thing it does. The other facts are stated/implied in the text.

Passage 4

13. D It is stated that princess Victoria and her governess where devoted to each other before the governess died in 1870 (**C** is incorrect). It says that she would have excelled in music and drawing if she had devoted more time to them (**B**) and that she started lessons at age 5 (**A**).

14. C Tea was only allowed in later years as a treat. The other options are stated in the passage.

15. A It states that Princess Victoria did not know 'that she was likely at any future time to be Queen'. **B** is wrong because she states that she always slept in her mother's room before becoming Queen, **C** is wrong because she said that she read books about history recommended by her uncle, the King of the Belgians, but this does not mean that it is Belgian history. **D** is incorrect because although the passage states she is 'especially fond of music' it does not directly compare music to reading.

16. D It says that her education was 'chiefly' but not solely with a governess. The other statements are stated in the passage.

Passage 5

17. A 'the limit of endurance is reached more quickly towards the end of pregnancy'

18. C The passage states that exertion may need to be prohibited to prevent miscarriage. However, this does not mean that exhaustion is the cause of miscarriage.

19. B The passage states that sometimes women find it necessary to stop exercise a week or two before delivery.

20. B The passage states that women who have laborious household duties may need to exercise less than women with sedentary lives, but not that they don't need to exercise at all.

Passage 6

21. C the passage states that it is the simplest form in which spaghetti is served, but does not generalize to pasta as a whole.

22. C the passage states that Italians eat macaroni unbroken, but their ability to do this 'is not the privilege of everybody'. However it does not explicitly state whether any people of other nationalities are able to eat macaroni this way or not.

23. B The passage states that macaroni must be cooked for 12-15 minutes, not spaghetti. The other statements are in the passage as truths.

24. C The passage states that macaroni should be drained in a colander which means that pasta can be drained in a colander and **C** is correct. Spaghetti (but not pasta as a generalization) should be but in the water at boiling point (**B**) and both macaroni and spaghetti (but not pasta as a generalization) should be cooked in salted boiling water (**A**). Mezzani is the preferable kind of macaroni, but again not generalized to all pasta, for butter and cheese (**D**).

Passage 7

25. D

A. The passage states that penguins cannot fly, so this statement is incorrect

B. The text states that penguins 'have been termed the true inhabitants of that country' – this is not synonymous with being 'the first land inhabitants of the Antarctic region' – to see this you must think about how they qualify their statement and you will see that it due to penguins being best adapted rather than the first animals to evolve to this environment.

C. They would but the text clearly states that 'they never do this'.

D. By exclusion and by default after reading the phrase 'there is no food of any description to be had inland.' – this is the correct option

26. C The passage states that bears and foxes live in the North Polar regions, so penguins in the Antarctic are safe, but it makes no statement as to whether penguins live in the North Polar regions, so one can't tell.

27. B Penguins do appear to be safe on land according to the text BUT there is no food for them on land – if both parts of the statement are not true it cannot be classed as true hence FALSE.

28. C The passage states that 'their four legs, in course of time, gave place to wide paddles or "flippers,"', so their legs did not evolve into fins; they evolved into flippers.

Passage 8

29. **D** The passage states that 'The breeding season of these birds begins in October or November', so this statement is incorrect.

30. **A** The passage states that 'Bonelli's eagles (Hieraetus fasciatus)' so one can imply that the Latin name for Bonelli's eagles is Hieraetus fasciatus

31. **C** The passage states that the pellets of the Benelli's eagle contains squirrels' skulls, but reveals no information about the kite's nest, so one cannot tell.

32. **C** The passage states that 'In India winter is the time of year at which the larger birds of prey, both diurnal and nocturnal, rear up their broods.', so one can imply that the larger birds of prey breed in winter.

Passage 9

33. **C** The passage states that 'it was 1896 and 1897 which were very serious years for the country' which suggests that these are the worst years for the country, not 1889

34. **C** The passage states that foreign capitalists were pouring in, and in this time there was a boom in land, so one can imply that foreign input influenced the boom in land.

35. **B** The passage states that 'The province of Buenos Aires, the largest in the country', so this statement is false.

36. **C** The passage makes no reference to when the monsoon in Buenos Aires is.

Passage 10

37. **B** 'Mosquito eggs are laid in water or in places where water is apt to accumulate, otherwise they will not hatch.' Hence the true statement would be that water is required for eggs to hatch rather than to be laid.

38. **B** 'Some of the eggs may remain over winter, but usually those laid in the summer hatch in thirty-six to forty-eight hours or longer according to the temperature.' This just requires some delicate reading of the phrase to understand that eggs generally hatch 36/48 hours after being laid, so you cannot directly compare the hatching times of summer eggs to winter eggs, as whichever were laid first will be hatched first, apart from the rare example where some eggs stay over Winter.

39. **C** There is no mention of whether the eyes can see or not when first hatched, so this answer is false.

40. **C** The feeding habits described are far too diverse to be summarised in one blanket statement – one point in exception to this statement is that 'Some mosquito larva are predaceous, feeding on the young of other species or on other insects'. Answer C however is mentioned directly in the text.

Passage 11

41. **A** The passage states that mites can transmit diseases from 'man to man' so this statement is correct

42. **B** The passage states that 'they may possibly aid in the dissemination of leprosy.', which is evidently more serious than blackheads, so this statement is false

43. **C** One would think this might be true considering that domestic animals are worse affected but there is no direct comparison, nor can it be inferred, so the answer is Can't Tell

44. **D** The passage states that 'The young have six legs, the adult eight' so one can infer that the mites develop two more legs as they mature

END OF SECTION

Section B: Decision Making

1. **B** Express the information in the following Venn diagram

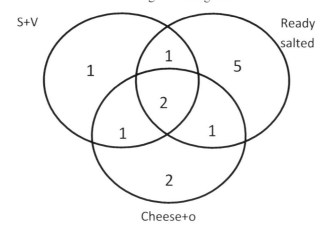

5+1+1+1+1+2+2=13

2. **D** The statements can be written as: S<T, T< J<L, L<A. Therefore, Sam is the shortest.

3. **C** We do not know whether Alexandra and Katie are dancers, so **A** and **B** are wrong. We do not know whether any dancers are ugly, so **D** is wrong.

4. **D** 1 – 1/6 = 5/6
   5/6 x 3/4 = 15/24
   15/24 x 1/3 = 5/24

5. **B** 50 pieces minus 5 spares = 45. Half the meat eaters want 2 pieces and half want 1 piece splitting the remainder into 30 pieces for half and 15 for the other half. This means there are 30 meat eaters in total, and 40 total guests (1/4 are vegetarian).

6. **C** If he travels north, then turns to his right, he will be facing east. If he turns right after that, he will be facing south. If he travels 5km and 5km south, he is due east of his starting point.

7. **B** The pattern is squares of even numbers starting at 12, so 18 squared = 324

8. **E** A takes 10 days and B takes 15 days.
   Calculating the LCM of these we get 30 days – think of this as the 'total work'
   If we then look at A in terms of 30/10 = 3 units of work
   And similarly B is 30/15 = 2 units of work
   So, finally 30/(3+2) = 6 days in total!

9. **E** Let the age of a = 2x and the age of b = 3x -> 2x=20 -> X=10
   Then present age of b is 3 * x= 3*10=30 years
   After 5 years, age of b = 30+5= 35 years

10. **C** 11+4=15
    15+22=37
    37-3=34

11. **C** 'Southampton is bigger than only Romsey.' -> therefore 4th- Southampton, 5th - Romsey
    Hence: London>Oxford>Cambridge>Southampton>Romsey

12. **C** If V is shorter only than R, R is the tallest and V is the 2nd tallest.
    J < M must mean that M is in the middle, followed by J and D therefore must be the shortest.
    R>V=160>M>J=135>D
    Hence M must lie between 135 and 160 -> can only be C

13. **C**
    A. All chemistry teachers are professionals-This cannot be determined based on the text and therefore cannot be assumed
    B. All professionals require a chemistry degree-This would imply that chemistry is the only university degree available which cannot be inferred from the questions
    C. All lawyers have a degree-This is true, considering that all lawyers are professionals and all professional jobs require a university degree
    D. All lawyers have a chemistry degree-there is no information in the text linking lawyers with chemistry
    E. All universities offer law as a degree course-this cannot be inferred from the information provided

14. **C** Joseph is a man and no man is a lion i.e. all men are NOT lions hence Joseph is not a lion.
    The statement is '1 to many' connection i.e. it puts Joseph in the bigger group of men BUT does not state that all men are or are not Joseph hence can't tell.

15. **B** Radius of circle = C
    Side of Square = S
    C= 2/3(S)
    S2=441 therefore S=21
    Hence C=14
    Perimeter=2*pi*14

16. **D** a+b=3430
    0.12a=0.28b
    Rearrange the second equation to a=7/3 b and substitute this into the first equation to get 10/3b=3430
    Hence b=1029 and a = 2401

17. **C** Son age now= S
    Father age now= F
    F- 4 = 4(S-4) -> F = 4S -12
    F+6/S+6 =5/2 -> 2F +12 = 5S+ 30
    8S -12 = 5S +30
    S=14
    F=44

18. **D** Since the narrator has no brother, his father's son is he himself.
    So, the man who is talking is the father of the man in the photograph.
    Thus, the man in the photograph is his son

19. **C** Total of 12 = 73*12=876
    E + A + O = (73.6*15) – 876 = 228
    68 +A + (A+6) =228
    2A = 154 -> A=77

20. E For this we need to know the area of a curved face of a cone is radius pi x radius x length (i.e. diagonal)

Diameter = 40m

Radius = 40/2=20m

H=21m

l=√ (20²+21²) = √(400+441) = √841 =29m

Surface area of curved face = 29x20 pi

Total surface area = 480pi + 100pi

Ratio = 480pi : 580pi = 24:29

21. D All the women have left for dinner so A and B cannot be correct and the sons went out to play so C is incorrect hence D is correct.

22. B Build up the order as you read the statements:

If A is placed below E -> E, A

C is placed above D -> C, D

B is placed below A -> E, A, B

and D is placed above E -> C, D, E, A, B

23. B Distance = 2/3S

distance=1-2/3S=1/3S

21/15 hr=2/3 S/4 + 1/3s /5

84=14/3S * 3

S= 6km

24. E The relative speed of the boys = 5.5kmph – 5kmph = 0.5 kmph

Distance between them is 8.5 km

Time= 8.5km / 0.5 kmph = 17 hrs

25. C 34*8*9 = 2448

This is 2/5 of the total work, so multiplying this by 3/2 gives us 3/5 = 3672

3672 / (6*9) = 68

The question asks for how many more men should be employed so we need an additional 34.

26. C Jack is a human. Therefore, Jack is not a machine-If Jack is a human, and machines are not human, then one can infer that Jack is not a machine

All machines are robots-Just because all robots are machines does not mean that all machines are robots, one cannot tell this based on the information

27. B

A.   'Frustrated people are prone to be drug addicts' -> does not mean all are drug addicts hence A is false

B.   'Some artists are frustrated -> frustrated people are prone to be drug addicts' -> so Yes some artists may be drug addicts TRUE

C.   Nothing is said about the relationship in this direction

D.   The artists who are frustrated may also be whimsical but not ALL frustrated people.

28. D X is married to A who is a lawyer, so B and C are either a doctor or a lawyer respectively. Y is not married to an engineer, and C is not a doctor so Y must be married to B who is a doctor, and Z must be married to C who is an engineer, so the correct answer is D, as none of these options are available.

29. A

    A. Vigorously learning the times tables in primary school provides an unnecessary level of stress. This is correct, primary school children have enough to worry about with 11+ exams at the end of year 6, forcing them to maintain high standards in the times tables will only add to this stress from an earlier age, and could lead to detrimental long-term effects.

    B. Literacy is a more important skill than mathematics-This is false; both literacy and mathematics are equally important in terms of educational needs.

    C. Knowing the times tables provides no benefit for children in later life-This is false; times tables forms an important part of mental arithmetic, which is an important life skill regardless of what career one goes in to.

    D. There is no need for mental arithmetic when calculators are so readily available-This is false; as mentioned above mental arithmetic is still an important life skill to have at hand, and the availability of calculators should not affect this.

END OF SECTION

Section C: Quantitative Reasoning

Data Set 1

1. **D** 8.1g. A drink made of 18g of powder and 200ml of whole milk has 9.6g fat. 100g powder has 8.4g fat, therefore 18g powder has 18/100 x 8.4 = 1.512g. The amount of fat in 200ml whole milk is 9.6-1.512 = 8.1g.

2. **A** 56ml. The total fat content of 2.464g contains fat from the powder. Therefore, the milk/water mix contains 2.464-1.512 = 0.952g fat. Since there is no fat in water, this can only be from the semi skimmed milk. If 100ml semi skimmed milk contains 1.7g fat then to obtain 0.952, we need a volume of 0.952 x 100/1.7 is 56ml.

3. **C** 7.0 pints. The box contains 360g of powder. He therefore needs 20x200ml = 4000ml milk which is equivalent to 4000/570 = 7.0 pints.

4. **C** 164g. he box has 360g powder and therefore at 18g per drink there are 20 drinks in the box. For these 20 drinks John will need 20x200ml = 4 litres. The protein content of the whole box is 360/100 x 8.9g = 32.4g. The protein in the 4 litres of the semi skimmed milk is 40x3.3 = 132g. The total amount of protein is therefore 164g.

5. **C** 9kCal. (372 – 4x8.9 – 4x65.2)/8.4 = 9. OR (200-4x8..0-4x20.4)/9.6 = 9

Data Set 2

6. **E** £83.10. The Monday – Friday 9am-6pm stays will each cost:
   - 9am-1pm: 4hrs x 6 periods x 20p = £4.80 and - 1pm-6pm: 5hrs x 6 periods x 30p = £9.00
   Total cost = £13.80 per day.
   Saturday and Sunday 9am-6pm stays will each cost: 9hrs x 3 periods x 15p = £4.05 per day.
   6 overnight stays = £6.
   Total cost = 5 days x £13.80 + 2 days x £4.05 + £6 overnight = £83.10

7. **E** £1.30. The first period of 5 mins (5:55-6:00pm) will be charged at 30p and the second part (6:00-6:15) will be charged at £1. Hence the total is £1.30.

8. **D** 80p. A 37-minute stay will be charged as 40 minutes and will therefore cost 4 x 20 = 80p.

9. **D** £20. The week-day cost from 9am-6pm is £13.80 (see previous). However he can use the ticket from 8:30-9:00am, which would normally be charged at £1. The total for the day is £14.80. Over 5 days, this costs £74, hence a saving of £20.

10. **E** 50p. He will be charged a full 20p for the first 2 minutes and a full 30p for the next 2 minutes,

11. **E** Saturday 1:40pm.
    Cost of Friday 5pm to 6pm = 6 x 30 = £1.80.
    Cost over overnight stay (Friday to Saturday) = £1. Cumulative total = £2.80.
    This leaves £2.10 to spend on Saturday day time. The cost of 15p per period – he can spend 2:10/0.15 = 14 periods of 20 minutes i.e. 4 hrs 40 mins in total. He must therefore have left the car ark by Saturday 1:40pm.

Data Set 3

12. **B** 1 850 000 Rupees. From the table it shows that the price of 10 grams of gold in 2010 is 18500 Rupees. There are 1000 grams in a kilogram, so the answer is 18500 multiplied by 100=1 850 000 Rupees.

13. **E** £21765. If there are 85 rupees to £1, then 1 850 000 rupees, which is the price of 1kg of gold in 2010, then this corresponds to 1 850 000 divided by 85, which gives 21765 to the nearest pound.

14. **B** 590kg. £25000 corresponds to 2125000 Rupees, which in 1940 would have bought you 58962 lots of 10 grams of gold, which corresponds to 590kg to the nearest kilogram

15. **C** 763kg. £25000 in 1940 would've corresponded to 2750000 Rupees, which in 1940, would have bought you 76304 lots of 10 grams of gold, which corresponds to 763kg to the nearest kilogram

16. **A** 2. To calculate this, one must find the highest common factor between 38 and 46, which is 2. 2 is therefore the highest number of each coloured pen that can go into each pencil case without leaving a remainder.

Data Set 4

17. **C** December 2013. This is when the peak of the graph is.

18. **A** $1020. One can see reading off from the graph that the maximum value that Bitcoin reaches is $1020

19. **D** Between July 2013 and December 2013, the value of bitcoin rose from $100 to $1020, over the period of 5 months. If we simplify such that the rise is uniform over the 5 months, then the rise is (1020-100) divided by 6 = $184 per month

20. **B** $1 600 000. Jamil bought $1000 of Bitcoin shares when it was 50 cents a share. He would've been able to buy 2000 shares with this amount. He sold it in March 2013, when the value was $220 a share, meaning that Jamil turned over a revenue of $440 000, with an overall profit of $439 000. However, had he sold it in December 2013, when the value was $1020 a share, then he would've made a revenue of $2 040 000, with a net profit of $2 039 000. This means that had he sold it in December 2013, he would've made $1 600 000 more.

Data set 5

21. **C** The volume of a cone is defined is one-third of the height multiplied by the radius cubed multiplied by pi. As a result, the correct answer is 1 400 632 cm³, which is equivalent to 1.4m³

22. **E** The surface area of a cone is defined as $\pi r^2 + \pi rs$. As a result, the correct answer is 5850 cm², which is equivalent to 0.58m²

23. **B** If the angle between h and s=34°, and a right-angled triangle is formed between h, s and r, then the angle between s and r=180-(90+34) =56°

Data Set 6

24. **A** One can see that the data set with the highest number of deaths is ovary cancer

25. **A** 46250. The mean is the sum of all the groups divided by the number of groups there are. This gives a value of 46250 deaths

26. **D** 140 000. The range is taken as the largest value subtracted by the smallest value, which in this case gives a value of 140 000

Data Set 7

27. **C** 1.2 billion. Based on the graph, one can estimate the population to have been 1.2 billion in 1900.

28. **E** 80 million per year. One can see that the population in 1960 was 3 billion, and the population in 2010 was 7 billion. This is a rate of 4 billion over 50 years, which equates to a rate of increase of 80 million per year.

29. **C** 16.2 billion. If the population is 7 billion in 2010, and there is an increase in 80 million per year, then the population in 2125 will be 7 billion+(80 multiplied by 115) =16.2 billion

Data set 8

30. **A** 37.5%. There is a drop of 1350 accounts, which out of 3600 represents a 37.5% drop

31. **D** $202.5 million. If each client contributes $150 000 to the firm, then the firm would have lost 1350 multiplied by $150 000 which equals $202.5 million

32. **E** 2025. The minimum number of employees that must be fired would be if all the employees had the maximum salary of $100 000, so $202.5 million divided by 100000, which gives 2025.

Data Set 9

33. **D** 6 miles. If the bus takes 7 minutes to travel between these two stops, and the bus is travelling at 40 miles per hour, then the distance between the 2 stops is 40 multiplied by (7 divided by 60) which equals 4.66 miles

34. **C** 66mph. The bus takes 31 minutes to travel the 34 miles, which means that the average speed of the bus is 34 divided by (31 divided by 60) which = 66 miles per hour

35. **C** 5.3 miles per hour. If she misses the bus by 2 minutes, then she will be at the Weeping Cross Lane at 0735. If she needs to be at the station by 0752, then she has 17 minutes to walk 1.5 miles, which means she must walk at 5.3 miles per hour

36. **D** 0.264. If skittles only come in one of these 4 colours, then the probability that they draw a purple skittle will be (53-(12+13+14))/53, which equals 0.264.

END OF SECTION

Section D: Abstract Reasoning

Rules:

Set 1: In Set A, all objects with a right angle are striped. All objects with no right angle are black. The number of white circles is equal to the number of right angles present within the striped objects within the shape. In Set B, all objects with a right angle are black and all objects with no right angle are striped. The number of white circles is equal to the number of angles which are not right angles within the shapes.

Set 2: The number of sides in increasing by 1 in order, with even-sided shapes being shaded and odd-sided shapes being clear.

Set 3: The black circle is moving around the corners of the square, with the arrow being in the opposite corner to the black circle

Set 4: The letter on the right is 16 letters of the alphabet after the letter on the left.

Set 5: The shape on the right is the mirror image of the shape on the left, with a vertical line of symmetry.

Set 6: The shape on the right has 3 times the number of sides as the shape on the left

Set 7: The number of acute angles is odd and the number of obtuse angles is even in Set A; The number of acute angles is even and the number of obtuse angles is odd in Set B

Set 8: In Set A, if there is an upwards pointing arrow then there is an even number of circles. In Set B, if there is an upwards pointing arrow then there is an odd number of circles.

Set 9: In Set A, the number of edges adds up to a multiple of 3. In set B, the number of edges = 8.

Set 10: In Set A, arrows point towards rectangles and away from circles. In Set B, arrows point towards triangles and away from rectangles.

Set 11: In Set A, there are twice as many hexagons as triangles. In Set B, there are twice as many triangles as hexagons.

Set 12: In Set A, there is always a rectangle cut by one circle. In Set B, there is always a rectangle cut by two circles.

Set 13: In Set A, all shapes contain 9 arrowheads. In Set B, all shapes contain 8 arrowheads.

Set 14: In Set A, there are four curved edges. In Set B, there are 3 curved edges.

Set 15: In Set A, there are 10 shapes if white shapes count as two. In Set B, there are 8 shapes if black shapes count as two

END OF SECTION.

Section E: Situational Judgement Test

Scenario 1

1. **Very inappropriate.** It is highly inappropriate for a medical student to be drunk on a ward. As a fellow medical student, it John's professional duty to act in this situation.
2. **Very inappropriate.** Although he should act in this situation, the head of the medical school is not his first port of call.
3. **Very appropriate.** It would be most appropriate to speak to Tom first and see what is going on and advise him to go home. As John doesn't know what the context of the situation is, it is best to speak directly to Tom first.
4. **Inappropriate but not awful.** It would be more suitable to speak to a medical supervisor or to Tom first than to fellow medical students as a first port of call.
5. **Appropriate but not ideal.** Speaking to his supervisor would be a good thing to do but ideally, he should have spoken to Tom about it first.

Scenario 2

6. **Very inappropriate.** Jacob has a very legitimate reason for not attending the session but if nobody is aware of it then they will think he is missing the session because he can't be bothered to go (when the opposite is the case)
7. **Very inappropriate.** Jacob's mother was recently diagnosed with cancer so putting himself through a session like this would be unnecessary and unhelpful for him, especially as he thinks it would be too challenging. He shouldn't put himself through it just for the sake of it.
8. **Very appropriate.** The doctor will be extremely understanding if Jacob explains why he doesn't want to attend the session.
9. **Appropriate but not ideal.** It would be ideal to inform the doctor running the session directly about why he cannot come, but is understandable if he would rather his friend explained the situation to the doctor.
10. **Appropriate but not ideal.** Jacob shouldn't have felt the need to attend given his circumstances. But it was good that he left when he found it too overwhelming.

Scenario 3

11. **Very inappropriate.** As team captain, Lucy has a responsibility to the team and cannot abandon them a week before the match.
12. **Inappropriate but not awful.** As it is a week before the match, the team has probably practiced a lot with Lucy playing for the whole game. Changing this a week before the match is not ideal.
13. **Very appropriate.** This way Lucy is not letting down her team in any way and is attending as much of the conference as she can.
14. **Appropriate but not ideal.** Lucy's current priority is her responsibility for her hockey team and as a third year, there is plenty of time to attend conferences and build up medical interests. However, attending this conference in some capacity may have benefitted Lucy.
15. **Very inappropriate.** Lying to and abandoning her team members, especially as she is team captain, would be wrong.

Scenario 4

16. **Inappropriate but not awful.** It would be inappropriate to escalate a situation like this straight to the medical school director. However, Tara is distressed by the situation and at least in this scenario she is responding to the situation.

17. **Very appropriate.** A situation like this should be discussed with Tara's supervisor, as they can provide her with practical information and advice as to how to proceed.

18. **Inappropriate but not awful.** Although it is good that Tara did not passively accept the situation at hand, speaking directly to the consultant is difficult because of the power dynamic. It would be better to approach this situation from a different angle.

19. **Very inappropriate.** If Tara is finding the consultant distressing, she should definitely act in this situation as this should not be the experience that she is having.

20. **Very inappropriate.** Many things can be done in this situation, including changing wards, but stopping attending altogether compromises Tara's medical education.

Scenario 5

21. **Very inappropriate.** Revision cannot take place in two days even if she didn't fail by much and it is important to try not to fail the retake.

22. **Very appropriate.** This is ideal as she can be as involved in the wedding as is feasible whilst giving herself time to focus on the retake.

23. **Inappropriate but not awful.** Planning for a wedding is close to a full-time job and realistically does not leave time for much revision. However, at least she is planning on revising before the wedding unlike in scenario 1.

24. **Very inappropriate.** The wedding date has probably been in place for months and the venue hired etc. – this is not feasible.

25. **Appropriate but not ideal.** The wedding can still go ahead and Alice can revise for her retake. However, Alice's sister probably envisioned Alice as her maid of honour so changing this would not be the ideal situation for her sister.

Scenario 6

26. **Inappropriate but not awful.** Luke should speak to the student first before reporting him. This may be more appropriate if he found him doing the same thing a second time.

27. **Very inappropriate.** The student is breaking confidentiality and it is your responsibility as a medical professional to act in this situation.

28. **Very appropriate.** Speaking to a senior such as his supervisor is appropriate, especially if he is unsure of the most appropriate course of action in this situation.

29. **Very appropriate.** I is appropriate to speak directly to the student and explaining that what he did is breaking confidentiality.

30. **Very inappropriate.** The problem is not that he is in the hospital, it is that he is in the hospital lift and therefore a public place. The student should not be speaking about patients in any public place.

Scenario 7

31. **Fairly important.** This is a very good opportunity to practice on real patients and have a shot at the exam. However there are other ways to practice on real patients.

32. **Very important.** Some patients may have a compromised immune system and it is not worth risking their health.

33. **Very important.** This is Eva's only opportunity for a mock exam before finals.

34. **Of minor importance.** Although this is true it is still irresponsible to risk the health of patients who have medical conditions. Also they may have a serious condition that is stable at the moment.

35. **Very important.** Eva should obey the rules given by the medical school.

36. **Very important.** Eva has opportunities to practice for the exam, so even though the mock will be useful, she can still do well in her exam without it.

Scenario 8

37. **Very important.** It would not be fair for Lucy if she was not reimbursed, as it would cost her more in petrol to go to the further GP placement.

38. **Fairly important.** It would be kind to prevent Alice from making a long commute when it is much quicker and easier for Lucy by car.

39. **Very important.** She would be getting something equally valuable in return for swapping with Alice.

40. **Of minor importance.** Lucy would have to leave 20 minutes earlier than she would have originally left, which is not a big compromise.

41. **Of minor importance.** Alice may have already asked the other students who can drive and they may have refused, may not have their cars at university etc. Lucy does not know what the circumstances of the other students are.

42. **Of no importance.** Lucy chose her placement based on length of commute, not quality of placement/teaching. Therefore, by swapping with Alice, there is no reason to think that the quality of her placement would be decreased.

Scenario 9

43. **Fairly important.** If he learns better by reading a textbook, it is not as bad that he is missing some lectures. However, even if he learns better by reading a textbook, the lectures may be a helpful addition to his learning.

44. **Very important.** It would be worse to be tired all day and unable to concentrate in all the lectures, then to miss half of the lectures and have full concentration for the other half of the lectures.

45. **Of minor importance.** Even if the other students do not find the lectures useful, he may find them of use.

46. **Very important.** If the lectures are available online, Nick can catch up easily.

47. **Very important.** Again, if the lectures are available online, Nick will not have a problem catching up.

48. **Very important.** Insomnia is a medical condition and it is impacting Nick's studies. The medical student may be able to help him find a solution e.g. recording the lectures.

Scenario 10

49. **Very important.** If the consultant has other availabilities, Lorna need not compromise on either hockey or the teaching.

50. **Of no importance.** As Lorna is interested in paediatrics and likes her consultant, she is keen on the teaching for reasons unrelated to exams.

51. **Very important.** Again, if hockey practice were on for more than one evening a week, she could miss Thursday evenings but still have full attendance on other days in the week.

52. **Very important.** As Lorna is considering paediatrics, this teaching is extremely valuable for her.

53. **Fairly important.** It would not be good to let down the hockey team if she has made commitments. However, as she isn't captain, if she wants to prioritise the teaching over her extra-curricular activity, that would be valid.

54. **Fairly important.** It is healthy to have interests outside of studies and without hockey, Lorna wouldn't have this. However if she is committed to the teaching, Lorna can find a different extracurricular activity that does not take place on a Thursday evening.

Scenario 11

55. **Of minor importance.** Chris's friendship with Joe will undoubtedly have an effect on how Chris decides to act in this situation. However, plagiarism is wrong and Chris should not feel guilty about wanting to do act in this situation.

56. **Very important.** Plagiarism is wrong, and if it is directly stated that it is against university rules, Chris should act in this situation.

57. **Very important.** The medical student in the year above may have put a lot of effort into this essay and sent it to Joe and Chris to help them, but not for it to be replicated without any work on their part.

58. **Of no importance.** Whether or not the essay was good, what Joe did was wrong.

59. **Of no importance.** Again, what Joe did was wrong, even though the essay only counts for 5% of the grade.

60. **Of no importance.** Whether or not plagiarism is inevitable, if Chris has caught Joe doing it, he should act on this.

Scenario 12

61. **Very important.** By helping the surgeon, Chen may be assisting in saving this girl's life.

62. **Fairly important.** The procedure may be psychologically traumatic, but Chen is likely to be more traumatised if he does not assist and the girl subsequently dies. However, it is a lot to ask of a medical student.

63. **Very important.** Even though this wouldn't be a situation that ever occurred in the UK, the circumstances are different in Swaziland and so Chen should not feel guilty in helping the surgeon.

64. **Very important.** As the only person available to help, this increases Chen's responsibility and guilt if he doesn't.

65. **Of minor importance.** As previously stated, the UK and Swaziland are very different countries so Chen should feel justified in helping the surgeon.

Scenario 13

66. **Very important.** Although this publication could be useful, Jasmine should not compromise her exam results which are ultimately more important for her medical career.

67. **Fairly important.** A publication is useful for job application and CV, but as previously stated, it is not worth compromising exam results for.

68. **Of minor importance.** In this situation Jasmine's studies need to take priority since a manuscript can be submitted for publication at any time, her exam dates cannot be moved.

69. **Very important.** With further extracurricular activities to juggle her time for revision will be even more limited. Jasmine needs to prioritise and therefore knowing all of her possible commitments to make a reasoned decision.

END OF PAPER

Final Advice

Arrive well rested, well fed and well hydrated

The UKCAT is an intensive test, so make sure you're ready for it. You can choose when you sit the UKCAT so give yourself the best possible chance and choose a time when you know you do not have much else on. Ensure you get a good night's sleep before the exam (there is little point cramming) and don't miss breakfast. If you're taking water into the exam then make sure you've been to the toilet before so you don't have to leave during the exam. Make sure you're well rested and fed in order to be at your best!

Move on

If you're struggling, move on. Every question has equal weighting and there is no negative marking. In the time it takes to answer on hard question, you could gain three times the marks by answering the easier ones. Be smart to score points- especially in section 2 where some questions are far easier than others.

Afterword

Remember that the route to a high score is your approach and practice. Don't fall into the trap that "*you can't prepare for the UKCAT*"– this could not be further from the truth. With knowledge of the test, some useful time-saving techniques and plenty of practice you can dramatically boost your score.

Work hard, never give up and do yourself justice.

Good luck!

Acknowledgements

I would like to thank Rohan and the *UniAdmissions* Tutors for all their hard work and advice in compiling this book, and both my parents and Meg for their continued unwavering support.

Matthew

About UniAdmissions

UniAdmissions is an educational consultancy that specialises in supporting **applications to Medical School and to Oxbridge**.

Every year, we work with hundreds of applicants and schools across the UK. From free resources to our *Ultimate Guide Books* and from intensive courses to bespoke individual tuition – with a team of **300 Expert Tutors** and a proven track record, it's easy to see why UniAdmissions is the **UK's number one admissions company**.

To find out more about our support like intensive **UKCAT courses** and **UKCAT tuition** check out www.uniadmissions.co.uk/UKCAT

Your Free Books

Thanks for purchasing this Ultimate Guide Book. Readers like you have the power to make or break a book – hopefully you found this one useful and informative. If you have time, *UniAdmissions* would love to hear about your experiences with this book.

As thanks for your time we'll send you another ebook from our Ultimate Guide series absolutely <u>FREE</u>!

How to Redeem Your Free Ebook in 3 Easy Steps

1) Find the book you have either on your Amazon purchase history or your email receipt to help find the book on Amazon.

2) On the product page at the Customer Reviews area, click on 'Write a customer review'

Write your review and post it! Copy the review page or take a screen shot of the review you have left.

3) Head over to www.uniadmissions.co.uk/free-book and select your chosen free ebook! You can choose from:
- ➤ The Ultimate UKCAT Guide – 1250 Practice Questions
- ➤ The Ultimate BMAT Guide – 800 Practice Questions
- ➤ BMAT Mock Papers
- ➤ BMAT Past Paper Solutions
- ➤ The Ultimate Oxbridge Interview Guide
- ➤ The Ultimate Medical School Interview Guide
- ➤ The Ultimate Medical Personal Statement Guide
- ➤ The Ultimate Medical School Application Guide

Your ebook will then be emailed to you – it's as simple as that!

Alternatively, you can buy all the above titles at **www.uniadmisions.co.uk/our-books**

UKCAT Intensive Course

If you're looking to improve your UKCAT score in a short space of time, our **UKCAT intensive course** is perfect for you. It's a fully interactive seminar that guides you through sections 1, 2 and 3 of the UKCAT.

You are taught by our experienced UKCAT experts, who are Doctors or senior Oxbridge medical tutors who excelled in the UKCAT. The aim is to teach you powerful time-saving techniques and strategies to help you succeed for test day.

➢ Full Day intensive Course
➢ Copy of our acclaimed book "The Ultimate UKCAT Guide"
➢ Full access to extensive UKCAT online resources including:
➢ 6 complete mock papers
➢ 1250 practice questions
➢ Online on-demand lecture series
➢ Ongoing Tutor Support until Test date – never be alone again.

Timetable:

➢ **1000 – 1030:** Registration
➢ **1030 – 1100:** Introduction
➢ **1100 – 1200:** Verbal Reasoning
➢ **1200 – 1300:** Quantitative Reasoning

➢ **1300 – 1330:** Lunch
➢ **1330 – 1500:** Abstract Reasoning
➢ **1500 – 1600:** Decision Making
➢ **1600 – 1730:** SJT

The course is normally £195 but you can get **£ 10 off** by using the code *"BKTEN"* at checkout.

www.uniadmissions.co.uk/UKCAT-course

£10 VOUCHER:

BKTEN

Medicine Interview Course

If you've got an upcoming interview for medical school – this is the perfect course for you. You get individual attention throughout the day and are taught by Oxbridge tutors + senior doctors on how to approach the medical interview.

- ➤ Full Day intensive Course
- ➤ Guaranteed Small Groups
- ➤ 4 Hours of Small group teaching
- ➤ 2 x 30 minute individual Mock Interviews + Written Feedback
- ➤ Full MMI interview circuit with written feedback
- ➤ Ongoing Tutor Support until your interview – never be alone again

Timetable:

- ➤ **1000 - 1015:** Registration
- ➤ **1015 - 1030:** Talk: Key to interview Success
- ➤ **1030 - 1130:** Tutorial: Common Interview Questions
- ➤ **1145 - 1245:** 2 x Individual Mock Interviews
- ➤ **1245 - 1330:** Lunch
- ➤ **1330 - 1430:** Medical Ethics Workshop
- ➤ **1445 - 1545:** MMI Circuit
- ➤ **1600 - 1645:** Situational Judgement Workshop
- ➤ **1645 - 1730:** Debrief and Finish

The course is normally £395 but you can get £35 off by using the code "*BRK35*" at checkout.

www.uniadmissions.co.uk/medical-school-interview-course

£35 VOUCHER:

BRK35

Printed in Great Britain
by Amazon